White Rose Maths

White Rose Maths Edition

POWER MATHS

Year 2A
A Guide to Teaching for Mastery

Series Editor: Tony Staneff
Lead author: Josh Lury

Pearson

Contents

Introduction to the author team

Power Maths arises from the work of maths mastery experts who are committed to proving that, given the right mastery mindset and approach, **everyone can do maths**. Based on robust research and best practice from around the world, *Power Maths* was developed in partnership with a group of UK teachers to make sure that it not only meets our children's wide-ranging needs but also aligns with the National Curriculum in England.

Power Maths – White Rose Maths edition

This edition of *Power Maths* has been developed and updated by:

Tony Staneff, Series Editor and Author

Vice Principal at Trinity Academy, Halifax, Tony also leads a team of mastery experts who help schools across the UK to develop teaching for mastery via nationally recognised CPD courses, problem-solving and reasoning resources, schemes of work, assessment materials and other tools.

Josh Lury, Lead Author

Josh is a specialist maths teacher, author and maths consultant with a passion for innovative and effective maths education.

The first edition of *Power Maths* was developed by a team of experienced authors, including:

- **Tony Staneff and Josh Lury**
- **Trinity Academy Halifax** (Michael Gosling CEO, Emily Fox, Kate Henshall, Rebecca Holland, Stephanie Kirk, Stephen Monaghan and Rachel Webster)
- **David Board, Belle Cottingham, Jonathan East, Tim Handley, Derek Huby, Neil Jarrett, Stephen Monaghan, Beth Smith, Tim Weal, Paul Wrangles** – skilled maths teachers and mastery experts
- **Cherri Moseley** – a maths author, former teacher and professional development provider
- **Professors Liu Jian and Zhang Dan**, Series Consultants and authors, and their team of mastery expert authors: **Wei Huinv, Huang Lihua, Zhu Dejiang, Zhu Yuhong, Hou Huiying, Yin Lili, Zhang Jing, Zhou Da and Liu Qimeng**

  Used by over 20 million children, Professor Liu Jian's textbook programme is one of the most popular in China. He and his author team are highly experienced in intelligent practice and in embedding key maths concepts using a C-P-A approach.

- **A group of 15 teachers and maths co-ordinators**

  We consulted our teacher group throughout the development of *Power Maths* to ensure we are meeting their real needs in the classroom.

What is *Power Maths*?

Created especially for UK primary schools, and aligned with the new National Curriculum, *Power Maths* is a whole-class, textbook-based mastery resource that empowers every child to understand and succeed. *Power Maths* rejects the notion that some people simply 'can't do' maths. Instead, it develops growth mindsets and encourages hard work, practice and a willingness to see mistakes as learning tools.

Best practice consistently shows that mastery of small, cumulative steps builds a solid foundation of deep mathematical understanding. *Power Maths* combines interactive teaching tools, high-quality textbooks and continuing professional development (CPD) to help you equip children with a deep and long-lasting understanding. Based on extensive evidence, and developed in partnership with practising teachers, *Power Maths* ensures that it meets the needs of children in the UK.

Power Maths and Mastery

Power Maths makes mastery practical and achievable by providing the structures, pathways, content, tools and support you need to make it happen in your classroom.

To develop mastery in maths children must be enabled to acquire a deep understanding of maths concepts, structures and procedures, step by step. Complex mathematical concepts are built on simpler conceptual components and when children understand every step in the learning sequence, maths becomes transparent and makes logical sense. Interactive lessons establish deep understanding in small steps, as well as effortless fluency in key facts such as tables and number bonds. The whole class works on the same content and no child is left behind.

Power Maths

- Builds every concept in small, progressive steps
- Is built with interactive, whole-class teaching in mind
- Provides the tools you need to develop growth mindsets
- Helps you check understanding and ensure that every child is keeping up
- Establishes core elements such as intelligent practice and reflection

The *Power Maths* approach

Everyone can!

Founded on the conviction that every child can achieve, *Power Maths* enables children to build number fluency, confidence and understanding, step by step.

Child-centred learning

Children master concepts one step at a time in lessons that embrace a concrete-pictorial-abstract (C-P-A) approach, avoid overload, build on prior learning and help them see patterns and connections. Same-day intervention ensures sustained progress.

Continuing professional development

Embedded teacher support and development offer every teacher the opportunity to continually improve their subject knowledge and manage whole-class teaching for mastery.

Whole-class teaching

An interactive, whole-class teaching model encourages thinking and precise mathematical language and allows children to deepen their understanding as far as they can.

What's different in the new edition?

If you have previously used the first editions of *Power Maths*, you might be interested to know how this edition is different. All of the improvements described below are based on feedback from *Power Maths* customers.

Changes to units and the progression

⚡ The order of units has been slightly adjusted, creating closer alignment between adjacent year groups, which will be useful for mixed age teaching.

⚡ The flow of lessons has been improved within units to optimise the pace of the progression and build in more recap where needed. For key topics, the sequence of lessons gives more opportunities to build up a solid base of understanding. Other units have fewer lessons than before, where appropriate, making it possible to fit in all the content.

⚡ Overall, the lessons put more focus on the most essential content for that year, with less time given to non-statutory content.

⚡ The progression of lessons matches the steps in the new White Rose Maths schemes of learning.

Lesson resources

⚡ There is a Quick recap for each lesson in the Teacher Guide, which offers an alternative lesson starter to the Power Up for cases where you feel it would be more beneficial to surface prerequisite learning than general number fluency.

⚡ In the **Discover** and **Share** sections there is now more of a progression from 1 a) to 1 b). Whereas before, 1 b) was mainly designed as a separate question, now 1 a) leads directly into 1 b). This means that there is an improved whole-class flow, and also an opportunity to focus on the logic and skills in more detail. As a teacher, you will be using 1 a) to lead the class into the thinking, then 1 b) to mould that thinking into the core new learning of the lesson.

⚡ In the **Share** section, for KS1 in particular, the number of different models and representations has been reduced, to support the clarity of thinking prompted by the flow from 1 a) into 1 b).

⚡ More fluency questions have been built into the guided and independent practice.

⚡ Pupil pages are as easy as possible for children to access independently. The pages are less full where this supports greater focus on key ideas and instructions. Also, more freedom is offered around answer format, with fewer boxes scaffolding children's responses; squared paper backgrounds are used in the Practice Books where appropriate. Artwork has also been revisited to ensure the highest standards of accessibility.

New components

480 Individual Practice Games are available in *ActiveLearn* for practising key facts and skills in Years 1 to 6. These are designed in an arcade style, to feel like fun games that children would choose to play outside school. They can be accessed via the Pupil World for homework or additional practice in school – and children can earn rewards. There are Support, Core and Extend levels to allocate, with Activity Reporting available for the teacher. There is a Quick Guide on *ActiveLearn* and you can use the Help area for support in setting up child accounts.

There is also a new set of lesson video resources on the Professional Development tile, designed for in-school training in 10- to 20-minute bursts. For each part of the *Power Maths* lesson sequence, there is a slide deck with embedded video, which will facilitate discussions about how you can take your *Power Maths* teaching to the next level.

Your *Power Maths* resources

To help you teach for mastery, *Power Maths* comprises a variety of high-quality resources.

Pupil Textbooks

Discover, **Share** and **Think together** sections promote discussion and introduce mathematical ideas logically, so that children understand more easily.

Using a Concrete-Pictorial-Abstract approach, clear mathematical models help children to make connections and grasp concepts.

Appealing scenarios stimulate curiosity, helping children to identify the maths problem and discover patterns and relationships for themselves.

Friendly, supportive characters help children develop a growth mindset by prompting them to think, reason and reflect.

The coherent *Power Maths* lesson structure carries through into the vibrant, high-quality textbooks. Setting out the core learning objectives for each class, the lesson structure follows a carefully mapped journey through the curriculum and supports children on their journey to deeper understanding.

Pupil Practice Books

The Practice Books offer just the right amount of intelligent practice for children to complete independently in the final section of each lesson.

Practice questions are finely tuned to move children forward in their thinking and to reveal misconceptions.

The practice questions are for everyone – each question varies one small element to move children on in their thinking.

Calculations are connected so that children think about the underlying concept.

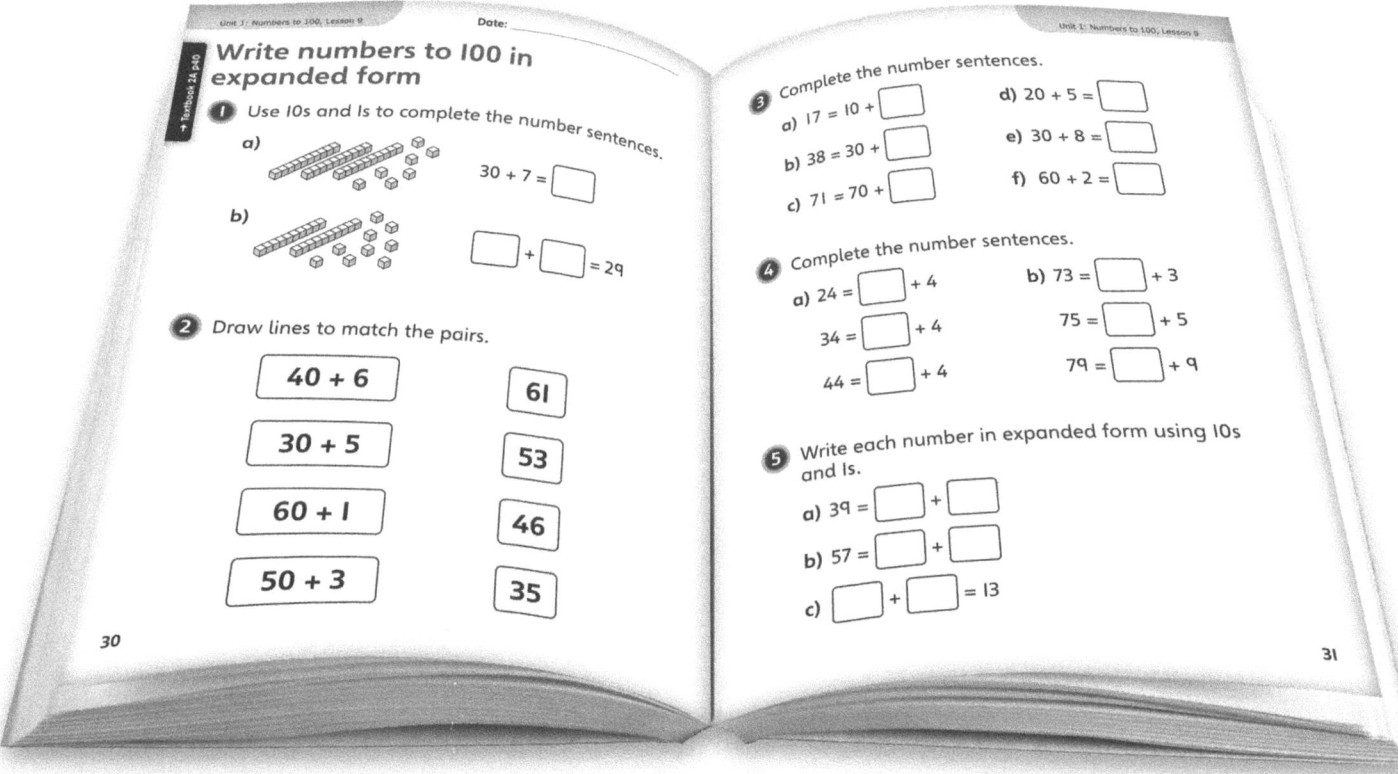

The *Power Maths* characters support and encourage children to think and work in different ways.

Challenge questions allow children to delve deeper into a concept.

Reflect questions reveal the depth of each child's understanding before they move on.

Think differently questions encourage children to use reasoning as well as their mathematical knowledge to reach a solution.

Online subscription

The online subscription will give you access to additional resources and answers from the Textbook and Practice Book.

eTextbooks

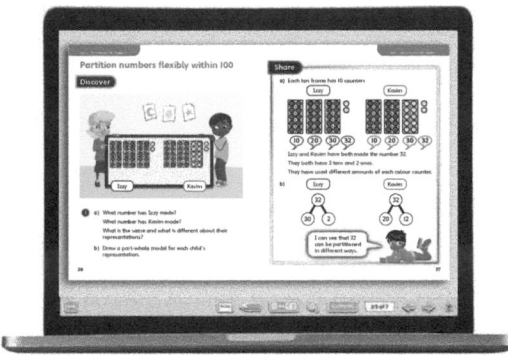

Digital versions of *Power Maths* Textbooks allow class groups to share and discuss questions, solutions and strategies. They allow you to project key structures and representations at the front of the class, to ensure all children are focusing on the same concept.

Teaching tools

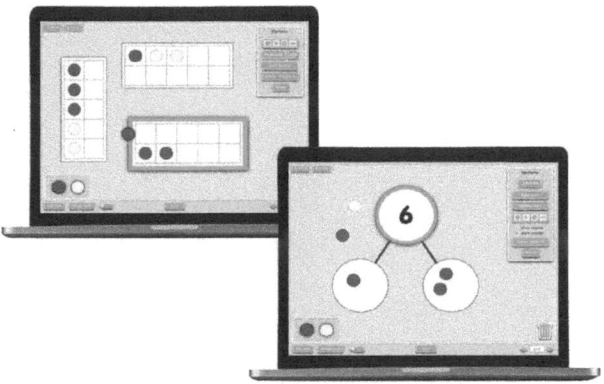

Here you will find interactive versions of key *Power Maths* structures and representations.

Power Ups

Use this series of daily activities to promote and check number fluency.

Online versions of Teacher Guide pages

PDF pages give support at both unit and lesson levels. You will also find help with key strategies and templates for tracking progress.

Unit videos

Watch the professional development videos at the start of each unit to help you teach with confidence. The videos explore common misconceptions in the unit, and include intervention suggestions as well as suggestions on what to look out for when assessing mastery in your students.

End of unit Strengthen and Deepen materials

The Strengthen activity at the end of every unit addresses a key misconception and can be used to support children who need it. The Deepen activities are designed to be low ceiling/high threshold and will challenge those children who can understand more deeply. These resources will help you ensure that every child understands and will help you keep the class moving forward together. These printable activities provide an optional resource bank for use after the assessment stage.

Individual Practice Games

These enjoyable games can be used at home or at school to embed key number skills (see page 6).

Professional Development videos and slides

These slides and videos of *Power Maths* lessons can be used for ongoing training in short bursts or to support new staff (see page 6).

The *Power Maths* teaching model

At the heart of *Power Maths* is a clearly structured teaching and learning process that helps you make certain that every child masters each maths concept securely and deeply. For each year group, the curriculum is broken down into core concepts, taught in units. A unit divides into smaller learning steps – lessons. Step by step, strong foundations of cumulative knowledge and understanding are built.

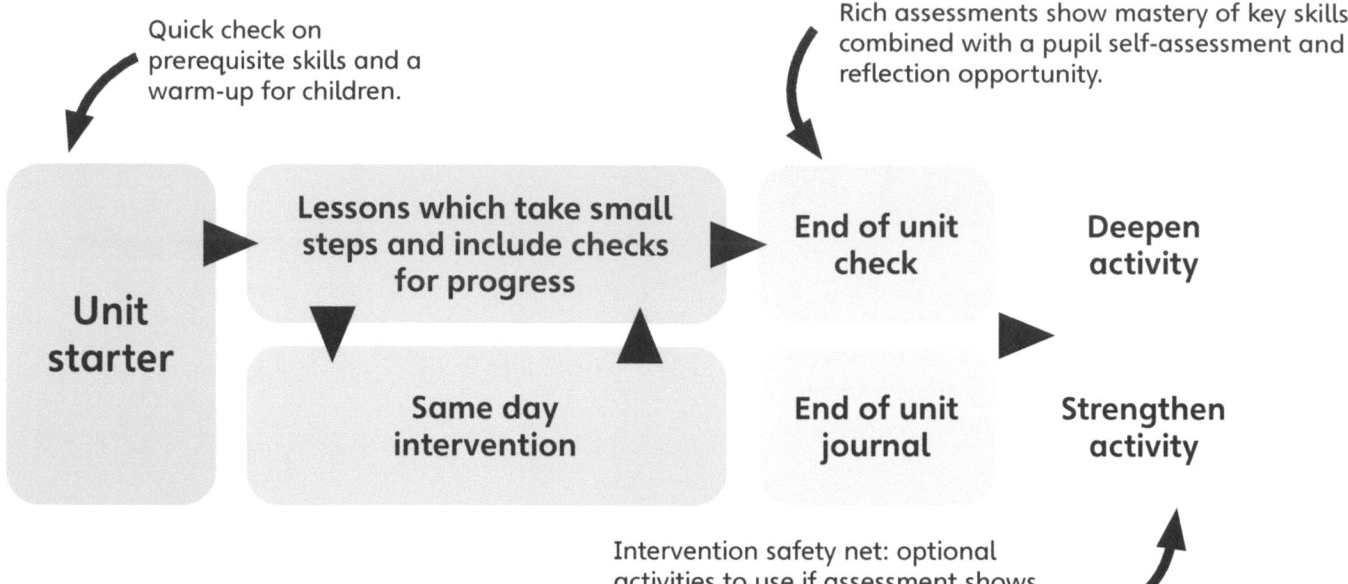

Quick check on prerequisite skills and a warm-up for children.

Rich assessments show mastery of key skills combined with a pupil self-assessment and reflection opportunity.

Unit starter

Lessons which take small steps and include checks for progress

Same day intervention

End of unit check

End of unit journal

Deepen activity

Strengthen activity

Intervention safety net: optional activities to use if assessment shows some children still have misconceptions.

Unit starter

Each unit begins with a unit starter, which introduces the learning context along with key mathematical vocabulary and structures and representations.

- The Textbooks include a check on readiness and a warm-up task for children to complete.
- Your Teacher Guide gives support right from the start on important structures and representations, mathematical language, common misconceptions and intervention strategies.
- Unit-specific videos develop your subject knowledge and insights so you feel confident and fully equipped to teach each new unit. These are available via the online subscription.

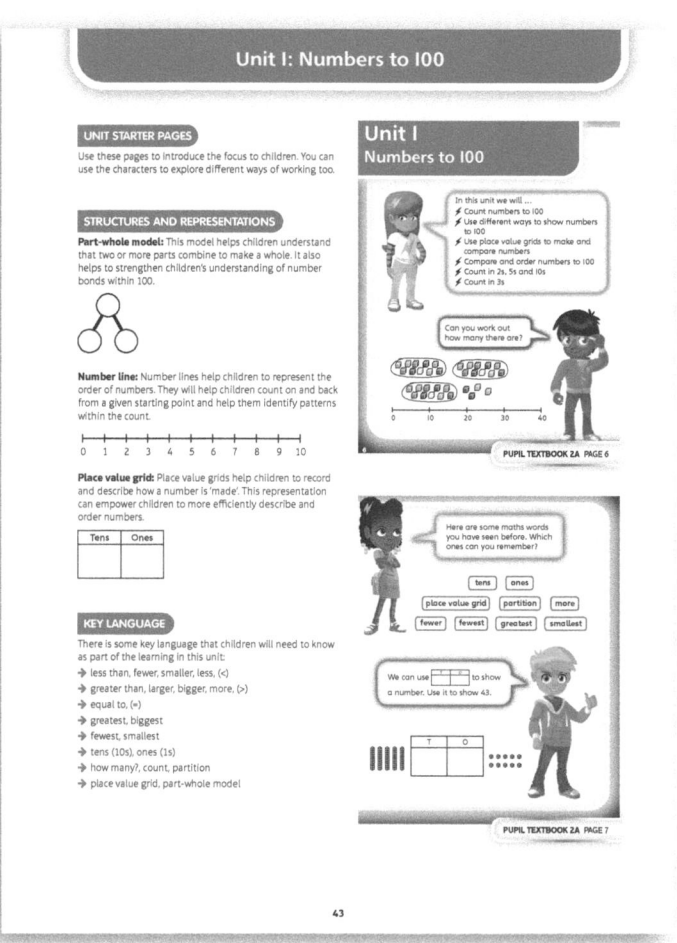

Lesson

Once a unit has been introduced, it is time to start teaching the series of lessons.

• Each lesson is scaffolded with Textbook and Practice Book activities and begins with a Power Up activity (available via online subscription) or the Quick recap activity in the Teacher Guide (see page 15).

• *Power Maths* identifies lesson by lesson what concepts are to be taught.

• Your Teacher Guide offers lots of support for you to get the most from every child in every lesson. As well as highlighting key points, tricky areas and how to handle them, you will also find question prompts to check on understanding and clarification on why particular activities and questions are used.

Same-day intervention

Same-day interventions are vital in order to keep the class progressing together. This can be during the lesson as well as afterwards (see page 28). Therefore, *Power Maths* provides plenty of support throughout the journey.

• Intervention is focused on keeping up now, not catching up later, so interventions should happen as soon as they are needed.

• Practice section questions are designed to bring misconceptions to the surface, allowing you to identify these easily as you circulate during independent practice time.

• Child-friendly assessment questions in the Teacher Guide help you identify easily which children need to strengthen their understanding.

End of unit check and journal

For each unit, the End of unit check in the Textbook lets you see which children have mastered the key concepts, which children have not and where their misconceptions lie. The Practice Books also include an End of unit journal in which children can reflect on what they have learned. Each unit also offers Strengthen and Deepen activities, available via the online subscription.

> The Teacher Guide offers different ways of managing the End of unit assessments as well as giving support with handling misconceptions.

> The End of unit check presents multiple-choice questions. Children think about their answer, decide on a solution and explain their choice.

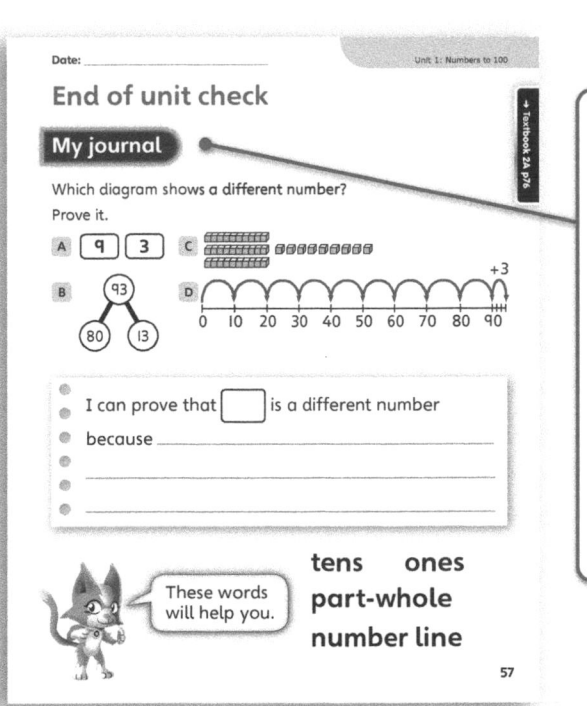

> The End of unit journal is an opportunity for children to test out their learning and reflect on how they feel about it. Tackling the 'journal' problem reveals whether a child understands the concept deeply enough to move on to the next unit.

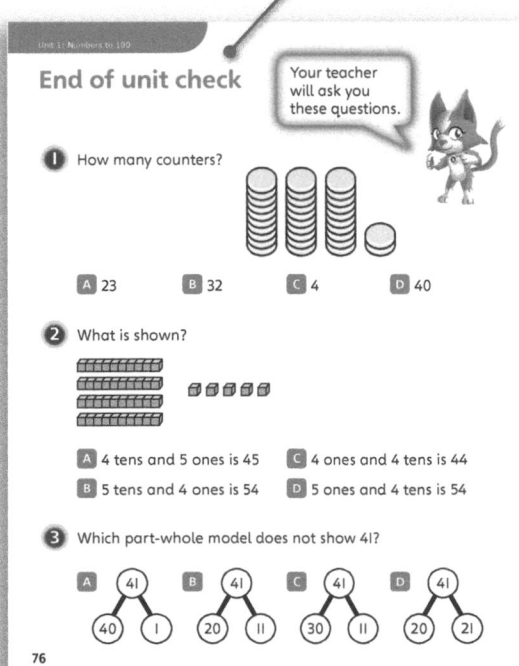

The *Power Maths* lesson sequence

At the heart of *Power Maths* is a unique lesson sequence designed to empower children to understand core concepts and grow in confidence. Embracing the National Centre for Excellence in the Teaching of Mathematics' (NCETM's) definition of mastery, the sequence guides and shapes every *Power Maths* lesson you teach.

Flexibility is built into the *Power Maths* programme so there is no one-to-one mapping of lessons and concepts and you can pace your teaching according to your class. While some children will need to spend longer on a particular concept (through interventions or additional lessons), others will reach deeper levels of understanding. However, it is important that the class moves forward together through the termly schedules.

Power Up ⏱ 5 minutes

Each lesson begins with a Power Up activity (available via the online subscription) which supports fluency in key number facts.

The whole-class approach depends on fluency, so the Power Up is a powerful and essential activity.

The Quick recap is an alternative starter, for when you think some or all children would benefit more from revisiting pre-requisite work (see page 15).

TOP TIP

If the class is struggling with the task, revisit it later and check understanding.

Power Ups reinforce the two key things that are essential for success: times-tables and number bonds.

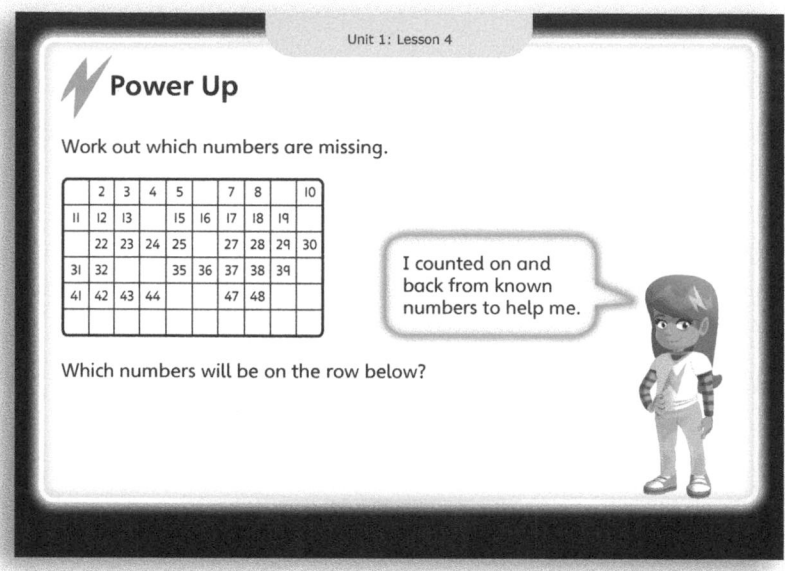

Discover ⏱ 10 minutes

A practical, real-life problem arouses curiosity. Children find the maths through story telling.

TOP TIP

Discover works best when run at tables, in pairs with concrete objects.

Question ❶ a) tackles the key concept and question ❶ b) digs a little deeper. Children have time to explore, play and discuss possible strategies.

Share 🕑 10 minutes

Teacher-led, this interactive section follows the **Discover** activity and highlights the variety of methods that can be used to solve a single problem.

TOP TIP
You can use the carpet area if you have this. Pairs sharing a textbook is a great format for **Share**!

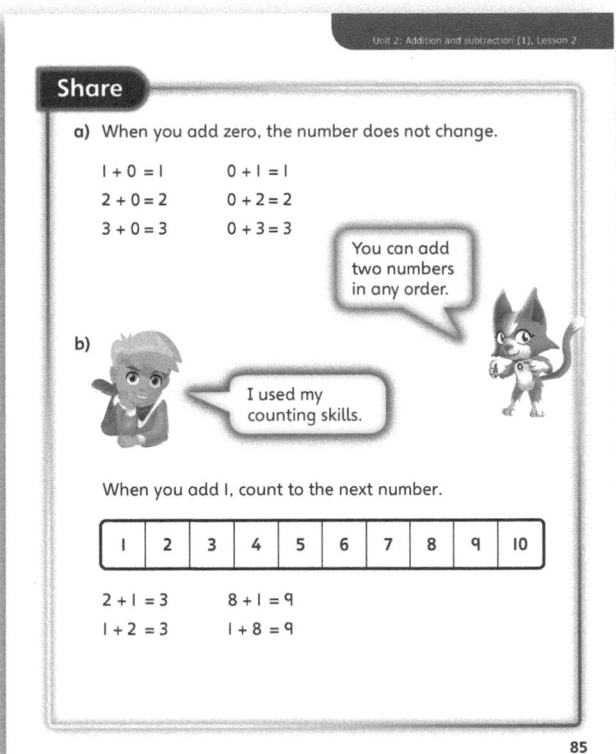

Your Teacher Guide gives target questions for children. The online toolkit provides interactive structures and representations to link concrete and pictorial to abstract concepts.

Bring children to the front to share and celebrate their solutions and strategies.

Think together

🕑 10 minutes

Children work in groups on the carpet or at tables, using their textbooks or eBooks.

TOP TIP
Make sure children have mini whiteboards or pads to write on if they are not at their tables.

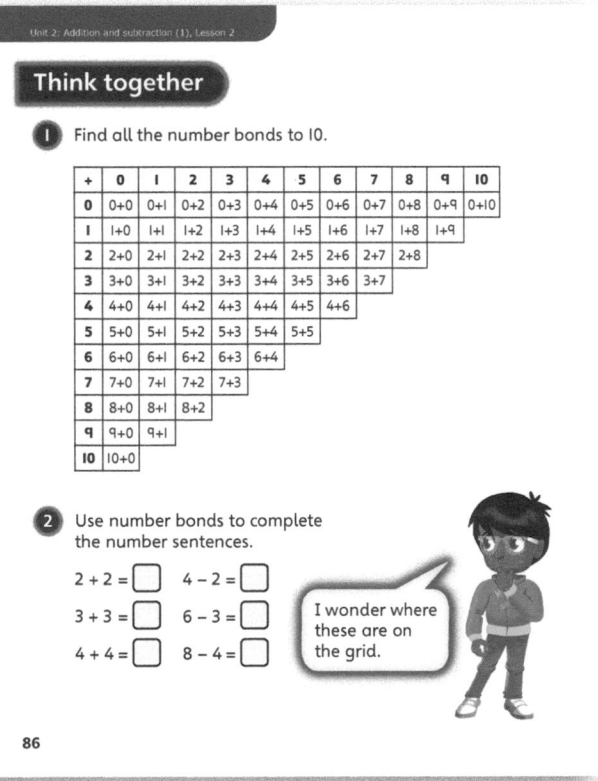

Using the Teacher Guide, model question ❶ for your class.

Question ❷ is less structured. Children will need to think together in their groups, then discuss their methods and solutions as a class.

Question ❸ – the openness of the **Challenge** question helps to check depth of understanding.

Practice ⏱ 15 minutes

Using their Practice Books, children work independently while you circulate and check on progress.

Questions follow small steps of progression to deepen learning.

TOP TIP
Some children could work separately with a teacher or assistant.

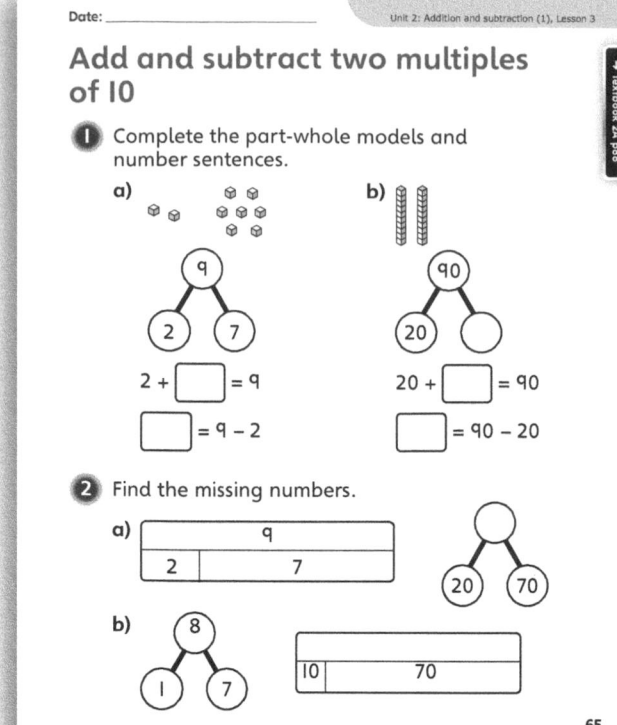

Date: _____

Add and subtract two multiples of 10

→ Textbook 2A p88

1 Complete the part-whole models and number sentences.

a)

9
2 7

$2 + \boxed{} = 9$

$\boxed{} = 9 - 2$

b)

90
20

$20 + \boxed{} = 90$

$\boxed{} = 90 - 20$

2 Find the missing numbers.

a)

9	
2	7

20 70

b)

8
1 7

10	70

65

Are some children struggling? If so, work with them as a group, using mathematical structures and representations to support understanding as necessary.

There are no set routines: for real understanding, children need to think about the problem in different ways.

Reflect ⏱ 5 minutes

'Spot the mistake' questions are great for checking misconceptions.

The **Reflect** section is your opportunity to check how deeply children understand the target concept.

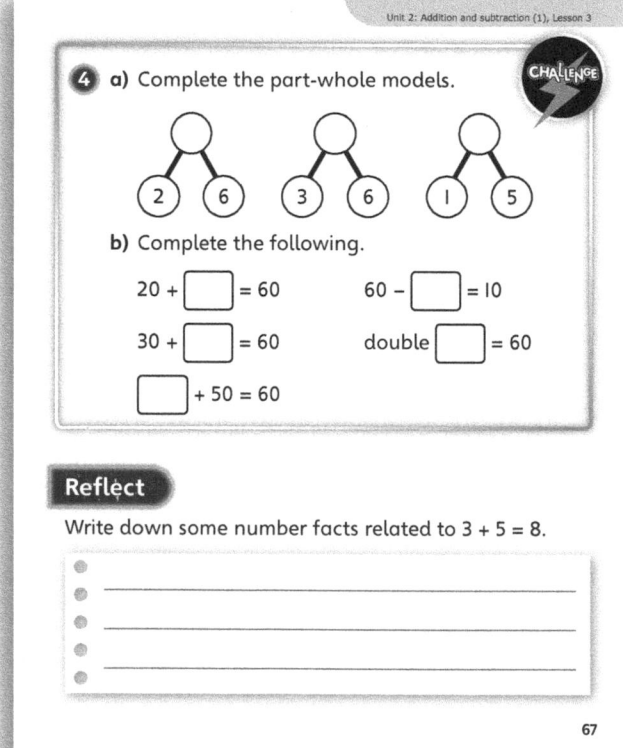

4 a) Complete the part-whole models. **CHALLENGE**

2 6   3 6   1 5

b) Complete the following.

$20 + \boxed{} = 60$ $60 - \boxed{} = 10$

$30 + \boxed{} = 60$ double $\boxed{} = 60$

$\boxed{} + 50 = 60$

Reflect

Write down some number facts related to $3 + 5 = 8$.

_____

_____

_____

_____

67

The Practice Books use various approaches to check that children have fully understood each concept.

Looking like they understand is not enough! It is essential that children can show they have grasped the concept.

Using the *Power Maths* Teacher Guide

Think of your Teacher Guides as *Power Maths* handbooks that will guide, support and inspire your day-to-day teaching. Clear and concise, and illustrated with helpful examples, your Teacher Guides will help you make the best possible use of every individual lesson. They also provide wrap-around professional development, enhancing your own subject knowledge and helping you to grow in confidence about moving your children forward together.

There is a Teacher Guide per year group for every term, with unit and lesson level guidance and support.

Never feel stuck! You will find ideas for introducing every unit and lesson and questions to encourage teacher reflection before and after each lesson.

Tips and advice on key elements such as C-P-A approaches, misconceptions, language, modelling growth mindsets and same day intervention.

Annotations for every Textbook and Practice Book page, providing prompts for key questions to ask to expose understanding and explanations as to why key questions have been chosen.

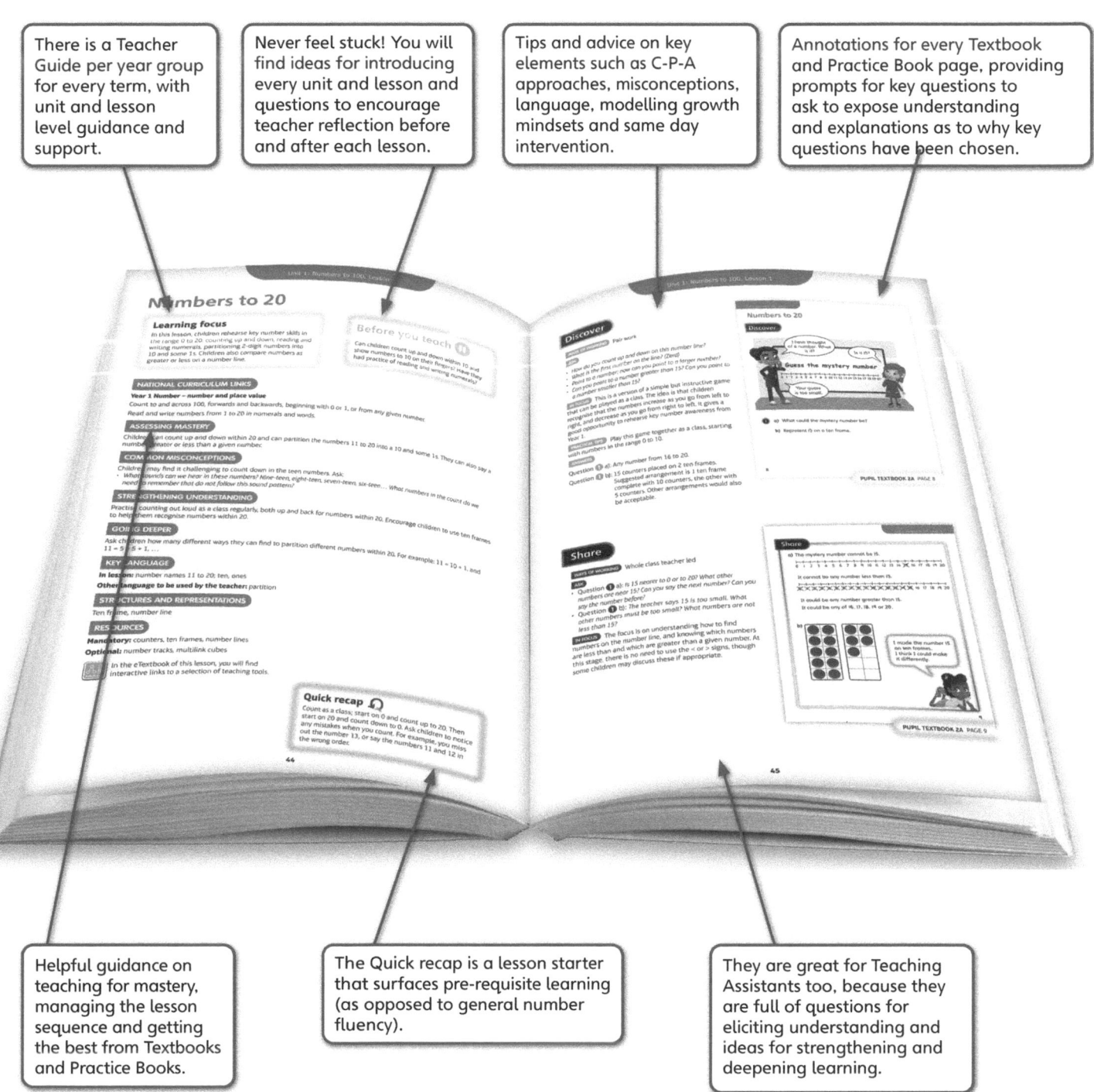

Helpful guidance on teaching for mastery, managing the lesson sequence and getting the best from Textbooks and Practice Books.

The Quick recap is a lesson starter that surfaces pre-requisite learning (as opposed to general number fluency).

They are great for Teaching Assistants too, because they are full of questions for eliciting understanding and ideas for strengthening and deepening learning.

At the end of each unit, your Teacher Guide helps you identify who has fully grasped the concept, who has not and how to move every child forward. This is covered later in the Assessment strategies section.

Power Maths Year 2, yearly overview

Textbook	Strand	Unit		Number of lessons
Textbook A / Practice Book A (Term 1)	Number – number and place value	1	Numbers to 100	17
	Number – addition and subtraction	2	Addition and subtraction (1)	13
	Number – addition and subtraction	3	Addition and subtraction (2)	12
	Geometry – properties of shape	4	Properties of shapes	12
Textbook B / Practice Book B (Term 2)	Measurement	5	Money	10
	Number – multiplication and division	6	Multiplication and division (1)	8
	Number – multiplication and division	7	Multiplication and division (2)	10
	Measurement	8	Length and height	5
	Measurement	9	Mass, capacity and temperature	8
	Statistics	10	Statistics	7
Textbook C / Practice Book C (Term 3)	Number – fractions	11	Fractions	15
	Geometry – position and direction	12	Position and direction	5
	Measurement	13	Time	8
	Number – addition and subtraction	14	Problem solving and efficient methods	12

Power Maths Year 2, Textbook 2A (Term I) overview

Strand	Unit		Lesson number	Lesson title	NC Objective 1	NC Objective 2
Number – number and place value	Unit 1	Numbers to 100	1	Numbers to 20	Count to and across 100, forwards and backwards, beginning with 0 or 1, or from any given number (Year 1)	Read and write numbers from 1 to 20 in numerals and words (Year 1)
Number – number and place value	Unit 1	Numbers to 100	2	Count in 10s	Count, read and write numbers to 100 in numerals; count in multiples of twos, fives and tens (Year 1)	
Number – number and place value	Unit 1	Numbers to 100	3	Count in 10s and 1s	Recognise the place value of each digit in a two-digit number (tens, ones)	Identify, represent and estimate numbers using different representations, including the number line
Number – number and place value	Unit 1	Numbers to 100	4	Recognise 10s and 1s	Recognise the place value of each digit in a two-digit number (tens, ones)	Identify, represent and estimate numbers using different representations, including the number line
Number – number and place value	Unit 1	Numbers to 100	5	Build a number from 10s and 1s	Recognise the place value of each digit in a two-digit number (tens, ones)	Identify, represent and estimate numbers using different representations, including the number line
Number – number and place value	Unit 1	Numbers to 100	6	Use a place value grid	Recognise the place value of each digit in a two-digit number (tens, ones)	Identify, represent and estimate numbers using different representations, including the number line
Number – number and place value	Unit 1	Numbers to 100	7	Partition numbers to 100	Recognise the place value of each digit in a two-digit number (tens, ones)	Identify, represent and estimate numbers using different representations, including the number line
Number – number and place value	Unit 1	Numbers to 100	8	Partition numbers flexibly within 100	Recognise the place value of each digit in a two-digit number (tens, ones)	Identify, represent and estimate numbers using different representations, including the number line
Number – number and place value	Unit 1	Numbers to 100	9	Write numbers to 100 in expanded form	Recognise the place value of each digit in a two-digit number (tens, ones)	Read and write numbers to at least 100 in numerals and in words
Number – number and place value	Unit 1	Numbers to 100	10	10s on a number line to 100	Identify, represent and estimate numbers using different representations, including the number line	
Number – number and place value	Unit 1	Numbers to 100	11	10s and 1s on a number line to 100	Identify, represent and estimate numbers using different representations, including the number line	Recognise the place value of each digit in a two-digit number (tens, ones)
Number – number and place value	Unit 1	Numbers to 100	12	Estimate numbers on a number line	Identify, represent and estimate numbers using different representations, including the number line	
Number – number and place value	Unit 1	Numbers to 100	13	Compare numbers (1)	Compare and order numbers from 0 up to 100; use <, > and = signs	Identify, represent and estimate numbers using different representations, including the number line
Number – number and place value	Unit 1	Numbers to 100	14	Compare numbers (2)	Compare and order numbers from 0 up to 100; use <, > and = signs	
Number – number and place value	Unit 1	Numbers to 100	15	Order numbers	Compare and order numbers from 0 up to 100; use <, > and = signs	
Number – number and place value	Unit 1	Numbers to 100	16	Count in 2s, 5s and 10s	Count in steps of 2, 3, and 5 from 0, and in tens from any number, forward and backward	
Number – number and place value	Unit 1	Numbers to 100	17	Count in 3s	Count in steps of 2, 3, and 5 from 0, and in tens from any number, forward and backward	

Strand	Unit		Lesson number	Lesson title	NC Objective 1	NC Objective 2
Number – addition and subtraction	Unit 2	Addition and subtraction (1)	1	Fact families	Recall and use addition and subtraction facts to 20 fluently, and derive and use related facts up to 100	
Number – addition and subtraction	Unit 2	Addition and subtraction (1)	2	Learn number bonds	Recall and use addition and subtraction facts to 20 fluently, and derive and use related facts up to 100	
Number – addition and subtraction	Unit 2	Addition and subtraction (1)	3	Add two multiples of 10	Recall and use addition and subtraction facts to 20 fluently, and derive and use related facts up to 100	
Number – addition and subtraction	Unit 2	Addition and subtraction (1)	4	Complements to 100 (tens)	Recall and use addition and subtraction facts to 20 fluently, and derive and use related facts up to 100	
Number – addition and subtraction	Unit 2	Addition and subtraction (1)	5	Add and subtract 1s	Add and subtract numbers using concrete objects, pictorial representations, and mentally, including: a two-digit number and ones	Solve problems with addition and subtraction: using concrete objects and pictorial representations, including those involving numbers, quantities and measures
Number – addition and subtraction	Unit 2	Addition and subtraction (1)	6	Add by making 10	Add and subtract numbers using concrete objects, pictorial representations, and mentally, including: a two-digit number and ones	Solve problems with addition and subtraction: using concrete objects and pictorial representations, including those involving numbers, quantities and measures
Number – addition and subtraction	Unit 2	Addition and subtraction (1)	7	Add using a number line	Add and subtract numbers using concrete objects, pictorial representations, and mentally, including: two two-digit numbers	Solve problems with addition and subtraction: applying their increasing knowledge of mental and written methods
Number – addition and subtraction	Unit 2	Addition and subtraction (1)	8	Add three 1-digit numbers	Add and subtract numbers using concrete objects, pictorial representations, and mentally, including: two two-digit numbers	Solve problems with addition and subtraction: applying their increasing knowledge of mental and written methods
Number – addition and subtraction	Unit 2	Addition and subtraction (1)	9	Add to the next 10	Add and subtract numbers using concrete objects, pictorial representations, and mentally, including: a two-digit number and ones	
Number – addition and subtraction	Unit 2	Addition and subtraction (1)	10	Add across a 10	Add and subtract numbers using concrete objects, pictorial representations, and mentally, including: a two-digit number and ones	Solve problems with addition and subtraction: using concrete objects and pictorial representations, including those involving numbers, quantities and measures
Number – addition and subtraction	Unit 2	Addition and subtraction (1)	11	Subtract across a 10	Add and subtract numbers using concrete objects, pictorial representations, and mentally, including: a two-digit number and ones	Solve problems with addition and subtraction: using concrete objects and pictorial representations, including those involving numbers, quantities and measures
Number – addition and subtraction	Unit 2	Addition and subtraction (1)	12	Subtract from a 10	Add and subtract numbers using concrete objects, pictorial representations, and mentally, including: two two-digit numbers	Solve problems with addition and subtraction: applying their increasing knowledge of mental and written methods
Number – addition and subtraction	Unit 2	Addition and subtraction (1)	13	Subtract a 1-digit number from a 2-digit number – across 10	Add and subtract numbers using concrete objects, pictorial representations, and mentally, including: a two-digit number and ones	Solve problems with addition and subtraction: using concrete objects and pictorial representations, including those involving numbers, quantities and measures
Number – addition and subtraction	Unit 3	Addition and subtraction (2)	1	10 more, 10 less	Count in steps of 2, 3, and 5 from 0, and in tens from any number, forward and backward	Solve problems with addition and subtraction: using concrete objects and pictorial representations, including those involving numbers, quantities and measures

Strand	Unit		Lesson number	Lesson title	NC Objective 1	NC Objective 2
Number – addition and subtraction	Unit 3	Addition and subtraction (2)	2	Add and subtract 10s	Add and subtract numbers using concrete objects, pictorial representations, and mentally, including: a two-digit number and tens	Solve problems with addition and subtraction: using concrete objects and pictorial representations, including those involving numbers, quantities and measures
Number – addition and subtraction	Unit 3	Addition and subtraction (2)	3	Add two 2-digit numbers – add 10s and add 1s	Add and subtract numbers using concrete objects, pictorial representations, and mentally, including: a two-digit number and tens	Solve problems with addition and subtraction: using concrete objects and pictorial representations, including those involving numbers, quantities and measures
Number – addition and subtraction	Unit 3	Addition and subtraction (2)	4	Add two 2-digit numbers – add more 10s then more 1s	Add and subtract numbers using concrete objects, pictorial representations, and mentally, including: a two-digit number and tens	Solve problems with addition and subtraction: using concrete objects and pictorial representations, including those involving numbers, quantities and measures
Number – addition and subtraction	Unit 3	Addition and subtraction (2)	5	Subtract a 2-digit number from a 2-digit number – not across 10	Add and subtract numbers using concrete objects, pictorial representations, and mentally, including: a two-digit number and tens	Solve problems with addition and subtraction: using concrete objects and pictorial representations, including those involving numbers, quantities and measures
Number – addition and subtraction	Unit 3	Addition and subtraction (2)	6	Subtract a 2-digit number from a 2-digit number – across 10	Add and subtract numbers using concrete objects, pictorial representations, and mentally, including: a two-digit number and tens	Solve problems with addition and subtraction: using concrete objects and pictorial representations, including those involving numbers, quantities and measures
Number – addition and subtraction	Unit 3	Addition and subtraction (2)	7	How many more? How many fewer?	Add and subtract numbers using concrete objects, pictorial representations, and mentally, including: a two-digit number and tens	Solve problems with addition and subtraction: using concrete objects and pictorial representations, including those involving numbers, quantities and measures
Number – addition and subtraction	Unit 3	Addition and subtraction (2)	8	Subtraction – find the difference	Solve problems with addition and subtraction: using concrete objects and pictorial representations, including those involving numbers, quantities and measures	
Number – addition and subtraction	Unit 3	Addition and subtraction (2)	9	Compare number sentences	Solve problems with addition and subtraction: using concrete objects and pictorial representations, including those involving numbers, quantities and measures	Recall and use addition and subtraction facts to 20 fluently, and derive and use related facts up to 100
Number – addition and subtraction	Unit 3	Addition and subtraction (2)	10	Missing number problems	Solve problems with addition and subtraction: using concrete objects and pictorial representations, including those involving numbers, quantities and measures	Recall and use addition and subtraction facts to 20 fluently, and derive and use related facts up to 100
Number – addition and subtraction	Unit 3	Addition and subtraction (2)	11	Mixed addition and subtraction	Solve problems with addition and subtraction: using concrete objects and pictorial representations, including those involving numbers, quantities and measures	Solve problems with addition and subtraction: applying their increasing knowledge of mental and written methods
Number – addition and subtraction	Unit 3	Addition and subtraction (2)	12	Two-step problems	Solve problems with addition and subtraction: using concrete objects and pictorial representations, including those involving numbers, quantities and measures	Solve problems with addition and subtraction: applying their increasing knowledge of mental and written methods
Geometry – properties of shape	Unit 4	Properties of shapes	1	Recognise 2D and 3D shapes	Compare and sort common 2D and 3D shapes and everyday objects.	

Strand	Unit		Lesson number	Lesson title	NC Objective 1	NC Objective 2
Geometry – properties of shape	Unit 4	Properties of shapes	2	Count sides on 2D shapes	Identify and describe the properties of 2D shapes, including the number of sides and line symmetry in a vertical line	
Geometry – properties of shape	Unit 4	Properties of shapes	3	Count vertices on 2D shapes	Identify and describe the properties of 2D shapes, including the number of sides and line symmetry in a vertical line	
Geometry – properties of shape	Unit 4	Properties of shapes	4	Draw 2D shapes	Identify and describe the properties of 2D shapes, including the number of sides and line symmetry in a vertical line	
Geometry – properties of shape	Unit 4	Properties of shapes	5	Lines of symmetry on shapes	Identify and describe the properties of 2D shapes, including the number of sides and line symmetry in a vertical line	
Geometry – properties of shape	Unit 4	Properties of shapes	6	Sort 2D shapes	Compare and sort common 2D and 3D shapes and everyday objects	
Geometry – properties of shape	Unit 4	Properties of shapes	7	Make patterns with 2D shapes	Order and arrange combinations of mathematical objects in patterns and sequences	
Geometry – properties of shape	Unit 4	Properties of shapes	8	Count faces on 3D shapes	Identify and describe the properties of 3D shapes, including the number of edges, vertices and faces	
Geometry – properties of shape	Unit 4	Properties of shapes	9	Count edges on 3D shapes	Identify and describe the properties of 3D shapes, including the number of edges, vertices and faces	
Geometry – properties of shape	Unit 4	Properties of shapes	10	Count vertices on 3D shapes	Identify and describe the properties of 3D shapes, including the number of edges, vertices and faces	
Geometry – properties of shape	Unit 4	Properties of shapes	11	Sort 3D shapes	Compare and sort common 2D and 3D shapes and everyday objects	
Geometry – properties of shape	Unit 4	Properties of shapes	12	Make patterns with 3D shapes	Order and arrange combinations of mathematical objects in patterns and sequences	

Mindset: an introduction

Global research and best practice deliver the same message: learning is greatly affected by what learners perceive they can or cannot do. What is more, it is also shaped by what their parents, carers and teachers perceive they can do. Mindset – the thinking that determines our beliefs and behaviours – therefore has a fundamental impact on teaching and learning.

Everyone can!

Power Maths and mastery methods focus on the distinction between 'fixed' and 'growth' mindsets (Dweck, 2007).[1] Those with a fixed mindset believe that their basic qualities (for example, intelligence, talent and ability to learn) are pre-wired or fixed: 'If you have a talent for maths, you will succeed at it. If not, too bad!' By contrast, those with a growth mindset believe that hard work, effort and commitment drive success and that 'smart' is not something you are or are not, but something you become. In short, everyone can do maths!

Key mindset strategies

A growth mindset needs to be actively nurtured and developed. *Power Maths* offers some key strategies for fostering healthy growth mindsets in your classroom.

It is okay to get it wrong

Mistakes are valuable opportunities to re-think and understand more deeply. Learning is richer when children and teachers alike focus on spotting and sharing mistakes as well as solutions.

Praise hard work

Praise is a great motivator, and by focusing on praising effort and learning rather than success, children will be more willing to try harder, take risks and persist for longer.

Mind your language!

The language we use around learners has a profound effect on their mindsets. Make a habit of using growth phrases, such as, 'Everyone can!', 'Mistakes can help you *learn*' and 'Just try for a little longer'. The king of them all is one little word, 'yet'... I can't solve this...yet!' Encourage parents and carers to use the right language too.

Build in opportunities for success

The step-by-small-step approach enables children to enjoy the experience of success. In addition, avoid ability grouping and encourage every child to answer questions and explain or demonstrate their methods to others.

[1]Dweck, C (2007) *The New Psychology of Success*, Ballantine Books: New York

The *Power Maths* characters

The *Power Maths* characters model the traits of growth mindset learners and encourage resilience by prompting and questioning children as they work. Appearing frequently in the Textbooks and Practice Books, they are your allies in teaching and discussion, helping to model methods, alternatives and misconceptions, and to pose questions. They encourage and support your children, too: they are all hardworking, enthusiastic and unafraid of making and talking about mistakes.

Meet the team!

Creative Flo is open-minded and sometimes indecisive. She likes to think differently and come up with a variety of methods or ideas.

Determined Dexter is resolute, resilient and systematic. He concentrates hard, always tries his best and he'll never give up – even though he doesn't always choose the most efficient methods!

'Let's try again.'

'Mistakes are cool!'

'Have I found all of the solutions?'

'Let's try it this way…'

'Can we do it differently?'

'I've got another way of doing this!'

'I'm going to try this!'

'I know how to do that!'

'Want to share my ideas?'

Curious Ash is eager, interested and inquisitive, and he loves solving puzzles and problems. Ash asks lots of questions but sometimes gets distracted.

'What if we tried this…?'

'I wonder…'

'Is there a pattern here?'

Miaow!

Sparks the Cat

Brave Astrid is confident, willing to take risks and unafraid of failure. She's never scared to jump straight into a problem or question, and although she often makes simple mistakes she's happy to talk them through with others.

Mathematical language

Traditionally, we in the UK have tended to try simplifying mathematical language to make it easier for young children to understand. By contrast, evidence and experience show that by diluting the correct language, we actually mask concepts and meanings for children. We then wonder why they are confused by new and different terminology later down the line! *Power Maths* is not afraid of 'hard' words and avoids placing any barriers between children and their understanding of mathematical concepts. As a result, we need to be deliberate, precise and thorough in building every child's understanding of the language of maths. Throughout the Teacher Guides you will find support and guidance on how to deliver this, as well as individual explanations throughout the pupil Textbooks.

Use the following key strategies to build children's mathematical vocabulary, understanding and confidence.

Precise and consistent

Everyone in the classroom should use the correct mathematical terms in full, every time. For example, refer to 'equal parts', not 'parts'. Used consistently, precise maths language will be a familiar and non-threatening part of children's everyday experience.

Full sentences

Teachers and children alike need to use full sentences to explain or respond. When children use complete sentences, it both reveals their understanding and embeds their knowledge.

Stem sentences

These important sentences help children express mathematical concepts accurately, and are used throughout the *Power Maths* books. Encourage children to repeat them frequently, whether working independently or with others. Examples of stem sentences are:

'4 is a part, 5 is a part, 9 is the whole.'

'There are .... groups. There are .... in each group.'

Key vocabulary

The unit starters highlight essential vocabulary for every lesson. In the Pupil books, characters flag new terminology and the Teacher Guide lists important mathematical language for every unit and lesson. New terms are never introduced without a clear explanation.

Symbolic language

Symbols are used early on so that children quickly become familiar with them and their meaning. Often, the *Power Maths* characters will highlight the connection between language and particular symbols.

The role of talk and discussion

When children learn to talk purposefully together about maths, barriers of fear and anxiety are broken down and they grow in confidence, skills and understanding. Building a healthy culture of 'maths talk' empowers their learning from day one.

Explanation and discussion are integral to the *Power Maths* structure, so by simply following the books your lessons will stimulate structured talk. The following key 'maths talk' strategies will help you strengthen that culture and ensure that every child is included.

Sentences, not words

Encourage children to use full sentences when reasoning, explaining or discussing maths. This helps both speaker and listeners to clarify their own understanding. It also reveals whether or not the speaker truly understands, enabling you to address misconceptions as they arise.

Working together

Working with others in pairs, groups or as a whole class is a great way to support maths talk and discussion. Use different group structures to add variety and challenge. For example, children could take timed turns for talking, work independently alongside a 'discussion buddy', or perhaps play different *Power Maths* character roles within their group.

Think first – then talk

Provide clear opportunities within each lesson for children to think and reflect, so that their talk is purposeful, relevant and focused.

Give every child a voice

Where the 'hands up' model allows only the more confident child to shine, *Power Maths* involves everyone. Make sure that no child dominates and that even the shyest child is encouraged to contribute – and praised when they do.

Assessment strategies

Teaching for mastery demands that you are confident about what each child knows and where their misconceptions lie; therefore, practical and effective assessment is vitally important.

Formative assessment within lessons

The **Think together** section will often reveal any confusions or insecurities; try ironing these out by doing the first **Think together** question as a class. For children who continue to struggle, you or your Teaching Assistant should provide support and enable them to move on.

▶ Performance in practice can be very revealing: check Practice Books and listen out both during and after practice to identify misconceptions.

▶ The **Reflect** section is designed to check on the all-important depth of understanding. Be sure to review how the children performed in this final stage before you teach the next lesson.

End of unit check – Textbook

Each unit concludes with a summative check to help you assess quickly and clearly each child's understanding, fluency, reasoning and problem solving skills. Your Teacher Guide will suggest ideal ways of organising a given activity and offer advice and commentary on what children's responses mean. For example, 'What misconception does this reveal?'; 'How can you reinforce this particular concept?'

For Year 1 and Year 2 children, assess in small, teacher-led groups, giving each child time to think and respond while also consolidating correct mathematical language. Assessment with young children should always be an enjoyable activity, so avoid one-to-one individual assessments, which they may find threatening or scary. If you prefer, the End of unit check can be carried out as a whole-class group using whiteboards and Practice Books.

End of unit check – Practice Book

The Practice Book contains further opportunities for assessment, and can be completed by children independently whilst you are carrying out diagnostic assessment with small groups. Your Teacher Guide will advise you on what to do if children struggle to articulate an explanation – or perhaps encourage you to write down something they have explained well. It will also offer insights into children's answers and their implications for next learning steps. It is split into three main sections, outlined below.

My journal is designed to allow children to show their depth of understanding of the unit. It can also serve as a way of checking that children have grasped key mathematical vocabulary. The question children should answer is first presented in the Textbook in the Think! section. This provides an opportunity for you to discuss the question first as a class to ensure children have understood their task. Children should have some time to think about how they want to answer the question, and you could ask them to talk to a partner about their ideas. Then children should write their answer in their Practice Book, using the word bank provided to help them with vocabulary.

The **Power check** allows pupils to self-assess their level of confidence on the topic by colouring in different smiley faces. You may want to introduce the faces as follows:

I am starting to understand. I need more practice

I don't yet know how to do this

I know how to do this, but I will keep practising

Each unit ends with either a Power play or a Power puzzle. This is an activity, puzzle or game that allows children to use their new knowledge in a fun, informal way.

Progress Tests

There are *Power Maths* Progress Tests for each half term and at the end of the year, including an Arithmetic test and Reasoning test in each case. You can enter results in the online markbook to track and analyse results and see the average for all schools' results. The tests use a 6-step scale to show results against age-related expectation.

How to ask diagnostic questions

The diagnostic questions provided in children's Practice Books are carefully structured to identify both understanding and misconceptions (if children answer in a particular way, you will know why). The simple procedure below may be helpful:

Ask the question, offering the selection of answers provided.

Children take time to think about their response.

Each child selects an answer and shares their reasoning with the group.

Give minimal and neutral feedback (for example, 'That's interesting', or 'Okay').

Ask, 'Why did you choose that answer?', then offer an opportunity to change their mind by providing one correct and one incorrect answer.

Note which children responded and reasoned correctly first time and everyone's final choices.

Reflect that together, we can get the right answer.

Keeping the class together

Traditionally, children who learn quickly have been accelerated through the curriculum. As a consequence, their learning may be superficial and will lack the many benefits of enabling children to learn with and from each other.

By contrast, *Power Maths'* mastery approach values real understanding and richer, deeper learning above speed. It sees all children learning the same concept in small, cumulative steps, each finding and mastering challenge at their own level. Remember that when you teach for mastery, EVERYONE can do maths! Those who grasp a concept easily have time to explore and understand that concept at a deeper level. The whole class therefore moves through the curriculum at broadly the same pace via individual learning journeys.

For some teachers, the idea that a whole class can move forward together is revolutionary and challenging. However, the evidence of global good practice clearly shows that this approach drives engagement, confidence, motivation and success for all learners, and not just the high flyers. The strategies below will help you keep your class together on their maths journey.

Mix it up

Do not stick to set groups at each table. Every child should be working on the same concept, and mixing up the groupings widens children's opportunities for exploring, discussing and sharing their understanding with others.

Recycling questions

Reuse the Textbook and Practice Book questions with concrete materials to allow children to explore concepts and relationships and deepen their understanding. This strategy is especially useful for reinforcing learning in same-day interventions.

Strengthen at every opportunity

The next lesson in a *Power Maths* sequence always revises and builds on the previous step to help embed learning. These activities provide golden opportunities for individual children to strengthen their learning with the support of Teaching Assistants.

Prepare to be surprised!

Children may grasp a concept quickly or more slowly. The 'fast graspers' won't always be the same individuals, nor does the speed at which a child understands a concept predict their success in maths. Are they struggling or just working more slowly?

Same-day intervention

Since maths competence depends on mastering concepts one by one in a logical progression, it is important that no gaps in understanding are ever left unfilled. Same-day interventions – either within or after a lesson – are a crucial safety net for any child who has not fully made the small step covered that day. In other words, intervention is always about keeping up, not catching up, so that every child has the skills and understanding they need to tackle the next lesson. That means presenting the same problems used in the lesson, with a variety of concrete materials to help children model their solutions.

We offer two intervention strategies below, but you should feel free to choose others if they work better for your class.

Within-lesson intervention

The **Think together** activity will reveal those who are struggling, so when it is time for practice, bring these children together to work with you on the first practice questions. Observe these children carefully, ask questions, encourage them to use concrete models and check that they reach and can demonstrate their understanding.

After-lesson intervention

You might like to use the **Think together** questions to recap the lesson with children who are working behind expectations during assembly time. Teaching Assistants could also work with these children at other convenient points in the school day. Some children may benefit from revisiting work from the same topic in the previous year group. Note also the suggestion for recycling questions from the Textbook and Practice Book with concrete materials on page 26.

The role of practice

Practice plays a pivotal role in the *Power Maths* approach. It takes place in class groups, smaller groups, pairs, and independently, so that children always have the opportunities for thinking as well as the models and support they need to practise meaningfully and with understanding.

Intelligent practice

In *Power Maths*, practice never equates to the simple repetition of a process. Instead we embrace the concept of intelligent practice, in which all children become fluent in maths through varied, frequent and thoughtful practice that deepens and embeds conceptual understanding in a logical, planned sequence. To see the difference, take a look at the following examples.

Traditional practice

- Repetition can be rote – no need for a child to think hard about what they are doing

- Praise may be misplaced

- Does this prove understanding?

Intelligent practice

- Varied methods – concrete, pictorial and abstract

- Equation expressed in different ways, requiring thought and understanding

- Constructive feedback

All practice questions are designed to move children on and reveal misconceptions.

Simple, logical steps build onto earlier learning.

C-P-A runs throughout – different ways of modelling and understanding the same concept.

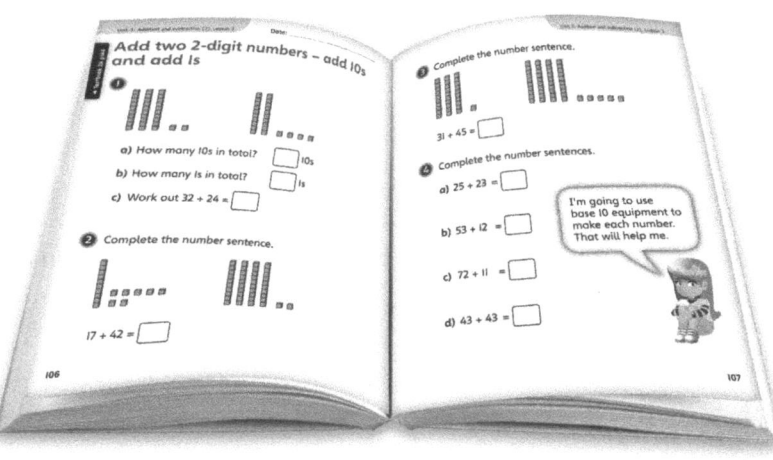

Conceptual variation – children work on different representations of the same maths concept.

Friendly characters offer support and encourage children to try different approaches.

A carefully designed progression

The Practice Books provide just the right amount of intelligent practice for children to complete independently in the final sections of each lesson. It is really important that all children are exposed to the practice questions, and that children are not directed to complete different sections. That is because each question is different and has been designed to challenge children to think about the maths they are doing. The questions become more challenging so children grasping concepts more quickly will start to slow down as they progress. Meanwhile, you have the chance to circulate and spot any misconceptions before they become barriers to further learning.

Homework and the role of parents and carers

While *Power Maths* does not prescribe any particular homework structure, we acknowledge the potential value of practice at home. For example, practising fluency in key facts, such as number bonds and times-tables, is an ideal homework task. You can share the Individual Practice Games for homework (see page 6), or parents and carers could work through uncompleted Practice Book questions with children at either primary stage.

However, it is important to recognise that many parents and carers may themselves lack confidence in maths, and few, if any, will be familiar with mastery methods. A Parents' and Carers' evening that helps them understand the basics of mindsets, mastery and mathematical language is a great way to ensure that children benefit from their homework. It could be a fun opportunity for children to teach their families that everyone can do maths!

Structures and representations

Unlike most other subjects, maths comprises a wide array of abstract concepts – and that is why children and adults so often find it difficult. By taking a concrete-pictorial-abstract (C-P-A) approach, *Power Maths* allows children to tackle concepts in a tangible and more comfortable way.

Non-linear stages

Concrete

Replacing the traditional approach of a teacher working through a problem in front of the class, the concrete stage introduces real objects that children can use to 'do' the maths – any familiar object that a child can manipulate and move to help bring the maths to life. It is important to appreciate, however, that children must always understand the link between models and the objects they represent. For example, children need to first understand that three cakes could be represented by three pretend cakes, and then by three counters or bricks. Frequent practice helps consolidate this essential insight. Although they can be used at any time, good concrete models are an essential first step in understanding.

Pictorial

This stage uses pictorial representations of objects to let children 'see' what particular maths problems look like. It helps them make connections between the concrete and pictorial representations and the abstract maths concept. Children can also create or view a pictorial representation together, enabling discussion and comparisons. The *Power Maths* teaching tools are fantastic for this learning stage, and bar modelling is invaluable for problem solving throughout the primary curriculum.

Abstract

Our ultimate goal is for children to understand abstract mathematical concepts, symbols and notation and of course, some children will reach this stage far more quickly than others. To work with abstract concepts, a child must be comfortable with the meaning of and relationships between concrete, pictorial and abstract models and representations. The C-P-A approach is not linear, and children may need different types of models at different times. However, when a child demonstrates with concrete models and pictorial representations that they have grasped a concept, we can be confident that they are ready to explore or model it with abstract symbols such as numbers and notation.

Use at any time and with any age to support understanding

Variation helps visualisation

Children find it much easier to visualise and grasp concepts if they see them presented in a number of ways, so be prepared to offer and encourage many different representations.

For example, the number six could be represented in various ways:

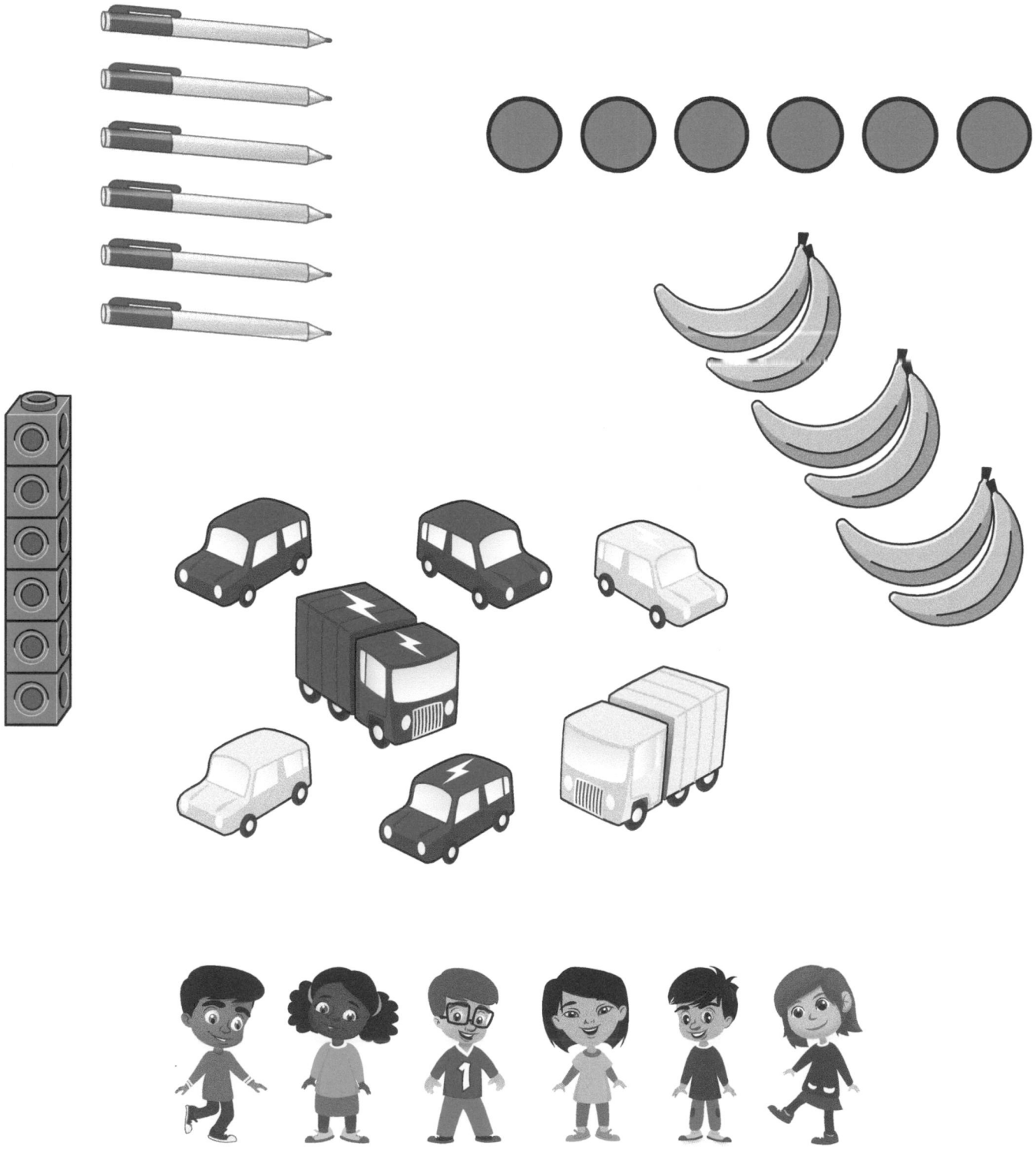

Practical aspects of *Power Maths*

One of the key underlying elements of *Power Maths* is its practical approach, allowing you to make maths real and relevant to your children, no matter their age.

Manipulatives are essential resources for both key stages and *Power Maths* encourages teachers to use these at every opportunity, and to continue the Concrete-Pictorial-Abstract approach right through to Year 6.

The Textbooks and Teacher Guides include lots of opportunities for teaching in a practical way to show children what maths means in real life.

Discover and Share

The **Discover** and **Share** sections of the Textbook give you scope to turn a real-life scenario into a practical and hands-on section of the lesson. Use these sections as inspiration to get active in the classroom. Where appropriate, use the **Discover** contexts as a springboard for your own examples that have particular resonance for your children – and allow them to get their hands dirty trying out the mathematics for themselves.

Unit videos

Every term has one unit video which incorporates real-life classroom sequences.

These videos show you how the reasoning behind mathematics can be carried out in a practical manner by showing real children using various concrete and pictorial methods to come to the solution. You can see how using these practical models, such as part-whole and bar models, helps them to find and articulate their answer.

Mastery tips

Mastery Experts give anecdotal advice on where they have used hands-on and real-life elements to inspire their children.

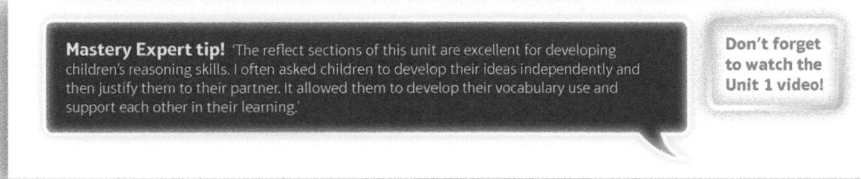

Mastery Expert tip! 'The reflect sections of this unit are excellent for developing children's reasoning skills. I often asked children to develop their ideas independently and then justify them to their partner. It allowed them to develop their vocabulary use and support each other in their learning.'

Don't forget to watch the Unit 1 video!

Concrete-Pictorial-Abstract (C-P-A) approach

Each **Share** section uses various methods to explain an answer, helping children to access abstract concepts by using concrete tools, such as counters. Remember, this isn't a linear process, so even children who appear confident using the more abstract method can deepen their knowledge by exploring the concrete representations. Encourage children to use all three methods to really solidify their understanding of a concept.

Pictorial representation – drawing the problem in a logical way that helps children visualise the maths

Concrete representation – using manipulatives to represent the problem. Encourage children to physically use resources to explore the maths.

Abstract representation – using words and calculations to represent the problem.

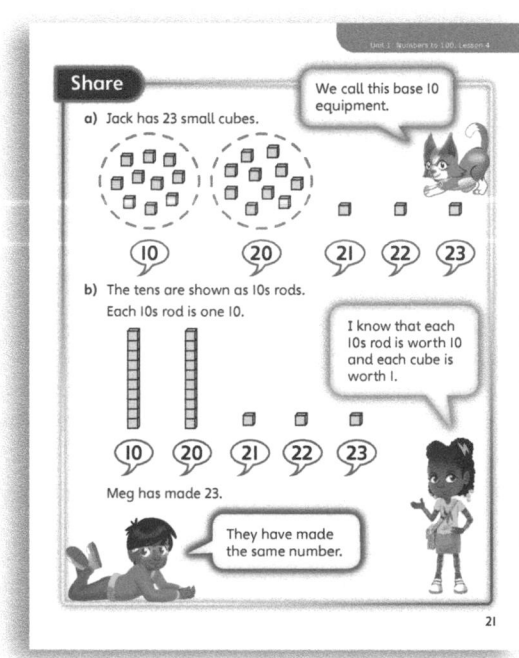

Practical tips

Every lesson suggests how to draw out the practical side of the **Discover** context.

You'll find these in the **Discover** section of the Teacher Guide for each lesson.

PRACTICAL TIPS Play this game together as a class, starting with numbers in the range 0 to 10.

Resources

Every lesson lists the practical resources you will need or might want to use. There is also a summary of all of the resources used throughout the term on page 41 to help you be prepared.

RESOURCES

Mandatory: counters, ten frames, number lines

Optional: number tracks, multilink cubes

Using *Power Maths* flexibly in Key Stage 1

Power Maths lessons have a coherent, regular structure that supports you in building up children's understanding in a series of small steps. This is something most classes will need to build up to, rather than running in from a standing start at the beginning of Year 1.

Start by using the Practice Books in small groups

In most Year 1 classes, it won't be realistic for the whole class to complete the Practice Book pages independently at the start of the year, but they will learn to do this gradually. For the Textbooks, children will need to get used to direct teaching and recording answers in their own books. And, of course, this will set them up well for the rest of Primary school.

Small teacher-led groups are likely to be the best approach for independent practice at the beginning of KS1. This format allows you to talk children through the question, discuss their ideas using manipulatives (often there will be manipulatives on the page as a hint), and guide them in representing their answer. (For instance, they can tell you the answer is 5, but they may need help writing 5 or knowing that they should colour in 5 apples.)

Go through the questions one-by-one with the group. You can mark their work/give feedback there and then. As children get used to the materials, the next stage could be for the small group to work through the questions at their own pace. The style of questions in *Power Maths* is quite regular, so children will get better at knowing what they need to do.

To facilitate small group work, you are likely to need some other activities as a carousel. A good way to do this is by turning a question from the Textbook into a game (usually **Think together** question 3 will work well) and teaching this to children before you break into groups. For instance, look at the example below (pages 78–79 in Textbook 1A). You could teach children a game with a part-whole model where one child puts in the whole using counters and the other children have to put in the parts. Or they could try this with beanbags and hoops. Base the practice on the key learning from the lesson.

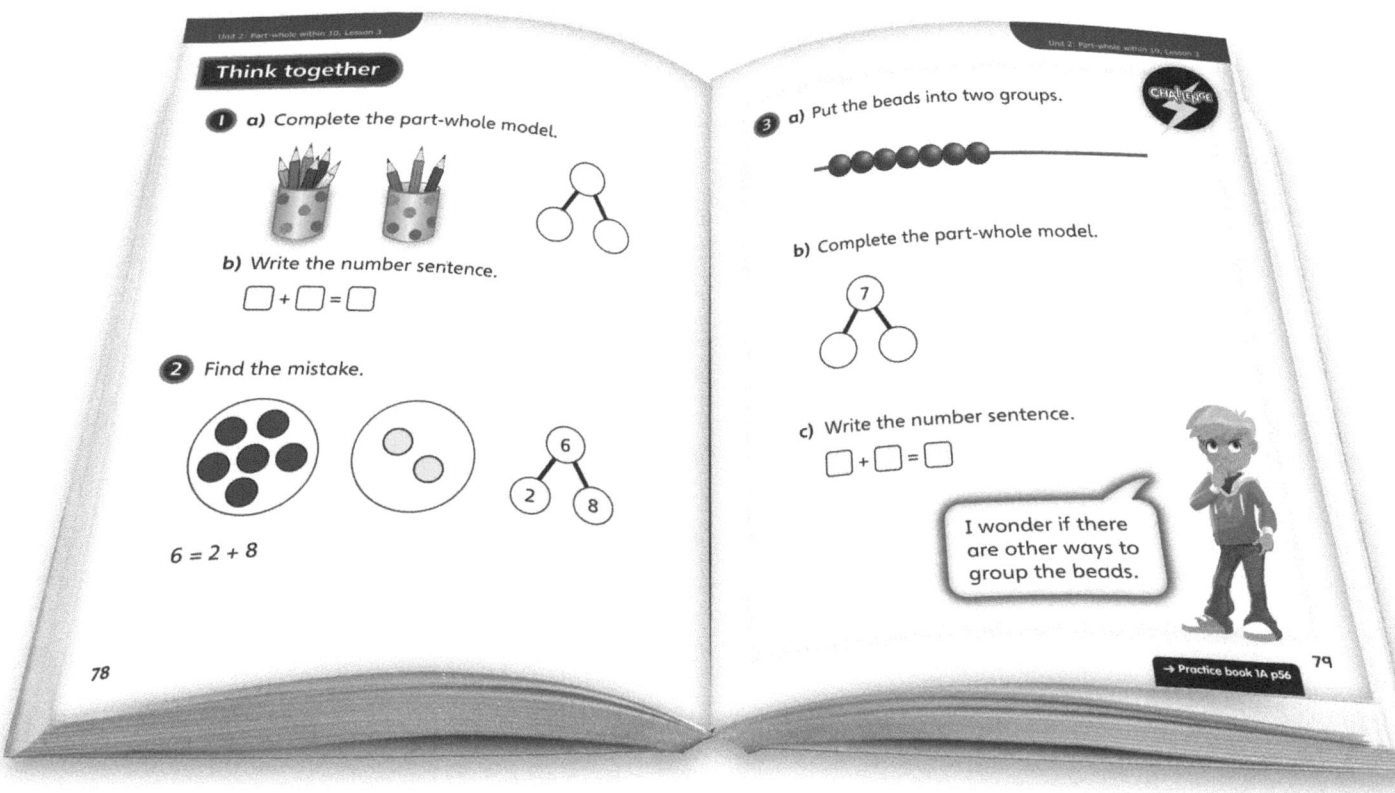

Are there any other ways to use the resources flexibly?

Don't be afraid to bring the **Discover** activity to life! Perhaps you could turn it into a game, or a role play. For instance, if the context is a teddy bear's picnic, you could share out fruit between teddies in the class. Or could you find a toy rocket to launch for the lesson below? (Textbook 1A page 32).

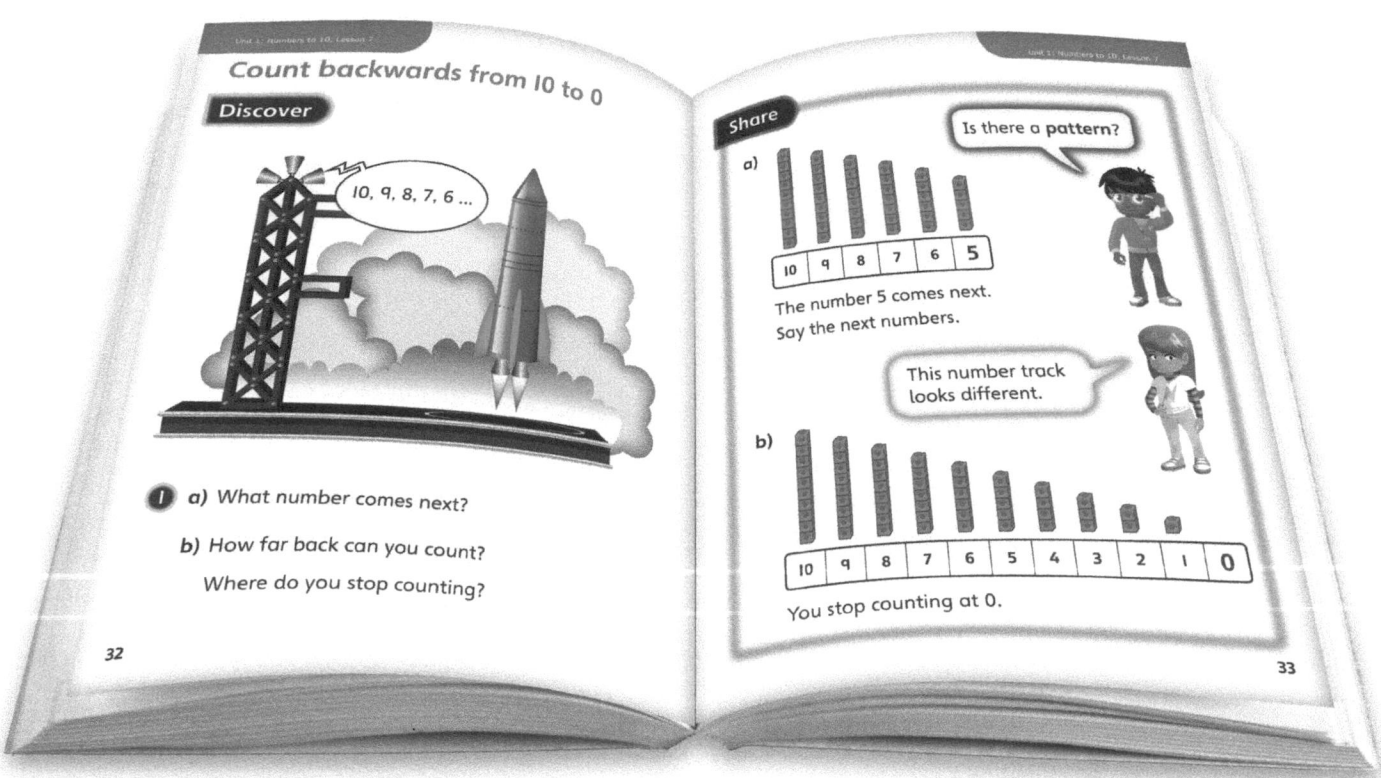

For some lessons you could consider a slightly different approach where you move backwards and forwards between the Textbook and Practice Book. If **Think together** question 1 links well with Practice Book question 1, you could do the **Think together** question together and then let children complete the Practice Book question, then the same for question 2, etc. This works better for some lessons than others, but it is one way of making practice more independent in short bursts, as a way of building up independence.

Don't forget, there isn't a *Power Maths* lesson for every lesson in the year. You can take more time where you need to, so that children's understanding is secure. In Key Stage 1, it will be all the more important to take your time, because children need to get used to the format as well as master the key learning. If using some of the ideas above means that a *Power Maths* lesson actually takes two lessons, e.g. for the first part of the year, then that's fine!

There are some further ideas for using the materials flexibly in the next section.

Working with children below age-related expectation

This section offers advice on using *Power Maths* with children who are significantly behind age-related expectation. Teacher judgement will be crucial in terms of where and why children are struggling, and in choosing the right approach. The suggestions can of course be adapted for children with special educational needs, depending on the specific details of those needs.

General approaches to support children who are struggling

Keeping the pace manageable

Remember, you have more teaching days than *Power Maths* lessons so you can cover a lesson over more than one day, and revisit key learning, to ensure all children are ready to move on. You can use the + and – buttons to adjust the time for each unit in the online planning. The NCETM's Ready-to-Progress criteria can be used to help determine what should be highest priority.

Same-day intervention

You could go over the Textbook pages or revisit the previous year's work if necessary (see Addressing gaps). Remember that same-day intervention can be within the lesson, as well as afterwards (see page 28). As children start their independent practice, you can work with those who found the first part of the lesson difficult, checking understanding using manipulatives.

Fluency sessions

Fit in as much practice as you can for number bonds and times-tables, etc., at other times of the day. If you can, plan a short 'maths meeting' for this in the afternoon. You might choose to use a Power Up you haven't used already.

Addressing gaps

Use material from the same topic in the previous year to consolidate or address gaps in learning, e.g. Textbook pages and Strengthen activities. The End of unit check will help gauge children's understanding.

Pre-teaching

Find a 5- to 10-minute slot before the lesson to work with the children you feel would benefit. The afternoon before the lesson can work well, because it gives children time to think in between. Recap previous work on the topic (addressing any gaps you're aware of) and do some fluency practice, targeting number facts etc. that will help children access the learning.

Focusing on the key concepts

If children are a long way behind, it can be helpful to take a step back and think about the key concepts for children to engage with, not just the fine detail of the objective for that year group (e.g. addition with a specific number of columns). Bearing that in mind, how could children advance their understanding of the topic?

Providing extra support within the lesson

Support in the Teacher Guide

First of all, use the Strengthen support in the Teacher Guide for guided and independent work in each lesson, and share this with Teaching Assistants, where relevant. As you read through the lesson content and corresponding Teacher Guide pages before the lesson, ask yourself what key idea or nugget of understanding is at the heart of the lesson. If children are struggling, this should help you decide what's essential for all children before they move on.

Annotating pages

You can annotate questions to provide extra scaffolding or hints if you need to, but aim to build up children's ability to access questions independently wherever you can. Children tend to get used to the style of the *Power Maths* questions over time.

Quick recap as lesson starter

The Quick recap for each lesson in the Teacher Guide is an alternative starter activity to the Power Up. You might choose to use this with some or all children if you feel they will need support accessing the main lesson.

Consolidation questions

If you think some children would benefit from additional questions at the same level before moving on, write one or two similar questions on the board. (This shouldn't be at the expense of reasoning and problem-solving opportunities: take longer over the lesson if you need to.)

Hard copy Textbooks

The Textbooks help children focus in more easily on the mathematical representations, read the text more comfortably, and revisit work from a previous lesson that you are building on, as well as giving children ownership of their learning journey. In main lessons, it can work well to use the e-Textbook for Discover and give out the books when discussing the methods in the Share section.

Reading support

It's important that all children are exposed to problem solving and reasoning questions, which often involves reading. For whole-class work you can read questions together. For independent practice you could consider annotating pages to help children see what the question is asking, and stem sentences to help structure their answer. A general focus on specific mathematical language and vocabulary will help children access the questions. You could consider pairing weaker readers with stronger readers, or read questions as a group if those who need support are on the same table.

Providing extra depth and challenge with *Power Maths*

Just as prescribed in the National Curriculum, the goal of *Power Maths* is never to accelerate through a topic but rather to gain a clear, deep and broad understanding. Here are some suggestions to help ensure all children are appropriately challenged as you work with the resources.

Overall approaches

First of all, remember that the materials are designed to help you keep the class together, allowing all children to master a concept while those who grasp it quickly have time to explore it in more depth. Use the Deepen support in the Teacher Guide (see below) to challenge children who work through the questions quickly. Here are some questions and ideas to encourage breadth and depth during specific parts of the lesson, or at any time (where no part of the lesson sequence is specified):

- **Discover**: 'Can you demonstrate your solution another way?'

- **Share**: Make sure every child is encouraged to give answers and engage with the discussion, not just the most confident.

- **Think together**: 'Can you model your answers using concrete materials? Can you explain your solution to a partner?'

- Practice: Allow all children to work through the full set of questions, so that they benefit from the logical sequence.

- **Reflect**: 'Is there another way of working out the answer? And another way?'
  'Have you found all the solutions?'
  'Is that always true?'
  'What's different between this question and that question? And what's the same?'

Note that the **Challenge** questions are designed so that all children can access and attempt them, if they have worked through the steps leading up to them. There may be some children in a given lesson who don't manage to do the **Challenge**, but it is not supposed to be a distinct task for a subset of the class. When you look through the lesson materials before teaching, think about what each question is specifically asking, and compare this with the key learning point for the lesson. This will help you decide which questions you feel it's essential for all children to answer, before moving on. You can at least aim for all children to try the **Challenge**!

Deepen activities and support

The Teacher Guide provides valuable support for each stage of the lesson. This includes Deepen tips for the guided and independent practice sections, which will help you provide extra stretch and challenge within your lesson, without having to organise additional tasks. If you have a Teaching Assistant, they can also make use of this advice. There are also suggestions for the lesson as a whole in the 'Going Deeper' section on the first page of the Teacher Guide section for that lesson. Every class is different, so you can always go a bit further in the direction indicated, if appropriate, and build on the suggestions given.

There is a Deepen activity for each unit. These are designed to follow on from the End of unit check, stretching children who have a firm understanding of the key learning from the unit. Children can work on them independently, which makes it easier for the teacher to facilitate the Strengthen activity for children who need extra support. Deepen activities could also be introduced earlier in the unit if the necessary work has been covered. The Deepen activities are on *ActiveLearn* on the Planning page for each unit, and also on the Resources page).

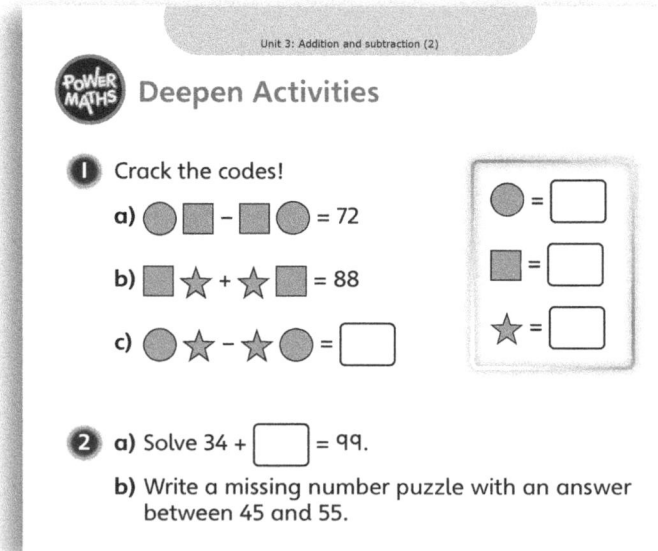

38

Using the questions flexibly to provide extra challenge

Sometimes you may want to write an extra question on the board or provide this on paper. You can usually do this by tweaking the lesson materials. The questions are designed to form a carefully structured sequence that builds understanding step by step, but, with careful thought about the purpose of each question, you can use the materials flexibly where you need to. Sometimes you might feel that children would benefit from another similar question for consolidation before moving on to the next one, or you might feel that they would benefit from a harder example in the same style. It should be quick and easy to generate 'more of the same' type questions where this is the case.

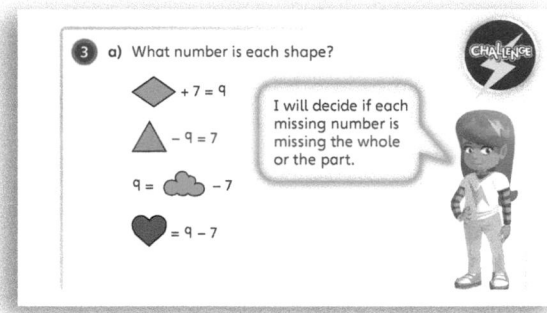

When you see a question like this one (from Unit 3, Lesson 10), it's easy to make harder examples to do afterwards if you need an extra challenge. For example, you could do something similar using multiples of 10, using 2-digit numbers with 1-digit numbers, or using 2-digit numbers. And, of course, more difficult examples will be 'across 10'.

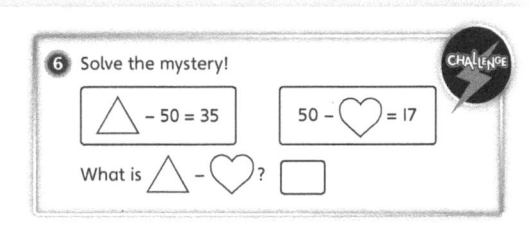

For this example (from Unit 3, Lesson 10), you could ask children to make up their own question(s) for a partner to solve. (In fact, for any of these examples you could ask early finishers to create their own question for a partner.)

Here's an example (from Unit 3, Lesson 12) where some of the sums and differences feature as questions in the lesson, but others don't. Clearly there are plenty of extra two-step problems you could ask using the same context. Children could work out the difference between Kasim's total and Amy and Ben's joint total. Or, if Kasim gives Ben 10 marbles, do they each have as many as Amy? A trickier one might be, how many more marbles does Kat need to have as many as Ben, Amy and Kasim combined.

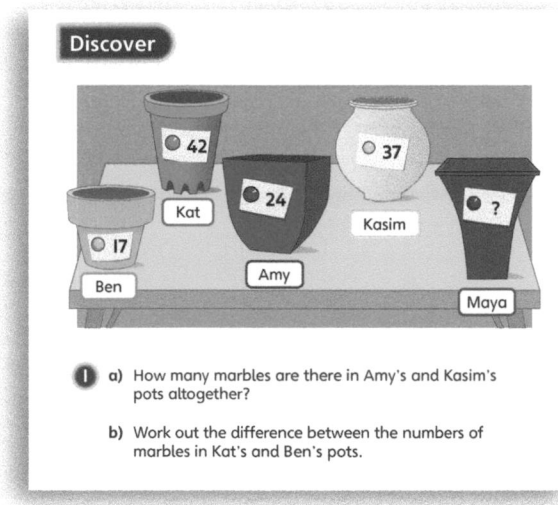

Besides creating additional questions, you should be able to find a question in the lesson that you can adapt into a game or open-ended investigation, if this helps to keep everyone engaged. It could simply be that, instead of answering 5 + 6 etc. on the page, they could build a robot with 5 cubes and 6 cylinders.

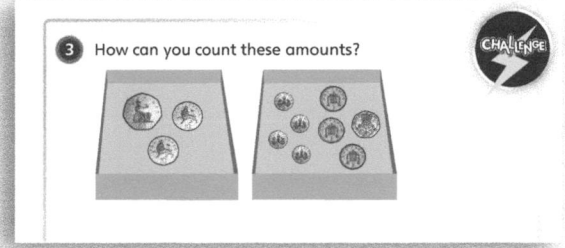

With a question like this (from Unit 5, Lesson 1), an extra challenge could be for children to play a game where they each take a handful of coins and then find the difference between the totals, or perhaps one child makes a total and their partner has to match it with different coins.

See the bullets above for some general ideas that will help with 'opening out' questions in the books, e.g. 'can you find all the solutions?' type questions.

Other suggestions

Another way of stretching children is through mixed ability pairs, or via other opportunities for children to explain their understanding in their own way. This is a good way of encouraging children to go deeper into the learning, rather than, for instance, tackling questions that are computationally more challenging but conceptually equivalent in level.

Using *Power Maths* with mixed age classes

Overall approaches

There are many variables between schools that would make it inadvisable to recommend a one-size-fits-all approach to mixed age teaching with *Power Maths*. These include how year groups are merged, availability of Teaching Assistants, experience and preference of teaching staff, range in pupil attainment across years, classroom space and layout, level of flexibility around timetables, and overall organisational structure (whether the school is part of a trust).

Some schools will find it best to timetable separate maths lessons for the different year groups. Others will aim to teach the class together as much as possible using the mixed age planning support on *ActiveLearn* (see the lesson exemplars for ways of organising lessons with strong/medium/weak correlation between year groups). There will also be ways of adapting these general approaches. For example, offset lessons where Year A start their lesson with the teacher, while Year B work independently on the practice from the previous lesson, and then start the next lesson with the teacher while Year A work independently; or teachers may choose to base their provision around the lesson from one year group and tweak the content up/down for the other group.

Key strategies for mixed age teaching

The mixed age teaching webinar on *ActiveLearn* provides advice on all aspects of mixed age teaching, including more detail on the ideas below.

Developing independence over time
Investing time in building up children's independence will pay off in the medium term.

Clear rationale
If someone asked, 'Why did you teach both Unit 3 and 4 in the same lesson/separate lessons?', what would your answer be?

Designing a lesson
1. Identify the core learning for each group
2. Identify any number skills necessary to access the core
3. Consider the flow of concepts and how one core leads to the other

Challenging all children
The questions are designed to build understanding step by step, but with careful thought about the purpose of each question you can tweak them to increase the challenge.

Multiple years combined
With more than two years together, teachers will inevitably need to use the resources flexibly if delivering a single lesson.

Enjoy the positives!

Comparison deepens understanding and there will be lots of opportunities for children, as well as misconceptions to explore. There is also in-built pre-teaching and the chance to build up a concept from its foundations. For teachers there is double the material to draw on! Mixed age teachers require a strong understanding of the progression of ideas across year groups, which is highly valuable for all teachers. Also, it is necessary to engage deeply with the lesson to see how to use the materials flexibly – this is recommended for all teachers and will help you bring your lesson to life!

List of practical resources

Year 2A Mandatory resources

Resource	Lesson
100 square	**Unit 1** Lesson 2
	Unit 3 Lesson 1
2D shapes	**Unit 4** Lessons 4, 7
2D shapes (regular and irregular)	**Unit 4** Lessons 2, 3
2D shapes (with labels)	**Unit 4** Lesson 6
2D and 3D shapes (with labels)	**Unit 4** Lesson 1
3D shapes	**Unit 4** Lessons 8, 9, 10, 11, 12
base 10 equipment	**Unit 1** Lessons 4, 5, 6, 7, 8, 9, 13 **Unit 2** Lessons 3, 5 **Unit 3** Lessons 1, 2, 11
calculation scaffolds (addition and subtraction)	**Unit 2** Lesson 1
construction materials (to create polyhedrons)	**Unit 4** Lesson 8
counters	**Unit 1** Lessons 1, 2, 8 **Unit 2** Lessons 7, 8, 12
counters (double-sided or two colours, if possible)	**Unit 2** Lesson 10
cubes and/or counters	**Unit 2** Lessons 1, 6 **Unit 3** Lessons 7, 9
dice	**Unit 1** Lesson 17
mirrors	**Unit 4** Lessons 5, 12
multilink cubes	**Unit 1** Lessons 3, 4, 13, 16
number grid	**Unit 1** Lesson 6
number lines	**Unit 1** Lessons 1, 10, 12, 16, 17 **Unit 3** Lessons 6, 7, 8
paper (plain)	**Unit 4** Lesson 4
paper (square dotted)	**Unit 4** Lesson 4
paper (squared)	**Unit 4** Lesson 4
part-whole model	**Unit 2** Lesson 1
pattern shapes	**Unit 4** Lessons 2, 3
place value chart	**Unit 3** Lesson 2
place value equipment	**Unit 1** Lesson 5 **Unit 2** Lesson 4 **Unit 3** Lessons 3, 4
place value grid	**Unit 1** Lesson 6

Resource	Lesson
rulers	**Unit 1** Lesson 10 **Unit 4** Lessons 2, 4
sorting circles	**Unit 1** Lesson 8
sorting hoops	**Unit 4** Lesson 2
sticks	**Unit 1** Lesson 17
straws (construction or art)	**Unit 4** Lessons 3, 9
string	**Unit 4** Lesson 3
ten frames	**Unit 1** Lessons 1, 2, 8 **Unit 2** Lessons 6, 7, 8, 9, 10, 11, 12, 13

Year 2A Optional resources

Resource	Lesson
2D shapes (laminated)	**Unit 4** Lesson 4
3D shapes (pictures of)	**Unit 4** Lesson 12
bag	**Unit 4** Lesson 11
base 10 equipment	**Unit 1** Lesson 16 **Unit 2** Lesson 1 **Unit 3** Lessons 8, 10
bead string	**Unit 2** Lessons 4, 7
bean bags	**Unit 3** Lesson 12
butterfly pictures	**Unit 4** Lesson 5
butterfly templates	**Unit 4** Lesson 5
classroom stationery objects	**Unit 3** Lesson 7
cones	**Unit 1** Lesson 3
countable objects	**Unit 1** Lesson 17
countable objects (can be grouped in 10s, e.g. paper straws or pencils)	**Unit 1** Lesson 4 **Unit 2** Lesson 11
countable resources (e.g. base 10 equipment, counters)	**Unit 3** Lesson 12
counters	**Unit 1** Lessons 7, 13, 16, 17 **Unit 2** Lesson 8 **Unit 3** Lessons 10, 11
cubes (interlocking)	**Unit 2** Lesson 7
digit cards	**Unit 2** Lesson 1
dry-wipe markers	**Unit 4** Lessons 2, 4, 9
dry-wipe pen	**Unit 4** Lesson 3
feely bag	**Unit 4** Lesson 8
hoops	**Unit 3** Lesson 12
isometric paper	**Unit 4** Lessons 3, 4
joining tubes	**Unit 4** Lesson 10
marshmallows	**Unit 4** Lesson 10
materials to make parts and wholes	**Unit 2** Lesson 1
multilink cubes	**Unit 1** Lessons 1, 14, 17
number bonds within 20	**Unit 2** Lesson 8
number cards	**Unit 1** Lessons 12, 15
number line (1 to 100)	**Unit 2** Lesson 10
number lines	**Unit 1** Lessons 11, 15 **Unit 3** Lesson 10
number sentence scaffold	**Unit 2** Lesson 3 **Unit 3** Lesson 7
number track (completed, increasing in 10s)	**Unit 3** Lesson 1
number tracks	**Unit 1** Lesson 1 **Unit 2** Lesson 2
paint (various colours)	**Unit 4** Lessons 5, 8
paper (squared)	**Unit 4** Lesson 5
paper (strips of)	**Unit 3** Lessons 11, 12

Resource	Lesson
paper cutouts of regular polygons	**Unit 4** Lesson 5
paper plates	**Unit 1** Lesson 8
peg boards	**Unit 4** Lessons 3, 5
place value equipment	**Unit 1** Lesson 4 **Unit 2** Lesson 13 **Unit 3** Lesson 5
place value grid (large)	**Unit 1** Lessons 6, 14
place value grids	**Unit 1** Lesson 15 **Unit 3** Lesson 11
printing materials (with 3D shapes)	**Unit 4** Lesson 1
Rangoli-style patterns	**Unit 4** Lesson 5
rekenreks	**Unit 2** Lesson 4 **Unit 3** Lesson 6
rulers	**Unit 1** Lesson 2
screen or bag (to hide 2D shapes)	**Unit 4** Lesson 2
sticks	**Unit 4** Lesson 2
sticky notes	**Unit 1** Lesson 12 **Unit 3** Lesson 10 **Unit 4** Lesson 2
sticky tack	**Unit 4** Lesson 4
straws	**Unit 1** Lesson 3 **Unit 4** Lesson 10
straws (art)	**Unit 4** Lessons 2, 4
ten frames	**Unit 1** Lesson 7
ten frames (pictures of completed)	**Unit 2** Lesson 10
tracing paper	**Unit 4** Lessons 5, 7

Getting started with *Power Maths*

As you prepare to put *Power Maths* into action, you might find the tips and advice below helpful.

STEP 1: Train up!

A practical, up-front full day professional development course will give you and your team a brilliant head-start as you begin your *Power Maths* journey. You will learn more about the ethos, how it works and why.

STEP 2: Check out the progression

Take a look at the yearly and termly overviews. Next take a look at the unit overview for the unit you are about to teach in your Teacher Guide, remembering that you can match your lessons and pacing to match your class.

STEP 3: Explore the context

Take a little time to look at the context for this unit: what are the implications for the unit ahead? (Think about key language, common misunderstandings and intervention strategies, for example.) If you have the online subscription, don't forget to watch the corresponding unit video.

STEP 4: Prepare for your first lesson

Familiarise yourself with the objectives, essential questions to ask and the resources you will need. The Teacher Guide offers tips, ideas and guidance on individual lessons to help you anticipate children's misconceptions and challenge those who are ready to think more deeply.

STEP 5: Teach and reflect

Deliver your lesson — and enjoy!

Afterwards, reflect on how it went… Did you cover all five stages? Does the lesson need more time? How could you improve it?

Unit I
Numbers to 100

Don't forget to watch the Unit 1 video!

Mastery Expert tip! 'The reflect sections of this unit are excellent for developing children's reasoning skills. I often asked children to develop their ideas independently and then justify them to their partner. It allowed them to develop their vocabulary use and support each other in their learning.'

WHY THIS UNIT IS IMPORTANT

This unit focuses on children's ability to read and understand numbers to 100. They will use their growing understanding of place value to help them sort, compare and order numbers.

Within this unit, children will revise their understanding of different representations of numbers and also meet other representations for the first time. They will use these representations to show a number's '10s' and '1s' and use this to help them compare and order. Children will use part-whole models and place value grids to show their partitioning of numbers and use these to support their reasoning when comparing and ordering.

Moving on from partitioning and ordering numbers, the children will begin to develop their ability to count forwards and backwards efficiently in steps of 2, 3, 5, and 10.

WHERE THIS UNIT FITS

→ **Unit 1: Numbers to 100**

→ Unit 2: Addition and subtraction (1)

This unit builds on children's work in Year 1 on numbers to 100. It is important that children can read and write numbers to 100 and recognise the place value of each digit in order to go onto addition and subtraction later in the term.

Before they start this unit, it is expected that children:
- know how to group objects into groups of 10
- count up and back in 1s.

ASSESSING MASTERY

Children who have mastered this unit will be able to confidently partition any 2-digit number in different ways and be able to recognise how many 10s and 1s make up any given 2-digit number. They will be able to use multiple concrete, pictorial and abstract representations of numbers, such as place value grids and part-whole models, to support their reasoning and justification of their ideas.

COMMON MISCONCEPTIONS	STRENGTHENING UNDERSTANDING	GOING DEEPER
Children may have more trouble counting backwards than forwards.	Give children games to play such as hide and seek where they will get to practise counting down (10, 9 … 3, 2, 1, coming!). Sing songs such as 'Ten Green Bottles'. Practise counting over the 10s bridges (for example, 42, 41, 40, 39 …).	Give children images of castles built with cubes (as in Lesson 17 – **Think together**). If you add three blocks each time, can they predict what the fourth castle in the pattern will be? What about the seventh? Can children explain how they know?
Children may place the 10s number in a place value grid, as opposed to just the digit representing how many 10s (50 instead of 5, for example).	Use digit cards to generate two numbers, say 5 and 2. Arrange the cards to make 52 and 25, discussing the value of each digit in both numbers. Reinforce the description: 5 tens and 2 ones is the same as 52; 2 tens and 5 ones is the same as 25.	

Unit I: Numbers to 100

UNIT STARTER PAGES

Use these pages to introduce the focus to children. You can use the characters to explore different ways of working too.

STRUCTURES AND REPRESENTATIONS

Part-whole model: This model helps children understand that two or more parts combine to make a whole. It also helps to strengthen children's understanding of number bonds within 100.

Number line: Number lines help children to represent the order of numbers. They will help children count on and back from a given starting point and help them identify patterns within the count.

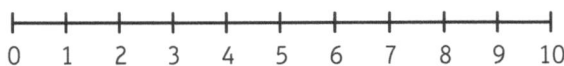

0 1 2 3 4 5 6 7 8 9 10

Place value grid: Place value grids help children to record and describe how a number is 'made'. This representation can empower children to more efficiently describe and order numbers.

T	O

KEY LANGUAGE

There is some key language that children will need to know as part of the learning in this unit:

→ less than, fewer, smaller, less, (<)
→ greater than, larger, bigger, more, (>)
→ equal to, (=)
→ greatest, biggest
→ fewest, smallest
→ tens (10s), ones (1s)
→ how many?, count, partition
→ place value grid, part-whole model

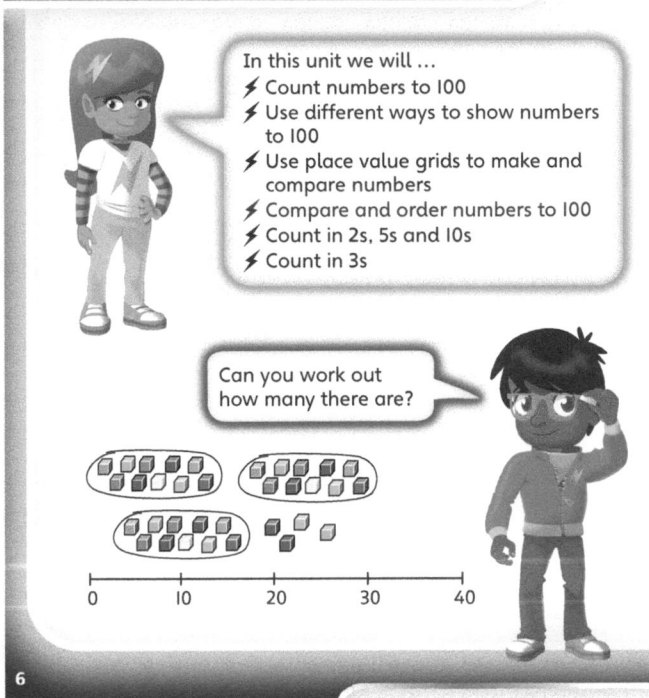

Unit I
Numbers to 100

In this unit we will …
⚡ Count numbers to 100
⚡ Use different ways to show numbers to 100
⚡ Use place value grids to make and compare numbers
⚡ Compare and order numbers to 100
⚡ Count in 2s, 5s and 10s
⚡ Count in 3s

Can you work out how many there are?

0   10   20   30   40

6

PUPIL TEXTBOOK 2A PAGE 6

Here are some maths words you have seen before. Which ones can you remember?

tens   ones
place value grid   partition   more
fewer   fewest   greatest   smallest

We can use [T|O] to show a number. Use it to show 43.

T	O

7

PUPIL TEXTBOOK 2A PAGE 7

Numbers to 20

Learning focus

In this lesson, children rehearse key number skills in the range 0 to 20: counting up and down, reading and writing numerals, partitioning 2-digit numbers into 10 and some 1s. Children also compare numbers as greater or less on a number line.

Before you teach ▮▮

Can children count up and down within 10 and show numbers to 10 on their fingers? Have they had practice of reading and writing numerals?

NATIONAL CURRICULUM LINKS

Year 1 Number – number and place value

Count to and across 100, forwards and backwards, beginning with 0 or 1, or from any given number.

Read and write numbers from 1 to 20 in numerals and words.

ASSESSING MASTERY

Children can count up and down within 20 and can partition the numbers 11 to 20 into a 10 and some 1s. They can also say a number greater or less than a given number.

COMMON MISCONCEPTIONS

Children may find it challenging to count down in the teen numbers. Ask:
• *What sounds can we hear in these numbers? Nine-teen, eight-teen, seven-teen, six-teen… What numbers in the count do we need to remember that do not follow this sound pattern?*

STRENGTHENING UNDERSTANDING

Practise counting out loud as a class regularly, both up and back for numbers within 20. Encourage children to use ten frames to help them recognise numbers within 20.

GOING DEEPER

Ask children how many different ways they can find to partition different numbers within 20. For example: 11 = 10 + 1, and 11 = 5 + 5 + 1, …

KEY LANGUAGE

In lesson: number names 11 to 20; ten, ones
Other language to be used by the teacher: partition

STRUCTURES AND REPRESENTATIONS

Ten frame, number line

RESOURCES

Mandatory: counters, ten frames, number lines

Optional: number tracks, multilink cubes

 In the eTextbook of this lesson, you will find interactive links to a selection of teaching tools.

Quick recap

Count as a class; start on 0 and count on to 20. Then start on 20 and count back to 0. Ask children to notice any mistakes when you count. For example, if you miss out the number 13, or say the numbers 11 and 12 in the wrong order.

46

Discover

ASK

- Question ❶ a): *How do you count up and down on this number line?*
- Question ❶ a): *What is the first number on the line?*
- Question ❶ a): *Point to a number; now can you point to a larger number?*
- Question ❶ b): *Can you point to a number greater than 15? Can you point to a number smaller than 15?*

IN FOCUS This is a version of a simple but instructive game that can be played as a class. The idea is that children recognise that the numbers increase as you go from left to right, and decrease as you go from right to left. It gives a good opportunity to rehearse key number awareness from Year 1.

PRACTICAL TIPS Play this game together as a class, starting with numbers in the range 0 to 10.

ANSWERS

Question ❶ a): Any number from 16 to 20.

Question ❶ b): 15 counters placed on 2 ten frames. Suggested arrangement is 1 ten frame complete with 10 counters, the other with 5 counters. Other arrangements would also be acceptable.

PUPIL TEXTBOOK 2A PAGE 8

Share

ASK

- Question ❶ a): *Is 15 nearer to 0 or to 20? What other numbers are near 15? Can you say the next number? Can you say the number before?*
- Question ❶ b): *The teacher says 15 is too small. What other numbers must be too small? What numbers are not less than 15?*

IN FOCUS The focus is on understanding how to find numbers on the number line, and knowing which numbers are less than and which are greater than a given number. At this stage, there is no need to use the < or > signs, though some children may discuss these if appropriate.

PUPIL TEXTBOOK 2A PAGE 9

Think together

WAYS OF WORKING Whole class teacher led (I do, We do, You do)

ASK

- Question **1**: *Are the numbers in order? Could you use counting skills to help?*
- Question **2**: *What is the same and what is different about each set of counters? How many counters fill one frame?*
- Question **3**: *What are the parts? What is the whole? What is the same and what is different about each part-whole model?*

IN FOCUS The key in question **1** is using counting as ordering and being able to recognise mistakes.

The visual representations in question **2** show partitions of numbers between 11 and 20.

Question **3** gives a more abstract way to consider partitions of 2-digit numbers.

STRENGTHEN Work with ten frames and counters to build numbers and count in 1s as each counter is added. Play a game with children in pairs, making different numbers from a 10 and some 1s.

DEEPEN Explore how many different partitions children can find for the number 11.

ASSESSMENT CHECKPOINT Assess whether children can explain how to partition any number up to 20.

ANSWERS

Question **1**: Missing numbers: 3, 13, 19

Question **2** a): 13

Question **2** b): 19

Question **2** c): 20

Question **3** a): 12

Question **3** b): 14

Question **3** c): 13

Question **3** d): 18

Question **3** e): 7

Question **3** f): 6

Question **3** g): 10

Question **3** h): Children could have written any number bonds to 19, e.g. 10, 9; 11, 8: 12, 7; 13, 6; 14, 5; 15, 4; 16, 3; 17, 2; 18, 1

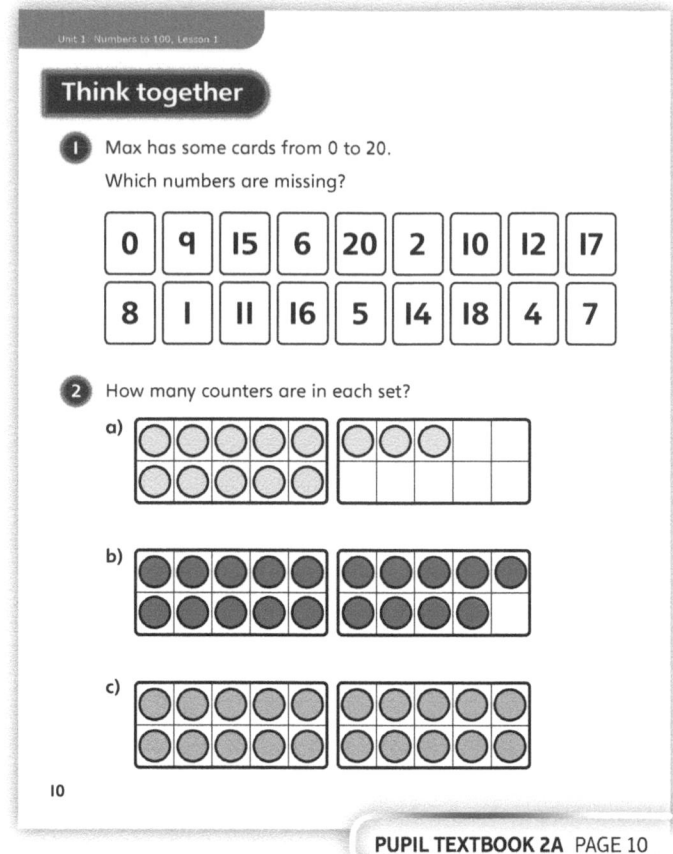

PUPIL TEXTBOOK 2A PAGE 10

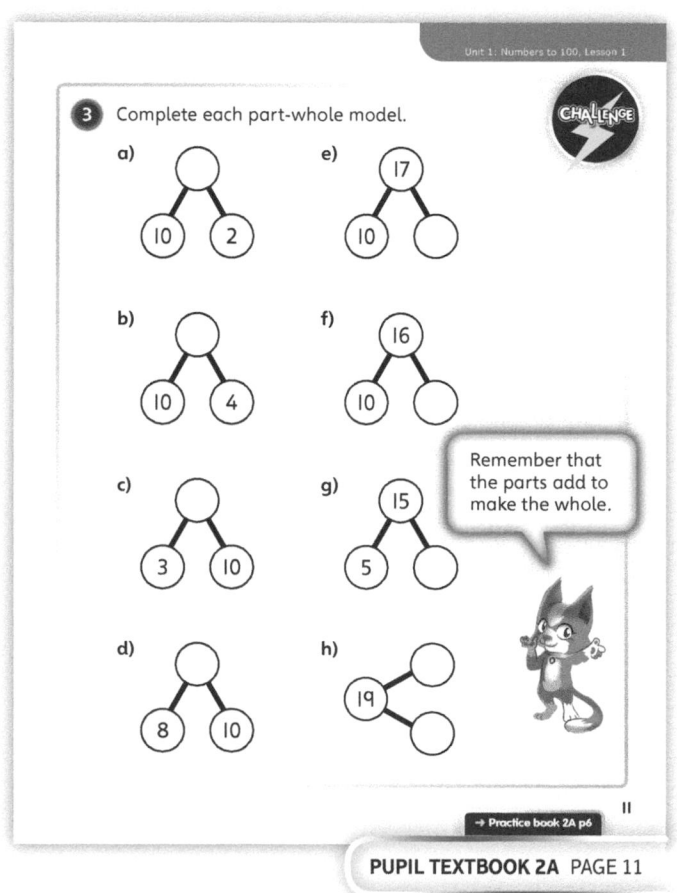

PUPIL TEXTBOOK 2A PAGE 11

Practice

WAYS OF WORKING Independent thinking

IN FOCUS Question ❶ requires children to write numbers and count in order. In question ❷, they are recognising numbers greater or less than a given number and in question ❸ they are recognising representations of 2-digit numbers as 10 and 1s.

Question ❹ uses part-whole models to show partitions of numbers into a 10 and some 1s. Question ❺ asks children to order numbers from smallest to greatest.

STRENGTHEN Focus on awareness of the counting pattern and number names. Play counting or number track games such as hopscotch, from 0 to 20.

DEEPEN Make visual patterns using counters, cubes or drawings to represent numbers 11 to 20.

ASSESSMENT CHECKPOINT Question ❶ should help you to determine if children can count accurately from 0 to 20. Questions ❷ and ❹ will show if children can compare two numbers within 20.

Assess whether children can explain the partitions of a number from 11 to 20.

ANSWERS Answers for the **Practice** part of the lesson can be found in the *Power Maths* online subscription.

Reflect

WAYS OF WORKING Pair work

IN FOCUS The **Reflect** part of the lesson requires children to order and compare numbers in the range 0 to 20.

ASSESSMENT CHECKPOINT Assess whether children can explain how they know which of two numbers is greater or less.

ANSWERS Answers for the **Reflect** part of the lesson can be found in the *Power Maths* online subscription.

After the lesson ⏸

- Return to the 'Guess the mystery number' game. Play it together as a class and check if children are demonstrating increased confidence.

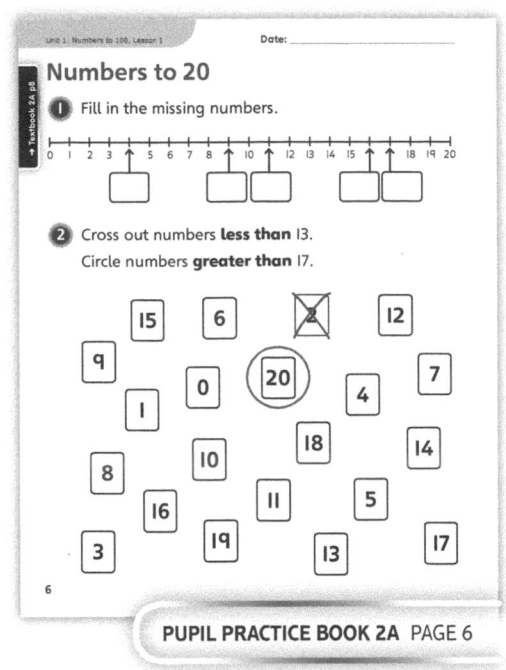

PUPIL PRACTICE BOOK 2A PAGE 6

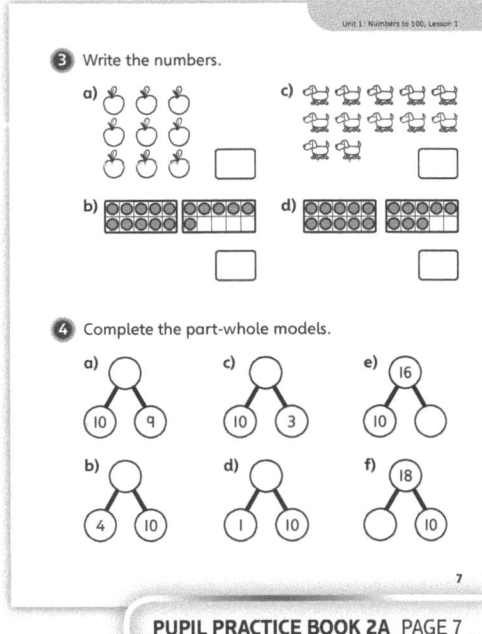

PUPIL PRACTICE BOOK 2A PAGE 7

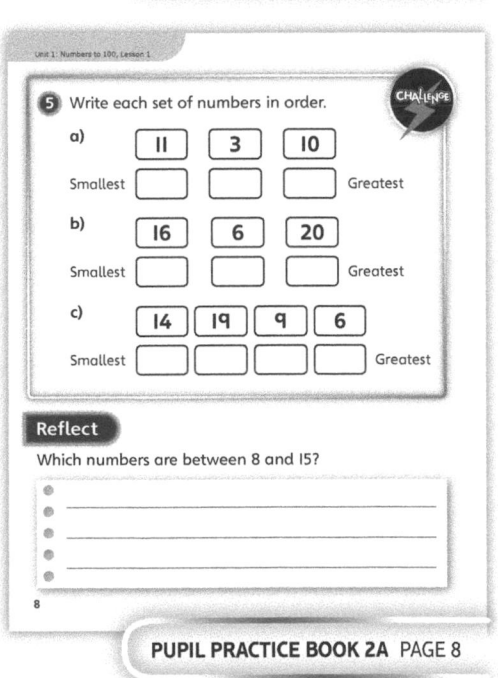

PUPIL PRACTICE BOOK 2A PAGE 8

49

Count in 10s

Learning focus

In this lesson, children count in multiples of 10 up to 100, using a variety of representations to support fluency and understanding. Children learn, for example, that 5 tens make 50, 7 tens make 70, and 90 is made up of 9 tens.

Before you teach

- How will you check children's confidence with numbers 20 to 100 and using the 100 square?
- Have children had opportunities to read and write the numbers 10, 20, 30, …, 100?

NATIONAL CURRICULUM LINKS

Year 1 Number – number and place value

Count, read and write numbers to 100 in numerals; count in multiples of twos, fives and tens.

ASSESSING MASTERY

Children can count in 10s up to 100. They can recognise representations of multiples of 10 and can say how many 10s make a given multiple of 10.

COMMON MISCONCEPTIONS

Children may mishear '-ty' numbers as '-teen' numbers, and so make counting mistakes such as 70, 80, 90, 20. Ask:
- *How many 10s are in this number? How many 10s will be in the next number?*

STRENGTHENING UNDERSTANDING

Find regular opportunities for children to quickly practise chanting the counting pattern in 10s as a class. For example, when lining up, go down the line with each child saying the next number in the count.

GOING DEEPER

Challenge children to find examples with patterns of counting in 10s, such as on a ruler. Encourage them to count in 10s using items such as coins.

KEY LANGUAGE

In lesson: counting words for 10, 20, 30; **tens (10s), ones (1s)**

Other language to be used by the teacher: pattern

STRUCTURES AND REPRESENTATIONS

Ten frame, finger numbers, 100 square

RESOURCES

Mandatory: 100 square, ten frames, counters

Optional: rulers

In the eTextbook of this lesson, you will find interactive links to a selection of teaching tools.

Quick recap

Look at a 100 square. Together, count up in 10s down the column from 10, 20, 30, and so on. Point to each of the numbers as you count.

Discover

Pair work

ASK

• Question ❶ a): *Can you see groups of 10?*
• Question ❶ a): *Can you make 10 using your fingers?*
• Question ❶ b): *How could you count? Can you think of different ways?*

IN FOCUS This requires children to recognise 10 as finger numbers, which is a form of subitising. It introduces counting in 10s as an efficient counting method.

PRACTICAL TIPS Act the activity out with the class, with some children each showing 10 fingers.

ANSWERS

Question ❶ a): 30

Question ❶ b): 40

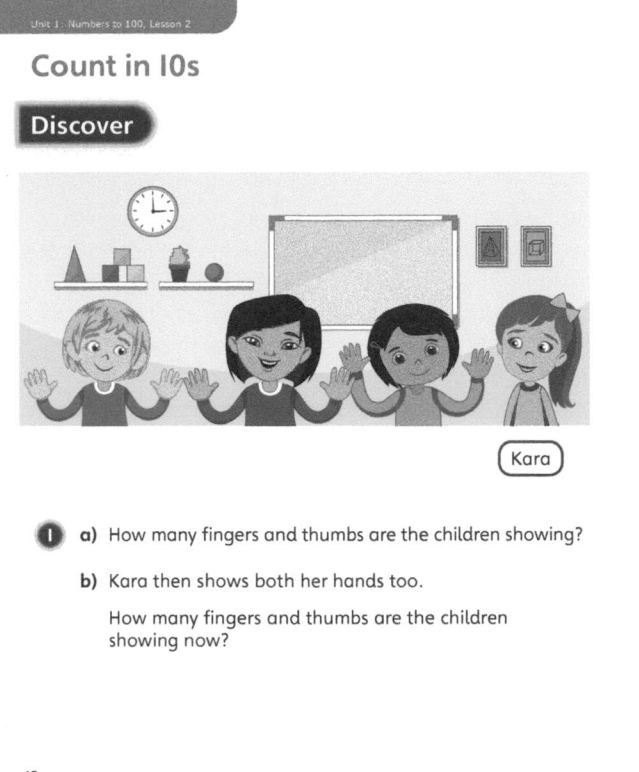

Count in 10s

Discover

Kara

❶ a) How many fingers and thumbs are the children showing?

b) Kara then shows both her hands too.

How many fingers and thumbs are the children showing now?

12

PUPIL TEXTBOOK 2A PAGE 12

Share

WAYS OF WORKING Whole class teacher led

ASK

• Question ❶ a): *How many fingers is each child showing? Can you show 10 fingers on your hands? Can you count the fingers in 10s?*
• Question ❶ b): *Can you keep counting 10 more?*

IN FOCUS The focus is on counting in 10s and also on counting 10 more than a given 10.

Share

a) Count 3 tens.

I counted in 1s.

10   20   30

You can count quicker if you count in 10s.

b) Count 4 tens.

10   20   30   40

13

PUPIL TEXTBOOK 2A PAGE 13

Think together

Think together

WAYS OF WORKING Whole class teacher led (I do, We do, You do)

ASK

- Question **1**: *How many counters does it take to fill each frame? Can you point to each frame and count them in 10s?*
- Question **2**: *Where is 20 on the number track? Where is 30 on the number track? How do you know? What do you notice about counting in 10s on a number track?*
- Question **3**: *Say each number aloud. Can you guess how many 10s makes each number? Are there any clues in the digits?*

IN FOCUS Question **1** uses ten frames to support the concept that each 10 is 10 ones. Children can use the number track in question **2** to help them consider the order of the counting pattern, as they identify each missing number. Question **3** reinforces the key learning that 10, 20, 30, and so on, are made from 1 ten, 2 tens, 3 tens, and vice versa.

STRENGTHEN Use a 100 square to support counting throughout the lesson.

DEEPEN Ask children to explore which numbers are missed out when counting in 10s.

ASSESSMENT CHECKPOINT Questions **1** and **2** assess whether children can count in 10s to 100, and question **3** assesses whether children can say how many 10s make a given number.

ANSWERS

Question **1** a): 50

Question **1** b): 60

Question **2**: 30, 40, 60, 80, 90

Question **3** a): 10 is 1 ten, 20 is 2 tens, 30 is 3 tens, 40 is 4 tens, 50 is 5 tens

Question **3** b): 70, 90, 100 is 10 tens

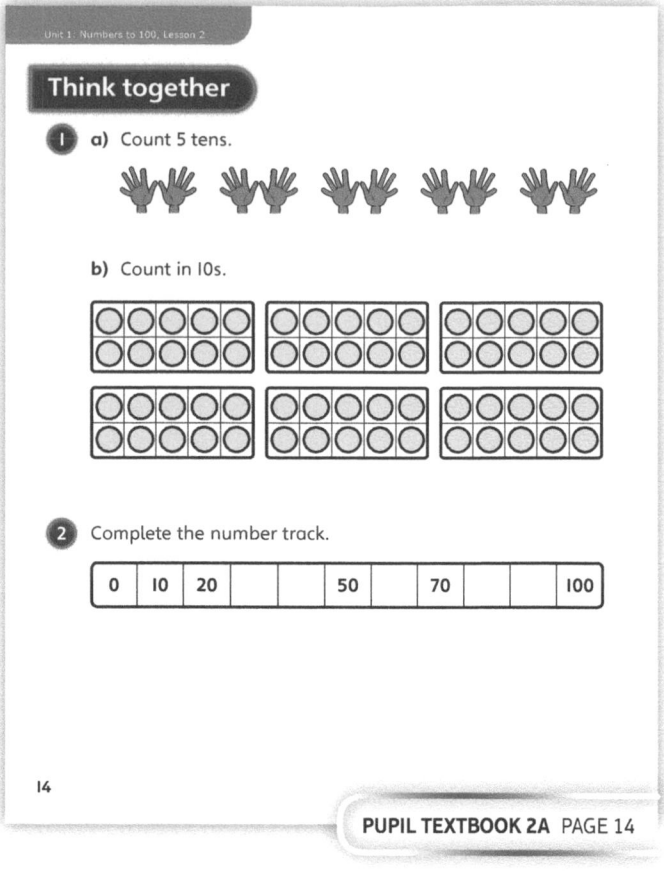

PUPIL TEXTBOOK 2A PAGE 14

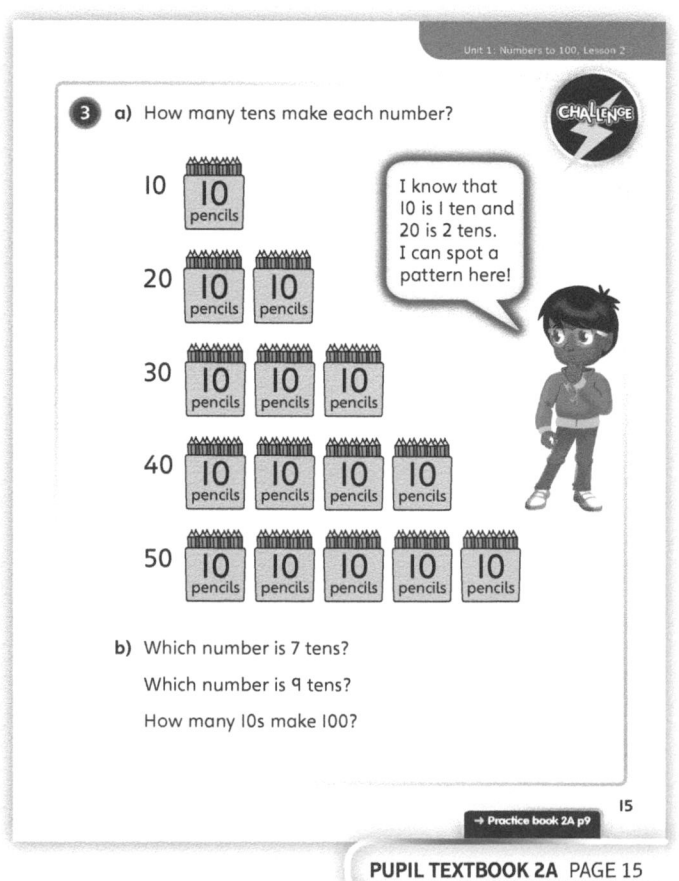

PUPIL TEXTBOOK 2A PAGE 15

Practice

WAYS OF WORKING Independent thinking

IN FOCUS Question ① rehearses the counting pattern of 10s. Question ② then moves on to counting in 10s using a number track. Question ③ uses the 100 square to support children in recognising the numbers in the 10s count as written numerals. In question ④ children are asked to count counters in 10s and write the correct number as a numeral.

STRENGTHEN Ask children to show 10 using fingers and then to repeat the gesture to show 20, then 30, then 40, and so on, counting up in 10s as they go.

DEEPEN Write a random selection of 2-digit numbers. Can children recognise by looking at the numerals which are in the 10s count pattern?

ASSESSMENT CHECKPOINT Questions ① and ② should show if children can count up in 10s to 100.

Assess whether children can explain what number is three 10s, and how many 10s make 40.

ANSWERS Answers for the **Practice** part of the lesson can be found in the *Power Maths* online subscription.

Reflect

WAYS OF WORKING Pair work

IN FOCUS The **Reflect** part of the lesson rehearses and extends fluency in counting in 10s. This will be an important skill for later lessons about numbers to 100.

ASSESSMENT CHECKPOINT Assess whether children can count accurately in 10s up to 100.

ANSWERS Answers for the **Reflect** part of the lesson can be found in the *Power Maths* online subscription.

After the lesson ⏸

- Ensure that children can count a set representation of 10s, and know to count 10 for each unit.
- Are children able to stop their 10s count at the right point to give the correct size of the group they are counting?

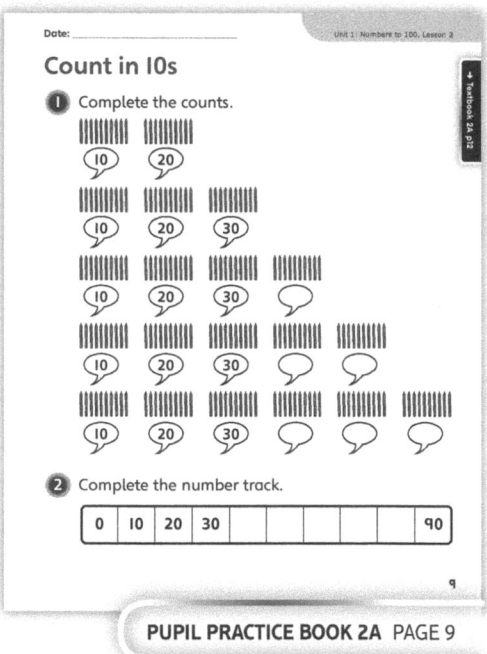

PUPIL PRACTICE BOOK 2A PAGE 9

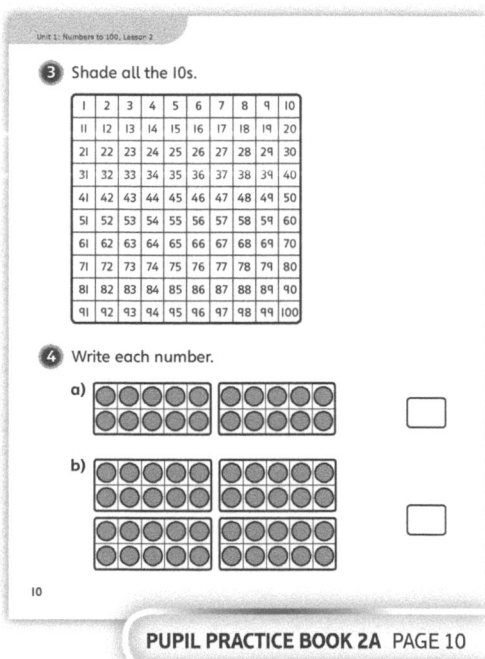

PUPIL PRACTICE BOOK 2A PAGE 10

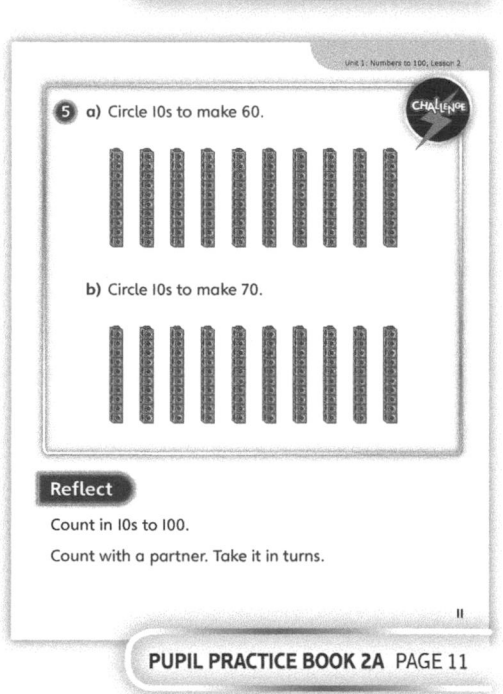

PUPIL PRACTICE BOOK 2A PAGE 11

Count in 10s and 1s

Learning focus

In this lesson, children learn to count in 10s, then count on in 1s, in order to count a certain number of objects or items, already arranged into 10s and 1s.

At this stage, the 10s and 1s are already made for children.

Before you teach ⏸

- Are children confident in counting up to 100 in 1s, counting on from any given number in 1s and counting up to 100 in 10s?

NATIONAL CURRICULUM LINKS

Year 1 Number – number and place value

Recognise the place value of each digit in a two-digit number (tens, ones).

Identify, represent and estimate numbers using different representations, including the number line.

ASSESSING MASTERY

Children can count in 10s, and then count on in 1s accurately.

COMMON MISCONCEPTIONS

Children may not know how to apply the counting principle of one number name for one item, or, in this case, one group of 10 items. Ask:
- *Did you make sure you counted every item? How can you check?*

STRENGTHENING UNDERSTANDING

Give children the opportunity to practise counting on in 1s from any given number. When counting in 10s and 1s, encourage them to take a pause after reaching the last 10, and then count on from there in 1s.

GOING DEEPER

Ask children to explore 2-digit numbers on a 100 square. Encourage them to discuss the patterns in the digits, and think about what the digits tell you about how to count up to this number in 10s and 1s.

KEY LANGUAGE

In lesson: more, tens (10s), ones (1s)

Other language to be used by the teacher: how many, pattern

STRUCTURES AND REPRESENTATIONS

Pictorial representations of 10s and 1s

RESOURCES

Mandatory: Multilink cubes

Optional: paper straws, cones

 In the eTextbook of this lesson, you will find interactive links to a selection of teaching tools.

Quick recap 🔄

Count up in 1s together, to count the number of children who are in the classroom today.

Discover

PUPIL TEXTBOOK 2A PAGE 16

WAYS OF WORKING Pair work

ASK

• Question ❶ a): *What do you notice about the cones? What is the same and what is different about each stack?*
• Question ❶ b): *Can you guess how many cones there are in total? Is 100 a good guess? Is 7 a good guess?*

IN FOCUS Children are required to recognise groups of 10, and count in 10s. They then count on in 1s from a given multiple of 10.

PRACTICAL TIPS Provide objects already grouped into 10s, for example bundles of paper straws or towers of multilink cubes.

ANSWERS

Question ❶ a): There are 40 stacked cones.

Question ❶ b): There are 43 cones altogether, 40 + 3.

Share

WAYS OF WORKING Whole class teacher led

ASK

• Question ❶ a): *How can we know that each stack has 10? How do you make sure you count the 10s and stop at the right number? Do you point as you count?*
• Question ❶ b): *Do you have to count all of them in 1s? How do you know when to change from counting in 10s to counting in 1s?*

IN FOCUS The focus is on counting sets of objects in 10s, applying the principle of counting one more 10 for each set of 10, and knowing when to switch from counting in 10s to counting in 1s.

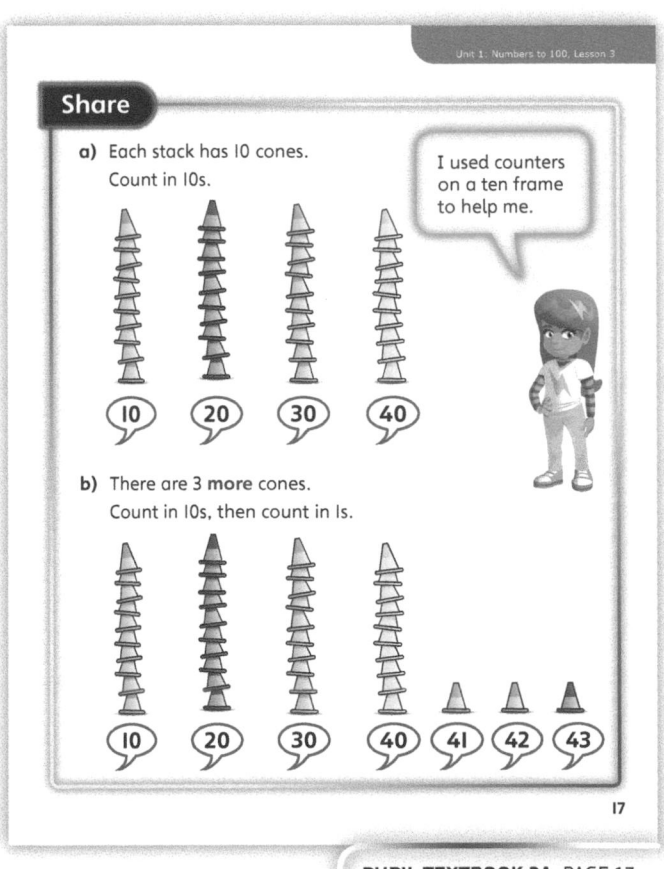

PUPIL TEXTBOOK 2A PAGE 17

Think together

Whole class teacher led (I do, We do, You do)

ASK

- Question **1**: *How many do you think there are in each different group?*
- Question **2**: *What mistakes could these children have made?*
- Question **3**: *How can you count when you are not able to see each item?*

IN FOCUS Question **1** gives fluency practice in counting in 10s then 1s. The mistakes shown in question **2** address a misconception, or counting error when grouping 10s. The image in question **3** introduces the concept of counting units of 10, without needing to see every individual item.

STRENGTHEN Recreate the questions with physical items for children to count, for example, bundles of paper straws or towers of 10 multilink cubes.

DEEPEN Discuss why it is important to count the 10s first, then the 1s. Ask: *What would happen if you counted the 1s first then the 10s?*

ASSESSMENT CHECKPOINT Questions **1**, **2** and **3** assess whether children can accurately count in 10s, then switch to counting in 1s.

ANSWERS

Question **1**: There are 32 stars.

Question **2**: Both Jo and Jim are incorrect. There are 26 seashells. They have both miscounted when grouping seashells into tens. Jo has one group of 11 and one group of 10 with 5 seashells in the middle. Jim has one group of 10 and one group of 9, with 7 in the middle.

Question **3**: There are 46 pencils, 58 pens and 42 rubbers.

PUPIL TEXTBOOK 2A PAGE 18

PUPIL TEXTBOOK 2A PAGE 19

Practice

WAYS OF WORKING Independent thinking

IN FOCUS Questions **1**, **2**, **3** and **6** require children to recognise the 10s, and count in 10s and 1s. Question **4** explores the idea of 2 tens as 20 ones. In question **5**, children are asked to count by first finding the groups of 10.

STRENGTHEN Provide physical objects for children to count that are already organised into groups of 10s and 1s.

DEEPEN Question **7** gives children an opportunity to practise counting in 10s and 1s up to 99. Use this as an opportunity for children to show how they are applying an awareness of 10s and 1s, and using pattern making skills.

ASSESSMENT CHECKPOINT Questions **1**, **2** and **3** assess whether children are able to count accurately in 10s and 1s to find an answer.

ANSWERS Answers for the **Practice** part of the lesson can be found in the *Power Maths* online subscription.

Reflect

WAYS OF WORKING Pair work

IN FOCUS The **Reflect** part of the lesson requires children to demonstrate counting in 10s and then 1s, in order to count a given number of objects.

ASSESSMENT CHECKPOINT Assess whether children can model the correct way to count the items accurately by counting in 10s, and then counting on in 1s.

ANSWERS Answers for the **Reflect** part of the lesson can be found in the *Power Maths* online subscription.

After the lesson ⏸

• Check that children are confident when switching the rhythm of the count from counting on in 10s to counting on in 1s.

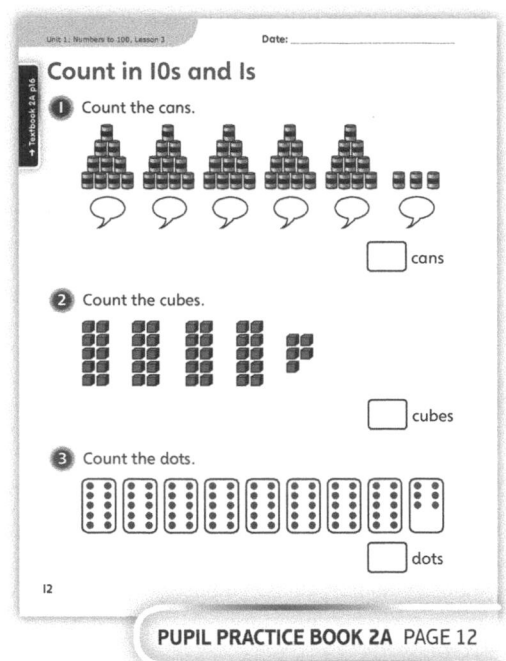

PUPIL PRACTICE BOOK 2A PAGE 12

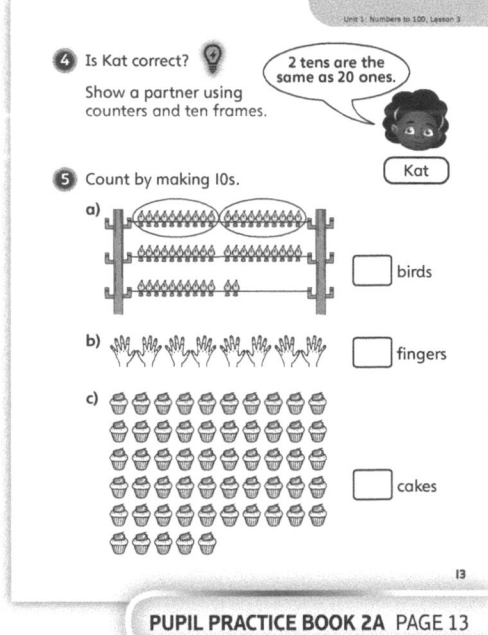

PUPIL PRACTICE BOOK 2A PAGE 13

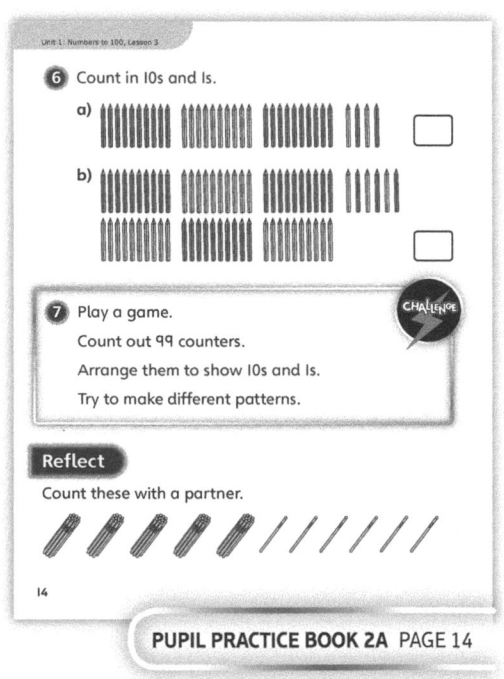

PUPIL PRACTICE BOOK 2A PAGE 14

Recognise 10s and 1s

Learning focus

In this lesson, children count in 10s and 1s, to and from 2-digit numbers.

Before you teach ⏸

- Are children confident about counting in 10s to 100, and about counting in 1s from different starting numbers to 100?

NATIONAL CURRICULUM LINKS

Year 1 Number – number and place value

Recognise the place value of each digit in a two-digit number (tens, ones).

Identify, represent and estimate numbers using different representations, including the number line.

ASSESSING MASTERY

Children can recognise that a 2-digit number is composed of 10s and 1s, and can count 10s and 1s to find a given amount.

COMMON MISCONCEPTIONS

Some children may have difficulty counting in different sized units and knowing when to switch from counting in 10s to counting on in 1s. Ask:
- *How many 10s are there? How will you check? Is this a 10 or a 1?*

STRENGTHENING UNDERSTANDING

Find regular opportunities for children to practise counting with numbers between 11 and 20. Start the count with a 10 and then ask children to continue 11, 12, 13, and so on.

GOING DEEPER

Challenge children to explore patterns in the numbers in a 100 square, using the language of 10s and 1s to explain their reasoning.

KEY LANGUAGE

In lesson: tens (10s), ones (1s)

Other language to be used by the teacher: count

STRUCTURES AND REPRESENTATIONS

Base 10 equipment, ten frame

RESOURCES

Mandatory: base 10 equipment, multilink cubes

Optional: objects in groups of 10, for example, paper straws, place value equipment

 In the eTextbook of this lesson, you will find interactive links to a selection of teaching tools.

Quick recap 🔄

Show 10s on your fingers by showing 10 fingers, then closing your hands, showing 10 fingers again, then closing your hands, and so on. Keep going, with children counting in 10s as you go.

Discover

WAYS OF WORKING Pair work

ASK

- Question ① a): *Can you count in 1s?*
- Question ① a): *Can you think of a quicker way to count?*
- Question ① b): *What do you notice about how some cubes have been joined together? Why do you think they are grouped like this?*

IN FOCUS Children are first required to count in 1s, then to demonstrate counting efficiently in 10s and then in 1s.

PRACTICAL TIPS Make available rods of 10 base 10 cubes to support children in their count.

ANSWERS

Question ① a): 23

Question ① b): 23

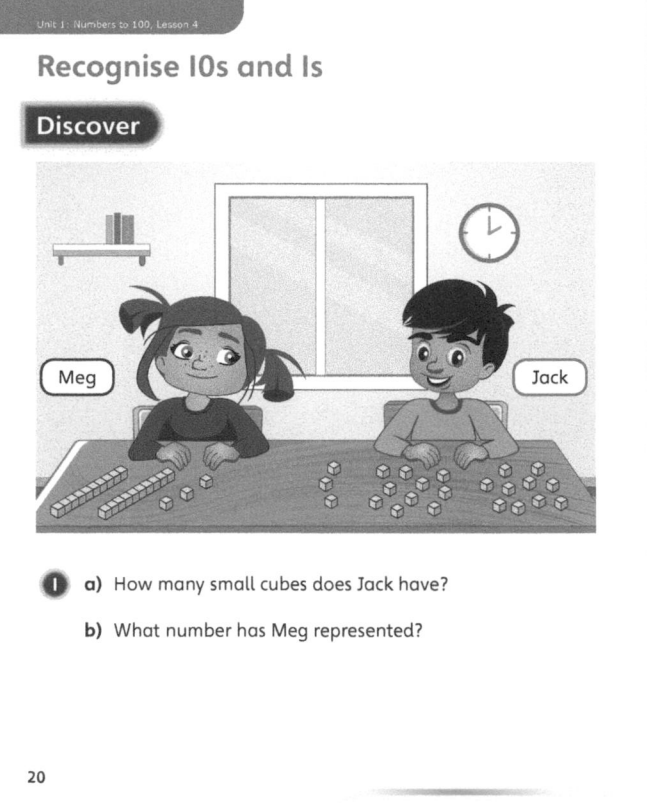

Share

WAYS OF WORKING Whole class teacher led

ASK

- Question ① a): *Can you see how it helps to group the cubes into 10s? Can you see how to count in 10s and then to count on in 1s?*
- Question ① b): *Have you seen 10s rods before? How could these make it easier to count?*

IN FOCUS In question ① a), 10 ones are grouped into 10s to help with the count. This is a key skill. Ensure that children have plenty of practice in switching from counting in 10s to counting on in 1s.

In question ① b), the groups of 10 are already formed. Children should aim to see the total without counting every cube in 1s.

Think together

Whole class teacher led (I do, We do, You do)

ASK

- Question ❶: *Where will you start counting? How will you know when to count on in 1s?*
- Question ❷: *How have the 10s and 1s been organised?*
- Question ❸: *Look at each representation; what is the same and what is different? Where are the 10s? Where are the 1s? What is a good way to count these?*

IN FOCUS In question ❶ children follow a given pattern, to further practise counting in 10s and then 1s. The first image in question ❷ is organised into 10s then 1s. Question ❷ b) needs reorganising. Children may point and count, or make the numbers themselves from 10s and 1s equipment. Question ❸ focuses on how the order of the 10s and 1s should not affect the total of the number represented.

STRENGTHEN Provide place value equipment for children to sort into 10s and 1s, and then count.

DEEPEN Ask children to show 30 with place value equipment. Instruct them to add 1, then say the number, add another 1 and say the new number. Challenge them to explore how the number changes as more and more 1s are added. Discuss the possibility of the existence of numbers such as 'thirty-ten', or 'thirty-eleven'.

ASSESSMENT CHECKPOINT Assess whether children can count a given number presented as 10s and 1s with base 10 and place value equipment.

ANSWERS

Question ❶: 34; Children count 10, 20, 30, 31, 32, 33, 34.

Question ❷ a): 64
b): 52

Question ❸: a), b) and c) are the same in that they all have 4 tens and 2 ones. They are different in that a) is shown in ten frames and b) and c) using multilink cubes. The cubes in b) and c) are oriented differently.

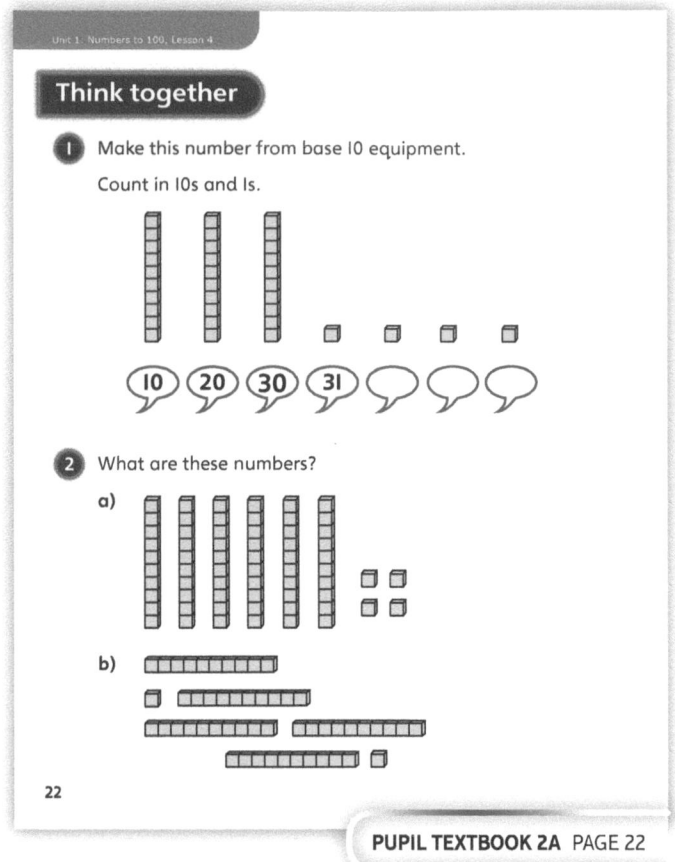

PUPIL TEXTBOOK 2A PAGE 22

PUPIL TEXTBOOK 2A PAGE 23

Practice

WAYS OF WORKING Independent thinking

IN FOCUS In questions ❶ and ❷ children are counting in 10s then 1s, with more scaffolding in question ❶. In question ❸ children are required to match different representations of numbers shown as 10s and 1s. Question ❹ explores the number system and where it bridges 10s boundaries, and question ❺ is a context-based word problem.

STRENGTHEN Adapt the **Reflect** activity so that children can explore the base 10 equipment more fully.

DEEPEN Challenge children to investigate question ❹ more fully. Ask: *What would the number 'eighty-twelve' be called?*

ASSESSMENT CHECKPOINT Assess whether children can explain how and why numbers are partitioned into 10s and 1s.

ANSWERS Answers for the **Practice** part of the lesson can be found in the *Power Maths* online subscription.

PUPIL PRACTICE BOOK 2A PAGE 15

Reflect

WAYS OF WORKING Pair work

IN FOCUS The **Reflect** part of the lesson provides a practical and fun way for children to practise counting in 10s and 1s.

ASSESSMENT CHECKPOINT Assess whether children are able to accurately and efficiently count the numbers shown in the pictures.

ANSWERS Answers for the **Reflect** part of the lesson can be found in the *Power Maths* online subscription.

PUPIL PRACTICE BOOK 2A PAGE 16

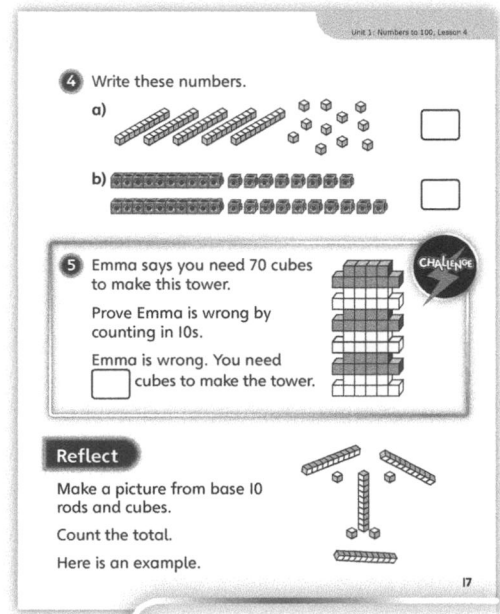

PUPIL PRACTICE BOOK 2A PAGE 17

After the lesson ⏸

- Ensure that children understand how to count first in 10s and then switch to counting on in 1s.

Build a number from 10s and 1s

Learning focus

In this lesson, children develop skills of counting in 10s and 1s, to form their own 10s and 1s, and represent and build an understanding of the number of 10s and 1s comprising a given 2-digit number.

Before you teach

- Are children familiar with handling and arranging 10s and 1s equipment?
- Have you given opportunities to count in 10s up to 100 and to count on in 1s from a given multiple of 10?
- Can you discuss as a class how and why a 10s rod represents a group of 10 ones?

NATIONAL CURRICULUM LINKS

Year 1 Number – number and place value

Recognise the place value of each digit in a two-digit number (tens, ones).

Identify, represent and estimate numbers using different representations, including the number line.

ASSESSING MASTERY

Children can draw or make representations of 10s and 1s to show a given 2-digit number.

COMMON MISCONCEPTIONS

Some children may not be confident with counting skills within 100, especially accurately counting in 10s then 1s. If this skill is not in place, the learning will be difficult to access. Ask:
- *Can you count up to 50 in 10s? Can you count from 50 to 60 in 1s?*

STRENGTHENING UNDERSTANDING

Give plenty of opportunities for children to handle and build confidence with 10s and 1s place value equipment. Introduce free-play activities such as building bridges, vehicles or creatures using the equipment, and model how to talk about the 10s rods and 1s cubes.

GOING DEEPER

Ask children to explore how to predict the number of 10s and 1s needed to make a given 2-digit number, by looking at the digits. This is in preparation for a more formal discussion of place value in the next lesson.

KEY LANGUAGE

In lesson: tens (10s), ones (1s)

STRUCTURES AND REPRESENTATIONS

Base 10 equipment; hand-drawn pictures of 10s and 1s (line for a 'ten', dot for a 'one')

RESOURCES

Mandatory: base 10 equipment, place value equipment

 In the eTextbook of this lesson, you will find interactive links to a selection of teaching tools.

Quick recap

Write some different 2-digit numbers on the board. Ask children to read the numbers out loud. Now say some 2-digit numbers and ask children to write them as numerals.

Discover

WAYS OF WORKING Pair work

ASK

- Question ① a): *What do you know about this equipment?*
- Question ① b): *Can you show me a 10? Can you show me a 1?*

IN FOCUS Children will need to recognise how to make a multiple of 10 from 10s rods, by placing a 10, then counting the next 10, and continuing until they reach 50. They will then be required to make a 2-digit number from 10s and 1s.

PRACTICAL TIPS Play a 'Show me' game together as a class. For example, *Show me 2 tens. Show me 1 ten and 2 ones,* and so on.

ANSWERS

Question ① a): Danny can use five 10s (five rods).

b): Maya can use five 10s and two 1s.

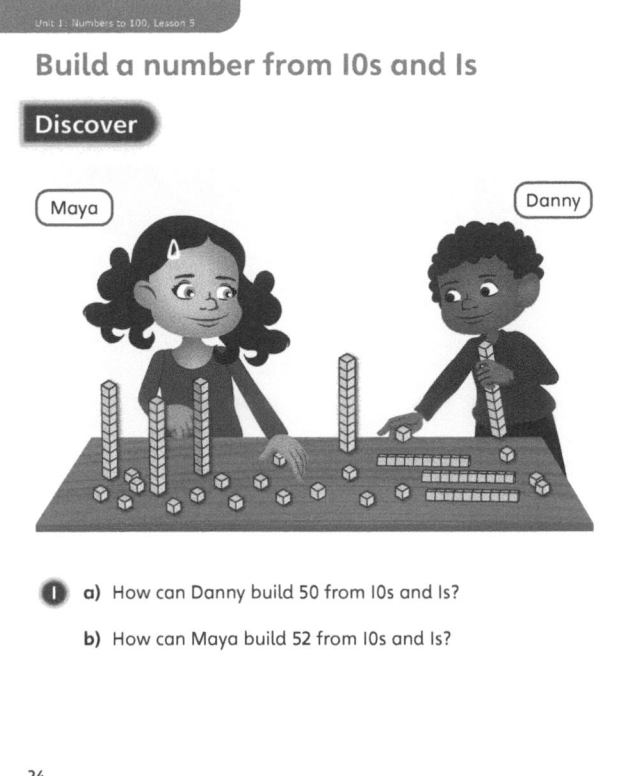

Build a number from 10s and 1s

Discover

Maya   Danny

① a) How can Danny build 50 from 10s and 1s?

b) How can Maya build 52 from 10s and 1s?

24

PUPIL TEXTBOOK 2A PAGE 24

Share

WAYS OF WORKING Whole class teacher led

ASK

- Question ① a): *How many 10s are there? Can you count in 10s? Can you show me how to keep adding more 10s and counting in 10s until you reach 50?*
- Question ① b): *Can we keep counting in 10s? How do we count on in 1s? What number does this show in total?*

IN FOCUS In question ① a), children are counting in 10s and placing 10s until they reach the target number. In question ① b), they are counting on from a multiple of 10, adding 1s to make the given 2-digit number. Make sure children also know how to quickly draw 10s rods as simple but clear neat lines, and 1s as clear dots.

Share

a) Count in 10s.

Each rod is worth 10. I remember this from the last lesson!

(10) (20) (30) (40) (50)

Danny does not need any 1s to make 50.

b) Count in 10s, then count in 1s.

(10) (20) (30) (40) (50) (51) (52)

Maya needs 5 tens and 2 ones to make 52.

25

PUPIL TEXTBOOK 2A PAGE 25

Think together

Whole class teacher led (I do, We do, You do)

ASK

- Question ❶: *Can you predict how many 10s you will need to place or draw?*
- Question ❷: *Do you notice how the number changes as you add more 1s?*
- Question ❸: *Can you predict the number that 6 tens and 4 ones will make? Can you make or draw the number and count to check?*

IN FOCUS In question ❶ children are counting in 10s, and then adding more 1s. For question ❷, children will need to notice the change in a number when adding a 1, then another 1, and another 1, and so on. Question ❸ is starting to prepare children to understand the place value of each digit more fully, ready for the next lesson.

STRENGTHEN Play 'Count as you draw'. Model how to draw a 10 line then count it as '10', draw another line and count '20', and so on.

DEEPEN Discuss what happens when you add or draw another 1 to a number such as 29, 39, or even 99.

ASSESSMENT CHECKPOINT Assess whether children can say that 54 is 5 tens and 4 ones, after drawing 10s and 1s to make 54 or making the number using equipment.

ANSWERS

Question ❶ a): Children should draw or make 70 with 7 tens.

Question ❶ b): Children should draw or make 72 with 7 tens and 2 ones.

Question ❷: Children should either draw or make 40 with 4 tens and then add on 1s to make 41, 42, 43, 44 and 45.

Question ❸ a): 64

Question ❸ b): 87

PUPIL TEXTBOOK 2A PAGE 26

PUPIL TEXTBOOK 2A PAGE 27

Practice

WAYS OF WORKING Independent thinking

IN FOCUS In question ❶ children draw 10s rods to show multiples of 10. In question ❷ they draw 10s to show a multiple of 10, and then draw 1s to make given 2-digit numbers. Question ❸ asks children to match numbers to 10s and 1s and question ❹ requires them to recognise and draw the number of 10s and 1s in 2-digit numbers. Question ❺ explores similarities and differences in 2-digit numbers.

STRENGTHEN Model how to quickly draw simple base 10 equipment, starting with 10s and then 1s.

DEEPEN Use question ❺ to prompt a discussion about how the 10s and 1s vary between different numbers. Link this to numbers on a 100 square.

ASSESSMENT CHECKPOINT Assess whether children can say the number of 10s and number of 1s that comprise a given 2-digit number.

ANSWERS Answers for the **Practice** part of the lesson can be found in the *Power Maths* online subscription.

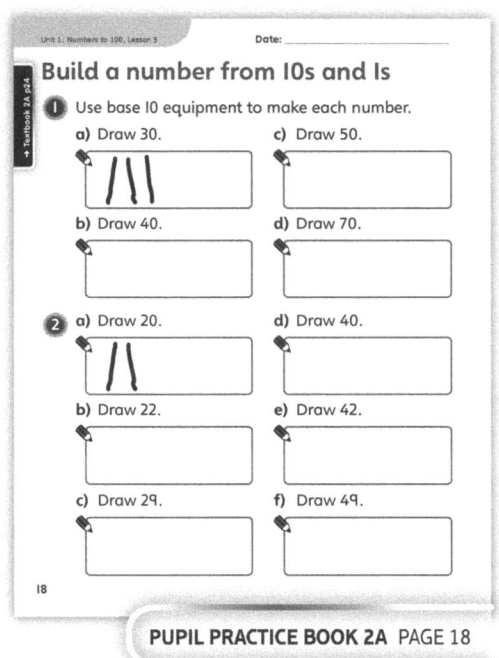

PUPIL PRACTICE BOOK 2A PAGE 18

Reflect

WAYS OF WORKING Pair work

IN FOCUS The **Reflect** part of the lesson requires children to explore and discuss variations in 10s and 1s of numbers that have similar digits.

ASSESSMENT CHECKPOINT Assess whether children can correctly recognise and describe how many 10s and how many 1s there are in given 2-digit numbers.

ANSWERS Answers for the **Reflect** part of the lesson can be found in the *Power Maths* online subscription.

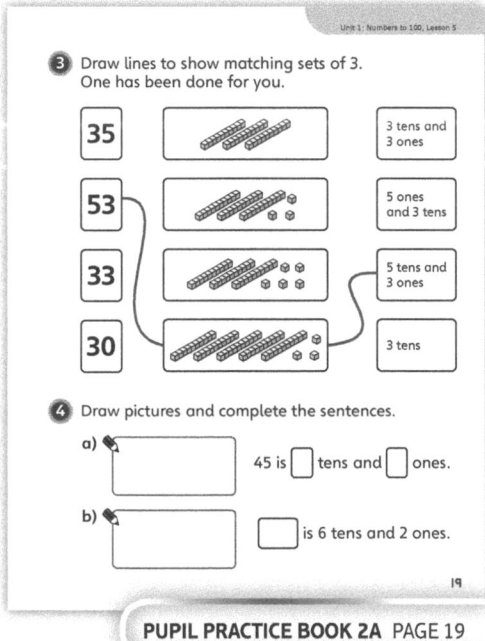

PUPIL PRACTICE BOOK 2A PAGE 19

After the lesson ⏸

- Write some 2-digit numbers on the board, or show them on a 100 square. Ensure that children can accurately represent each number using base 10 equipment or drawings of 10s and 1s.

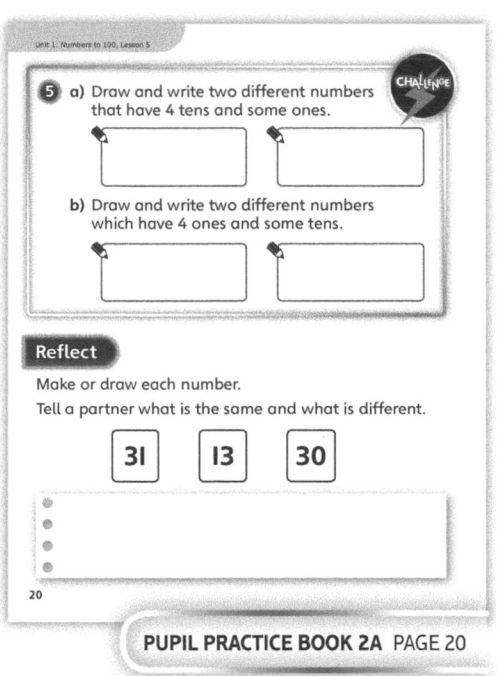

PUPIL PRACTICE BOOK 2A PAGE 20

Use a place value grid

Learning focus

In this lesson, children will learn to use a place value grid to show the value of digits within a 1- or 2-digit number.

Before you teach ⏸

- Which resources will best support recognising the 10s and 1s in a number?
- How will you link this lesson to the previous lesson on partitioning?

NATIONAL CURRICULUM LINKS

Year 1 Number – number and place value

Recognise the place value of each digit in a 2-digit number (tens, ones).

Identify, represent and estimate numbers using different representations, including the number line.

ASSESSING MASTERY

Children can show the value of the digits in a 1- or 2-digit number using a place value grid. They can convert different concrete and pictorial representations of numbers into 10s and 1s and place these correctly into the place value grid.

COMMON MISCONCEPTIONS

When filling in the place value grid, children may fill the 10s column with, for example, 60, instead of 6, when recording a number such as 63. Ask:

- *Can you show that number using cubes? How many 10s can you count? Are there 60 tens or 6?*

STRENGTHENING UNDERSTANDING

To strengthen understanding of the place value grid, provide children with pre-made representations of 10s and 1s. Using a large version of the place value grid, encourage children to make the number first, before putting the 10s and 1s in the correct columns. Once this is done children can count how many 10s and how many 1s there are and complete the columns with digit cards.

GOING DEEPER

Encourage children to find patterns. Ask: *Can you find all the numbers that have a 3 in the 1s column? Can you find all the numbers that have a 3 in the 10s column? What is the same and what is different about the numbers?*

KEY LANGUAGE

In lesson: place value grid, number, 10s, 1s, place value

Other language to be used by the teacher: digit, value

STRUCTURES AND REPRESENTATIONS

Place value grid, base 10 equipment, number line, digit cards

RESOURCES

Mandatory: place value grid, base 10 equipment, number grid

Optional: large place value grid

 In the eTextbook of this lesson, you will find interactive links to a selection of teaching tools.

Quick recap

Write these numbers on the board:

25, 20, 30

Ask children to choose the odd one out, and to explain why they chose that number.

Discover

Use a place value grid

WAYS OF WORKING Pair work

ASK

- Question ❶: *How do you know how many 10s a number has?*
- Question ❶: *How many numbers in the picture have 3 tens?*
- Question ❶ a): *Is Mr Taylor's number easier to find than 30?*
- Question ❶ b): *Can you show the numbers in a different way?*

IN FOCUS Use this part of the lesson to play a similar game in the classroom. Either you or the children could give the rest of the class clues about a number in the grid. Encourage children to give their reasoning when offering their suggestions.

PRACTICAL TIPS Use base 10 equipment to make each number, before discussing the 10s and 1s.

ANSWERS

Question ❶ a): 32

Question ❶ b): The 3 represents 3 tens so 30.
The 2 represents 2 ones so 2.

Discover

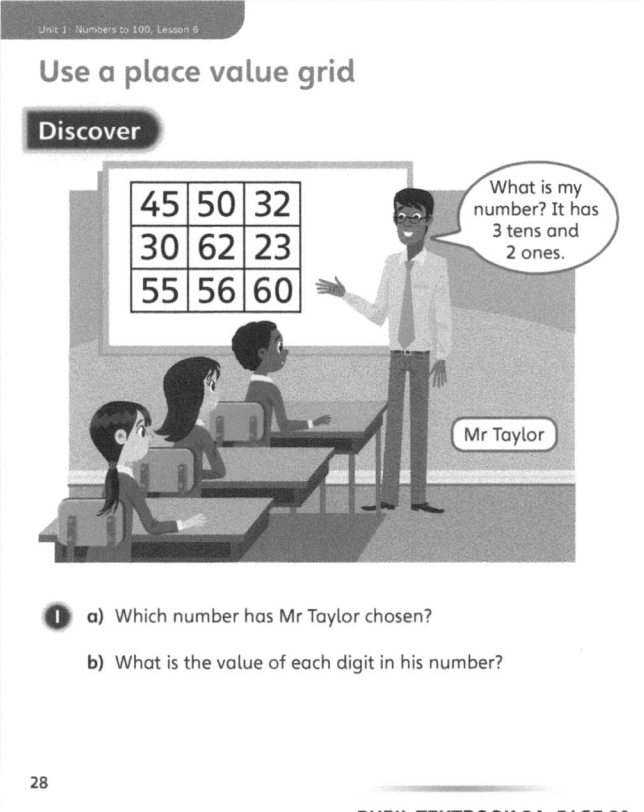

❶ a) Which number has Mr Taylor chosen?

b) What is the value of each digit in his number?

28

PUPIL TEXTBOOK 2A PAGE 28

Share

WAYS OF WORKING Whole class teacher led

ASK

- Question ❶ a): *Which numbers on the number grid have 3 tens? Which have 2 ones? Can you find the number that has 3 tens and 2 ones?*
- Question ❶ b): *Can you make the number from base 10 equipment, then say how many 10s and 1s there are?*

IN FOCUS In question ❶ children are recognising how the digits tell you how many 10s and how many 1s are in a given 2-digit number.

In question ❶ b) children start to use a place value grid to explore the place value of 2-digit numbers more formally.

Share

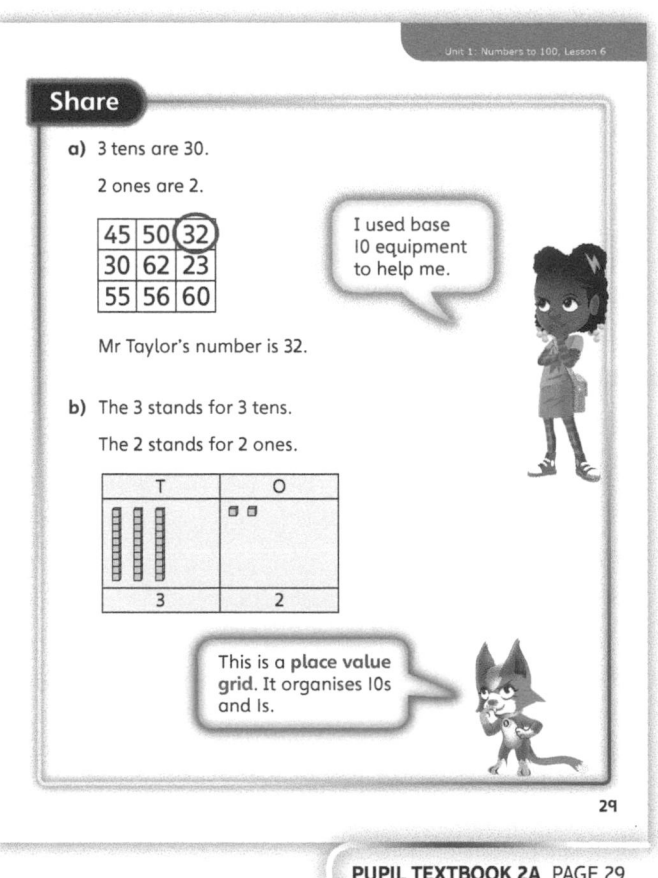

67

Think together

Whole class teacher led (I do, We do, You do)

ASK

- Question **1**: *How does using base 10 equipment in a place value grid help you to think about your answer?*
- Question **2**: *Can you predict the number of 10s and 1s, by thinking about the digits?*
- Question **3**: *Can you predict the digits of the numbers you can make by thinking about the 10s and 1s?*

IN FOCUS In question **1** children are starting to recognise place value through the use of base 10 equipment. In question **2**, they are using base 10 equipment and a place value grid to build their understanding. Question **3** explores using 0 as a place holder.

STRENGTHEN Provide base 10 equipment for children to make the numbers on their desks and explore this concept practically.

DEEPEN Use Dexter and Flo's conversation in question **3** to discuss what 0 in the 10s column represents. For example, ask: *How much is 0 tens? Is it possible to put 0 in the 10s column? Is Flo correct?* Encourage children to share their ideas. Ask: *Have you found any different solutions? How can you check your answers?* Give children the opportunity to swap their answers with a partner and check each other's work.

ASSESSMENT CHECKPOINT Assess whether children can correctly predict the value of each digit, and then justify their answer using base 10 equipment and place value grids.

ANSWERS

Question **1**:

T	O
|||||	☐☐☐☐☐

T	O

T	O
|||||	☐☐☐☐☐

The 5 in 45 stands for five 1s; the 5 in 50 stands for five 10s and the 5s in 55 stand for five 10s and five 1s.

Question **2** a): 41

Question **2** b): 65

Question **2** c): 30

Question **3**: Children could make these other numbers: 2, 5, 25, 50, 52.

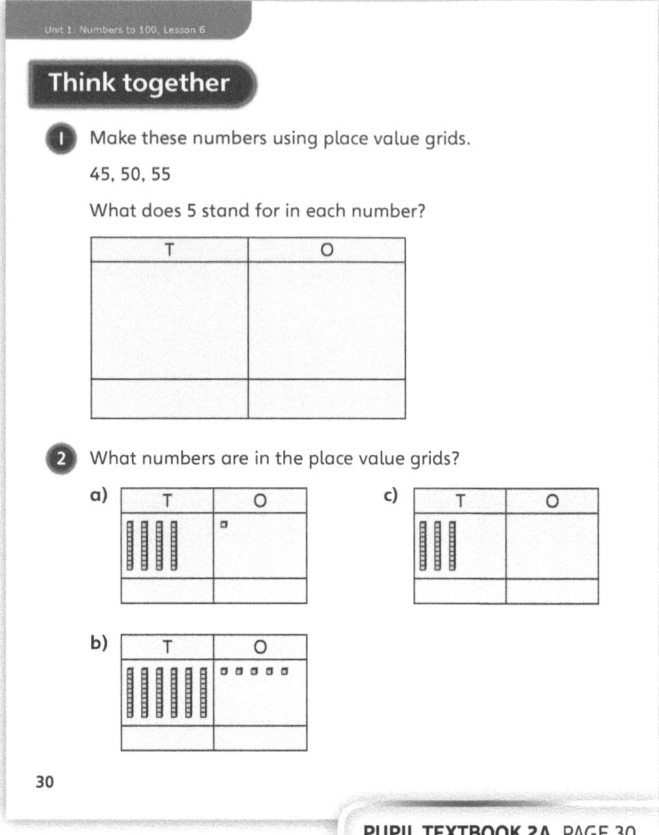

Unit 1: Numbers to 100, Lesson 6

Think together

1 Make these numbers using place value grids.

45, 50, 55

What does 5 stand for in each number?

T	O

2 What numbers are in the place value grids?

a)
T	O

c)
T	O

b)
T	O

30

PUPIL TEXTBOOK 2A PAGE 30

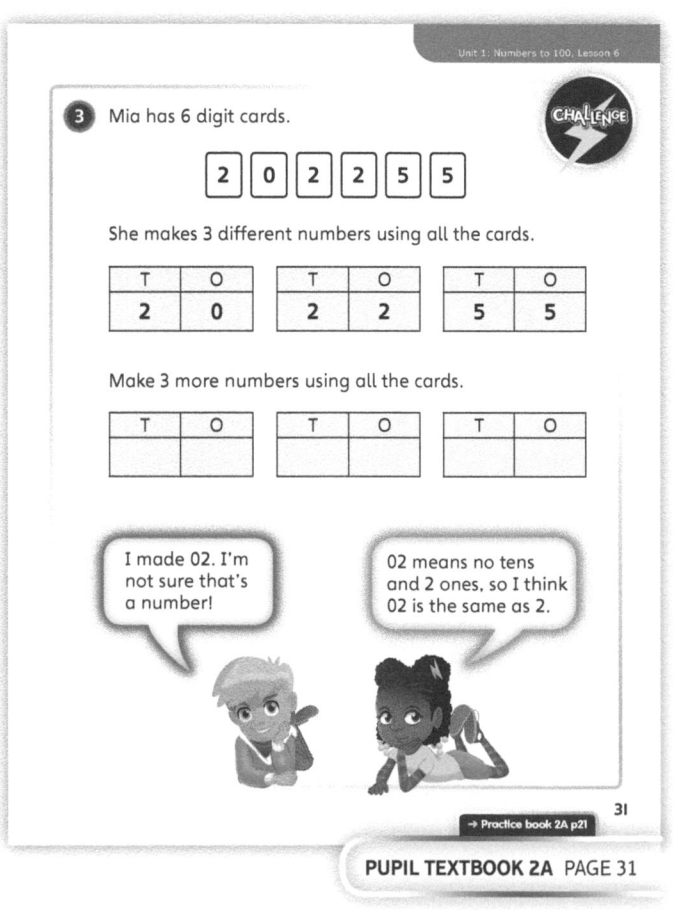

3 Mia has 6 digit cards.

2 0 2 2 5 5

She makes 3 different numbers using all the cards.

T	O
2	0

T	O
2	2

T	O
5	5

Make 3 more numbers using all the cards.

T	O

T	O

T	O

I made 02. I'm not sure that's a number!

02 means no tens and 2 ones, so I think 02 is the same as 2.

31

→ Practice book 2A p21

PUPIL TEXTBOOK 2A PAGE 31

Practice

WAYS OF WORKING Independent thinking

IN FOCUS In question ① children need to recognise that each digit in a number represents the number of 10s or 1s, depending on its position in a place value grid. In questions ② and ③ children recognise the place value of each digit in given 2-digit numbers and in question ③ they also confirm this using drawings. In question ④, children count place value equipment to complete place value grids. In question ⑤ children create numbers using different combinations of digits and in question ⑥ they recognise different possible answers based on certain criteria.

STRENGTHEN Encourage children to use base 10 equipment alongside their independent practice. Model how to predict, then use the base 10 equipment to confirm answers.

DEEPEN Challenge children to think about 0 as a place holder in the 1s column. Ask: *What is special about 0? Why can 0 be confusing for some people?*

ASSESSMENT CHECKPOINT Assess whether children can identify the value of each digit in a 2-digit number.

ANSWERS Answers for the **Practice** part of the lesson can be found in the *Power Maths* online subscription.

Reflect

WAYS OF WORKING Pair work

IN FOCUS The **Reflect** part of the lesson requires children to demonstrate the value of each '8' digit. Encourage children to use a place value grid to help them. Once children have worked independently on the problem, give them the opportunity to share their ideas with their partner.

ASSESSMENT CHECKPOINT Assess whether children can confidently explain the place value of each number, using a place value grid to support their thinking and justifications.

ANSWERS Answers for the **Reflect** part of the lesson can be found in the *Power Maths* online subscription.

After the lesson ⏸

- Check that children know that the '2' in 27 stands for 2 tens, or for 20.

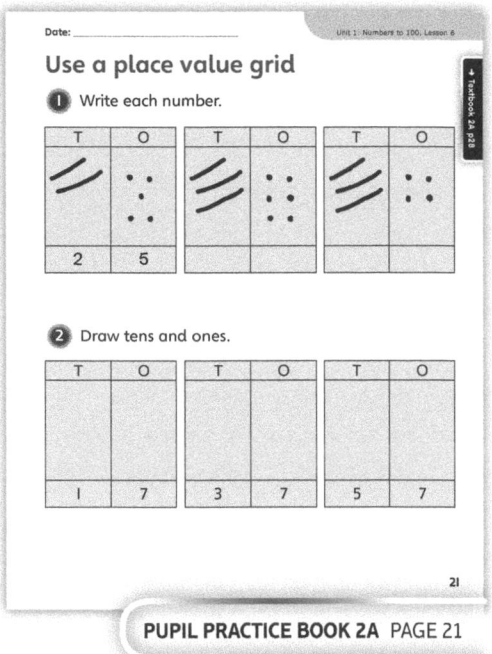

PUPIL PRACTICE BOOK 2A PAGE 21

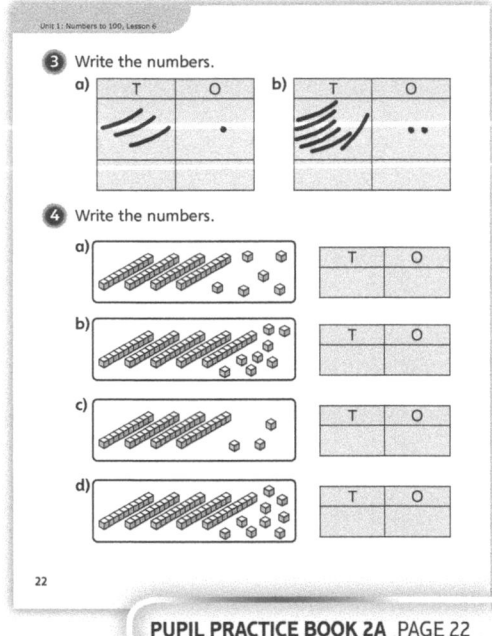

PUPIL PRACTICE BOOK 2A PAGE 22

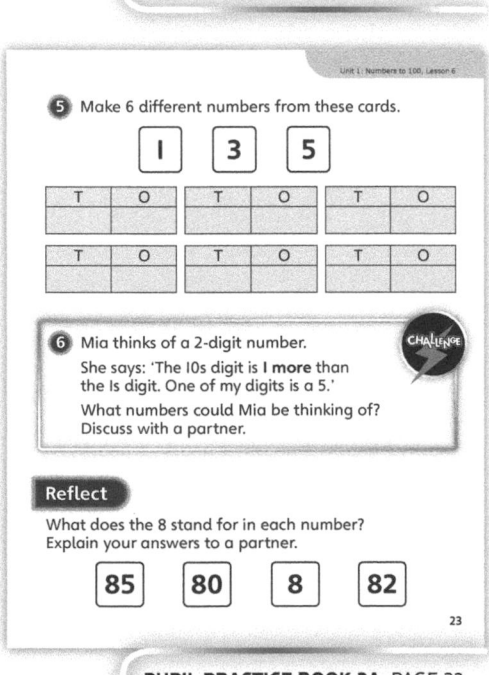

PUPIL PRACTICE BOOK 2A PAGE 23

Partition numbers to 100

Learning focus

In this lesson, children develop their understanding of the place value for 10s and 1s to partition 2-digit numbers.

Before you teach

- Can children build 2-digit numbers using base 10 equipment?
- Can children interpret representations of 2-digit numbers?
- Have children practised counting in 2-digit numbers from a given multiple of 10?

NATIONAL CURRICULUM LINKS

Year 1 Number – number and place value

Recognise the place value of each digit in a two-digit number (tens, ones).

Identify, represent and estimate numbers using different representations, including the number line.

ASSESSING MASTERY

Children can partition a given 2-digit number into 10s and 1s and they can combine partitions of 10s and 1s to form a 2-digit number.

COMMON MISCONCEPTIONS

Children may partition a number into parts incorrectly. For example, they partition 26 into 2 and 6, rather than 20 and 6. Ask:
- *What is the value of this digit? Does it show 2 ones or 2 tens? How do you know?*

STRENGTHENING UNDERSTANDING

Provide base 10 equipment for children to use alongside partitioning questions, so that they can check and confirm their partitioning.

GOING DEEPER

Challenge children to discuss unusual partitions, such as 50 being partitioned into 50 and 0.

KEY LANGUAGE

In lesson: partitioning, partition, part, whole

Other language to be used by the teacher: tens (10s), ones (1s)

STRUCTURES AND REPRESENTATIONS

Part-whole model, base 10 equipment

RESOURCES

Mandatory: base 10 equipment

Optional: ten frames, counters

 In the eTextbook of this lesson, you will find interactive links to a selection of teaching tools.

Quick recap

Show or draw some simple base 10 equipment that makes a 2-digit number. Ask children to say or write the number, and explain how they knew the correct digits.

Discover

WAYS OF WORKING Pair work

ASK

- Question ① a): *What are the children looking at? What do you think they have been doing?*
- Question ① b): *What number is in the circle?*
- Question ① b): *What do you think might go in the two circles below?*
- Question ① b): *Have the children used any equipment you recognise? What might they have used it for?*

IN FOCUS The picture should remind children of their previous learning as it references resources and equipment they met in the previous lesson. Question ① a) and b) develops the idea that an addition calculation can be written using the two numbers created from partitioning.

PRACTICAL TIPS Encourage children to explore a variety of different ways to build 10s and 1s, depending on the equipment that you have available.

ANSWERS

Question ① a): Children should make 56 using 5 tens and 6 ones.

Question ① b): 50 and 6.

Partition numbers to 100

Discover

① **a)** Make 56 using base 10 equipment.

   b) What numbers go into the part-whole model?

32

PUPIL TEXTBOOK 2A PAGE 32

Share

WAYS OF WORKING Whole class teacher led

ASK

- Question ① a): *How many 10s will you need? How many 1s? How can you tell this by reading the number 56?*
- Question ① b): *What is the value of the 5? What number does this 5 represent?*

IN FOCUS Question ① a) rehearses understanding of 2-digit numbers through 10s and 1s.

Question ① b) explores how to partition the number fully into parts using the part-whole model. Look out for the possible error of partitioning 56 into 5 and 6.

Share

a)

I made 56 as 10s and 1s.

56 is made up of 5 tens and 6 ones.

b)

56
50   6

The numbers 50 and 6 go in the part-whole model.

This is an example of **partitioning** a number.

33

PUPIL TEXTBOOK 2A PAGE 33

Think together

Think together

WAYS OF WORKING Whole class teacher led (I do, We do, You do)

ASK

- Question **1**: *How many 10s are there? What number does that make?*
- Question **2**: *Do you need to work out the parts or the whole?*
- Question **3**: *What is the same and what is different about each part-whole model?*

IN FOCUS In question **1**, children are partitioning a number into 10s and 1s. For question **2** they need to recognise how to combine 10s and 1s to make a whole. Question **3** addresses variations and potential misconceptions when partitioning using part-whole models.

STRENGTHEN Provide base 10 equipment that children can use to check their reasoning for each question.

DEEPEN Challenge children to investigate this statement: *All 2-digit numbers partition into two parts.* Can they explain what they find?

ASSESSMENT CHECKPOINT Use question **3** to assess whether children can identify and correct errors in partitioning.

ANSWERS

Question **1**: 30 and 5

Question **2** a): 71

Question **2** b): 44

Question **3** a): 40 has been incorrectly partitioned into 4 and 0. It should be 40 and 0.

Question **3** b): This partitioning is mathematically correct but it is more common for the 10s to be on the left-hand side and the 1s to be on the right-hand side.

Question **3** c): This is correct.

Question **3** d): The whole is in the wrong place: 26 should be on the left-hand side so that it partitions into 20 and 6. Alternatively, some children may recognise that 20 should be partitioned into 20 and 0.

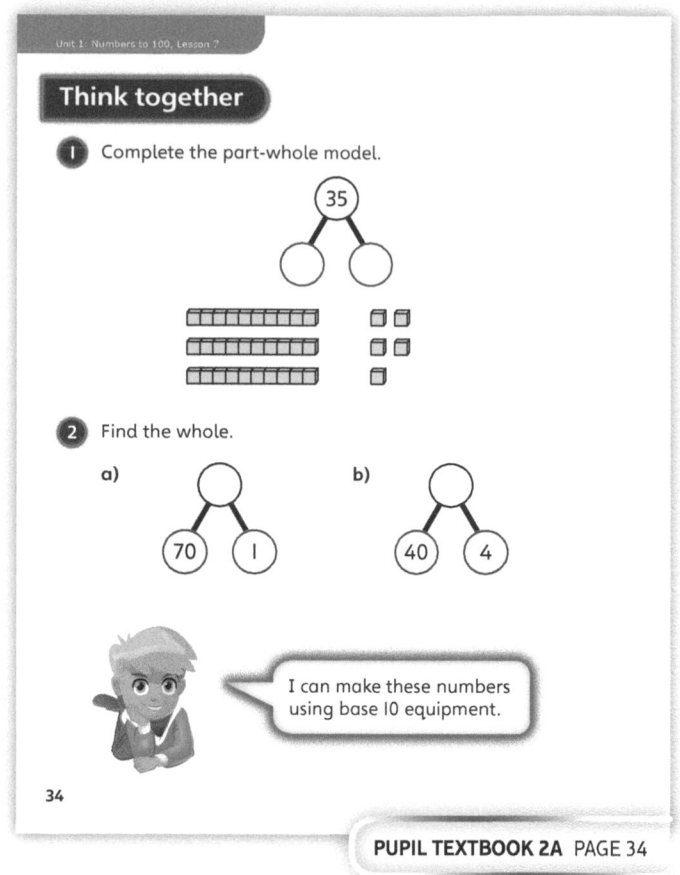

PUPIL TEXTBOOK 2A PAGE 34

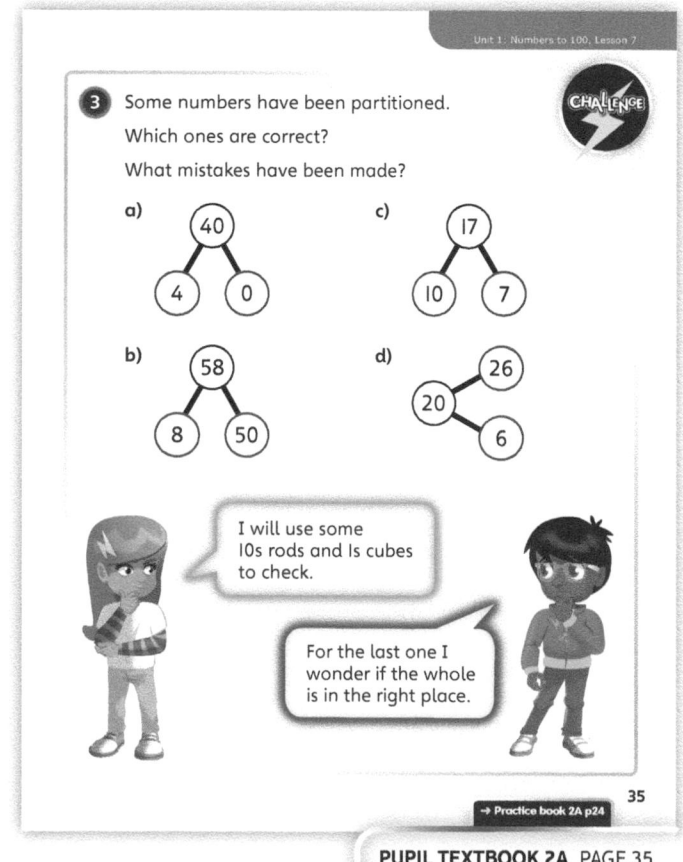

PUPIL TEXTBOOK 2A PAGE 35

Practice

WAYS OF WORKING Independent thinking

IN FOCUS In question ❶ children are partitioning 2-digit numbers into 10s and 1s, with base 10 equipment to scaffold their thinking. Question ❷ requires them to combine 10s and 1s into a 2-digit number, with a pictorial representation using items to scaffold this. In question ❸ children are partitioning 2-digit numbers into 10s and 1s and in question ❹ they are combining 10s and 1s into a 2-digit number. In question ❺ the 10s and 1s in 2-digit numbers are presented in different orders. Question ❻ challenges misconceptions about place value when partitioning.

STRENGTHEN Encourage children to use base 10 equipment alongside their independent work. Model how to use the equipment to confirm the number thinking.

DEEPEN Challenge children to explore more flexible partitions of different 2-digit numbers, beyond just 10s and 1s.

ASSESSMENT CHECKPOINT Assess whether children can accurately partition a given 2-digit number into 10s and 1s. Can they also accurately combine 10s and 1s to form a 2-digit number?

ANSWERS Answers for the **Practice** part of the lesson can be found in the *Power Maths* online subscription.

Reflect

WAYS OF WORKING Pair work

IN FOCUS The **Reflect** part of the lesson requires children to demonstrate an understanding of partitioning into 10s and 1s using a part-whole model.

ASSESSMENT CHECKPOINT Assess whether children can correctly identify the 10s and 1s in a 2-digit number and show this on a part-whole model. If you show them some part-whole models with 'deliberate' mistakes, can they discuss how to correct the errors?

ANSWERS Answers for the **Reflect** part of the lesson can be found in the *Power Maths* online subscription.

After the lesson

- Ensure that children can partition each of these numbers into 10s and 1s:

  13, 43, 37, 70.

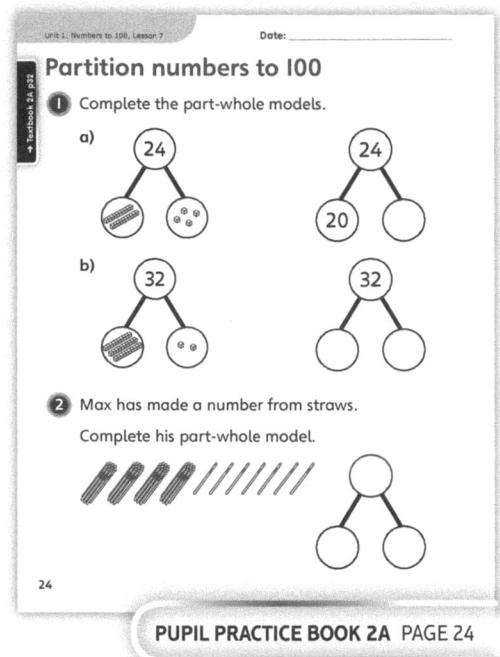

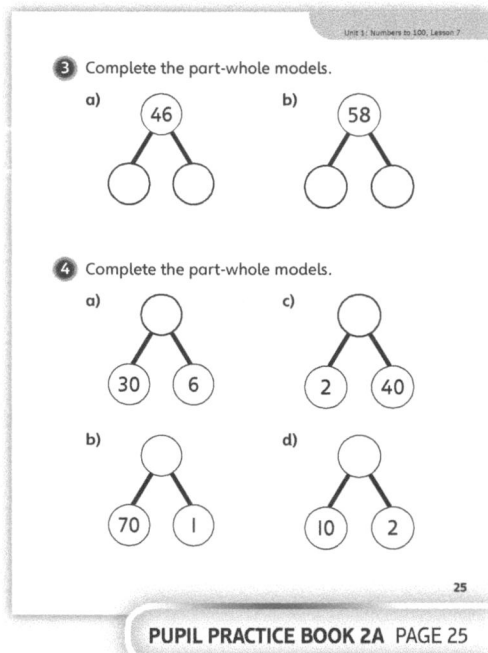

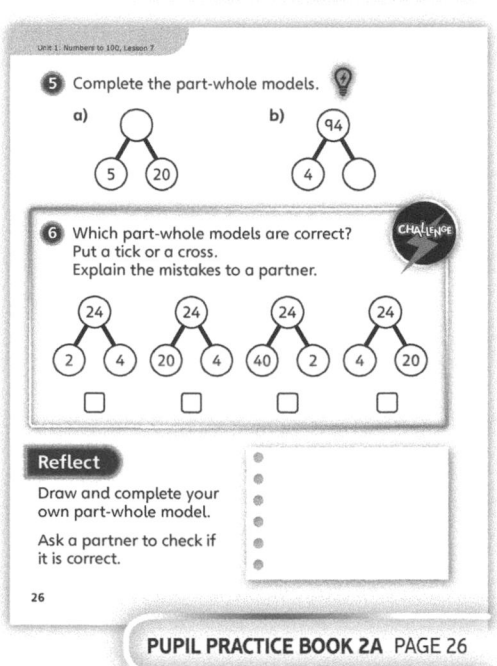

Partition numbers flexibly within 100

Learning focus

In this lesson, children are partitioning 2-digit numbers flexibly, by finding multiple partitions of 10s and 1s.

Before you teach ⏸

- Can children partition 2-digit numbers into 10s and 1s?
- Can children make 2-digit numbers using 10s and 1s equipment?
- Are children able to discuss the place value of different 2-digit numbers?

NATIONAL CURRICULUM LINKS

Year 1 Number – number and place value

Recognise the place value of each digit in a two-digit number (tens, ones).

Identify, represent and estimate numbers using different representations, including the number line.

ASSESSING MASTERY

Children can partition 2-digit numbers flexibly, finding multiple partitions of 10s and 1s.

This forms the foundation for exchange in calculations, and for working with numbers flexibly.

COMMON MISCONCEPTIONS

Children may expect that each 2-digit number has only one way of being partitioned. Ask:
- *Can you write this number in a part-whole model? Are there other numbers you could write in as the two parts?*

STRENGTHENING UNDERSTANDING

Provide base 10 equipment and two sorting circles or plates. Give children practice of physically moving the whole into the parts in different ways.

GOING DEEPER

Challenge children to find more than three different possible partitions for a given number.

KEY LANGUAGE

In lesson: partition, tens (10s), ones (1s), part-whole, part, whole

Other language to be used by the teacher: flexible

STRUCTURES AND REPRESENTATIONS

Ten frame, part-whole model

RESOURCES

Mandatory: base 10 equipment, ten frame, sorting circles, counters

Optional: paper plates

 In the eTextbook of this lesson, you will find interactive links to a selection of teaching tools.

Quick recap ↺

Show or make a 2-digit number using two ten frames. Ask children to separate it into 10s and 1s. Challenge them to use counters to make their own 2-digit numbers on ten frames.

Discover

WAYS OF WORKING Pair work

ASK

- Question ❶ a): *What is the same and what is different about Kasim's and Izzy's numbers?*
- Question ❶ a): *Can you make these numbers from base 10 equipment?*
- Question ❶ a): *What is the value of the digits in the total number?*

IN FOCUS Children are being shown that a given number can be partitioned in different ways. They need to understand how the parts combine to make the whole.

PRACTICAL TIPS Ask children to use base 10 equipment to make the number 32, and physically move the equipment into two different parts.

ANSWERS

Question ❶ a): Kasim has made 32 and Izzy has made 32.

Same: they both have the same total number of counters.

Different: Izzy has partitioned 32 into 30 and 2 whilst Kasim has partitioned 32 into 20 and 12.

Question ❶ b):

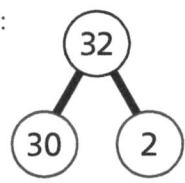 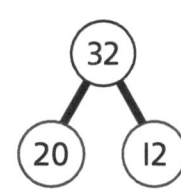

Share

WAYS OF WORKING Whole class teacher led

ASK

- Question ❶ a): *Can you see how each person has made some 10s and some 1s? Does the colour affect the number?*
- Question ❶ b): *Count the total for each part separately. What do you notice when you write the whole number?*

IN FOCUS In question ❶ a) children will find that counting in 10s and 1s is not affected by the change in colour of the counters used. Question ❶ b) shows that partitioning of the same number may be done in different ways.

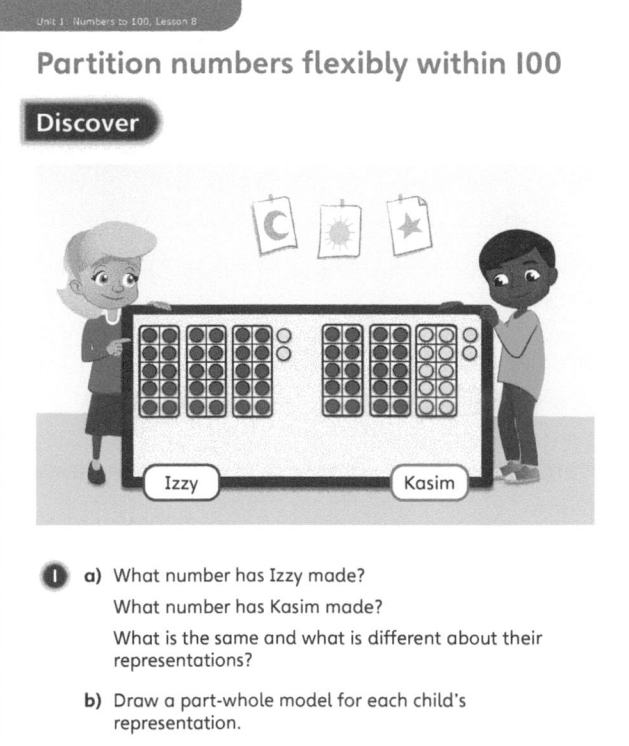

Unit 1: Numbers to 100, Lesson 8

Partition numbers flexibly within 100

Discover

❶ a) What number has Izzy made?

What number has Kasim made?

What is the same and what is different about their representations?

b) Draw a part-whole model for each child's representation.

36

PUPIL TEXTBOOK 2A PAGE 36

Share

a) Each ten frame has 10 counters

Izzy · Kasim

10 20 30 32 · 10 20 30 32

Izzy and Kasim have both made the number 32.

They both have 3 tens and 2 ones.

They have used different amounts of each colour counter.

b) Izzy · Kasim

32 · 32

30 2 · 20 12

I can see that 32 can be partitioned in different ways.

37

PUPIL TEXTBOOK 2A PAGE 37

Think together

Whole class teacher led (I do, We do, You do)

ASK

- Question ❶: *What are the parts? What is the whole?*
- Question ❷: *Could you make 67 from base 10 equipment to help you?*
- Question ❸: *Do you think there are more than three different ways? Could you work systematically?*

IN FOCUS Question ❶ builds on the use of ten frames to explore different partitions of the same number. Question ❷ has the part-whole model representing the partitions in different ways. In question ❸ children need to recognise that there are multiple partitions for a given number.

STRENGTHEN Provide base 10 equipment so that children can physically split numbers into different parts. They could use cups, circles or plates to hold the parts.

DEEPEN Use question ❸ as a model to challenge children to find more than five different partitions of given 2-digit numbers.

ASSESSMENT CHECKPOINT Assess whether children can use base 10 equipment to show how to make different partitions of a 2-digit number.

ANSWERS

Question ❶:

41
40   1

41
30   11

Question ❷:

67
60   7

67
50   17

Question ❸: There are multiple ways of partitioning 75, for example, 1 and 74, 70 and 5, 60 and 15, 40 and 35, etc.

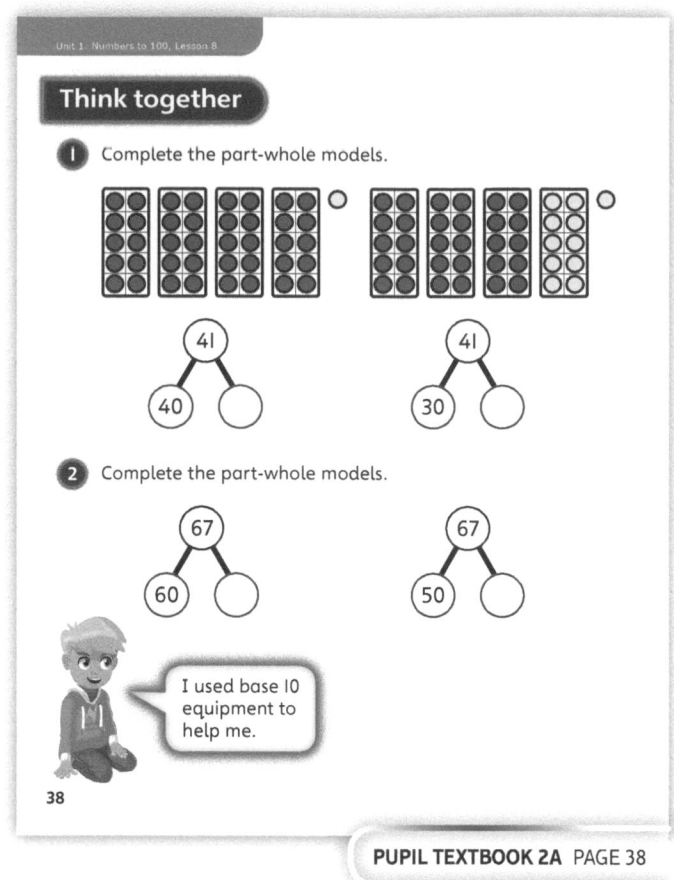

PUPIL TEXTBOOK 2A PAGE 38

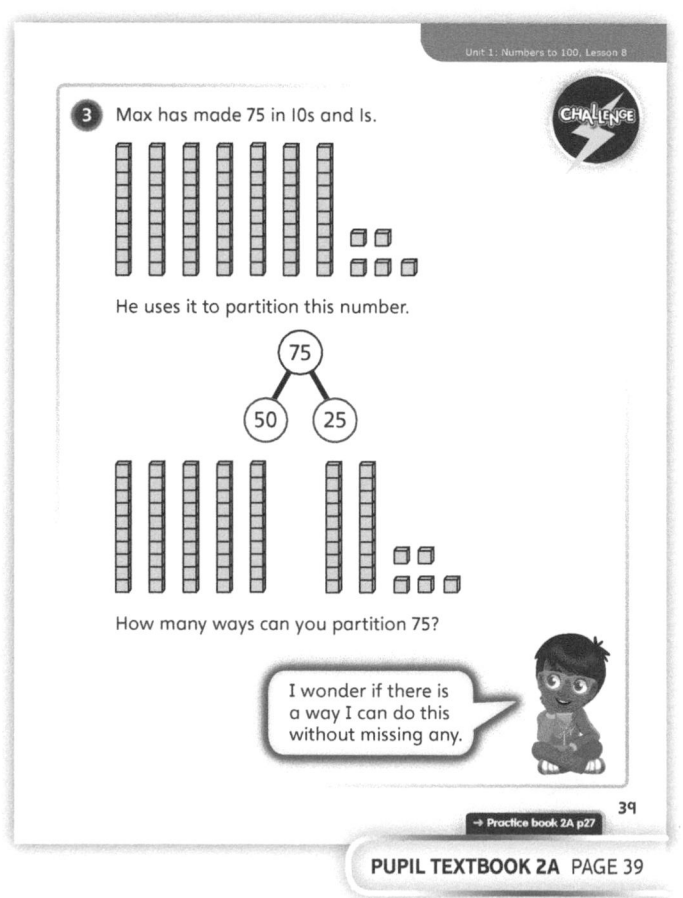

PUPIL TEXTBOOK 2A PAGE 39

Practice

Independent thinking

IN FOCUS Questions **1** and **2** involve partitioning supported by ten frames and base 10 equipment. In questions **3**, **4** and **5** children are partitioning in part-whole models without visual scaffolds.

STRENGTHEN Explore the use of base 10 equipment throughout. It is important to develop understanding of the concept, rather than increasing the speed or fluency of mental methods at this stage.

DEEPEN Some children may be ready to discuss partitions that have 10s and 1s in both parts.

ASSESSMENT CHECKPOINT Assess whether children can find two different partitions of 52.

ANSWERS Answers for the **Practice** part of the lesson can be found in the *Power Maths* online subscription.

Reflect

WAYS OF WORKING Pair work

IN FOCUS The **Reflect** part of the lesson requires children to explore flexible partitioning of a given written number, without the visual scaffolds of ten frames or base 10 equipment.

ASSESSMENT CHECKPOINT Assess whether children can justify their partitioning by using drawings, ten frames or base 10 equipment.

ANSWERS Answers for the **Reflect** part of the lesson can be found in the *Power Maths* online subscription.

After the lesson

• Ensure that children can use drawings or equipment to justify any partitioning that they do.

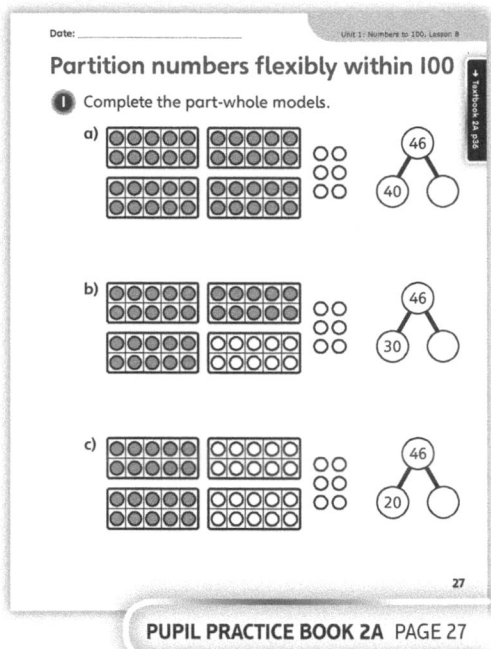

PUPIL PRACTICE BOOK 2A PAGE 27

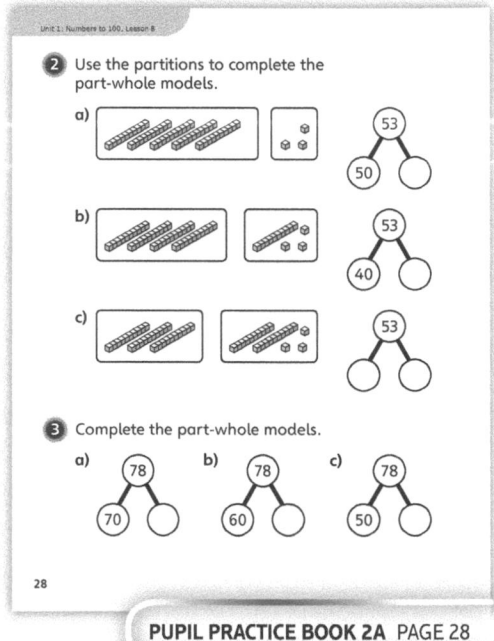

PUPIL PRACTICE BOOK 2A PAGE 28

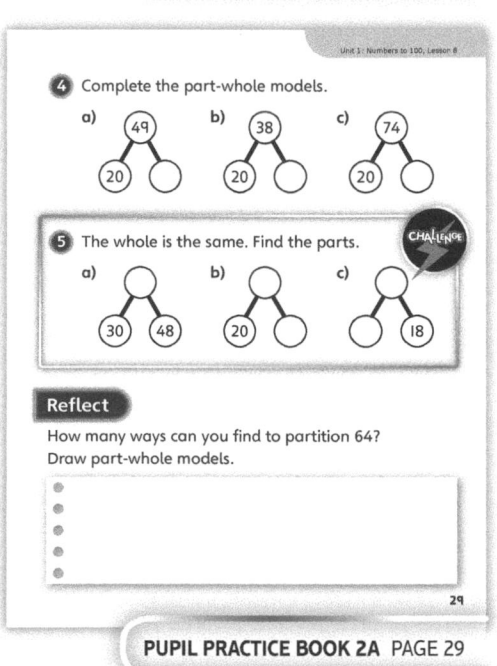

PUPIL PRACTICE BOOK 2A PAGE 29

Write numbers to 100 in expanded form

Learning focus

In this lesson, children build on their partitioning skills to write a 2-digit number as an addition of 10s and 1s, for example 43 = 40 + 3.

Before you teach

- Have children had the opportunity to make or draw 2-digit number representations?
- Are they able to partition 2-digit numbers into 10s and 1s?
- Can they combine 10s and 1s to make a 2-digit number?

NATIONAL CURRICULUM LINKS

Year 1 Number – number and place value

Recognise the place value of each digit in a two-digit number (tens, ones).

Read and write numbers to at least 100 in numerals and in words.

ASSESSING MASTERY

Children can complete a given addition involving 10s and 1s, to total a 2-digit number.

COMMON MISCONCEPTIONS

Children may forget to apply the link between partitioning and addition and may see addition as a process separate from place value. Ask:

- *How many 10s? How many 1s? How many altogether? Does that remind you of any other calculations you have done?*

STRENGTHENING UNDERSTANDING

Model how to link a written calculation with the numbers represented using base 10 equipment.

GOING DEEPER

Challenge children to explore more flexible partitions using expanded additions, for example, 43 = 30 + 13.

KEY LANGUAGE

In lesson: addition, tens (10s), ones (1s), partition, partitioning

Other language to be used by the teacher: expanded, calculation, number sentence

STRUCTURES AND REPRESENTATIONS

Base 10 equipment

RESOURCES

Mandatory: base 10 equipment

 In the eTextbook of this lesson, you will find interactive links to a selection of teaching tools.

Quick recap

Ask children to work out the missing numbers in these additions:

10 + 3 = ?   10 + 5 = ?

10 + ? = 19   ? + 10 = 11

Discover

Pair work

ASK

- Question ❶: *Do you ever solve puzzles? What puzzles do you like to solve?*
- Question ❶ a): *Can you see the shapes here? What do you think they are for?*
- Question ❶ a): *Why are there '=' and '+' signs? Do you think we have to calculate?*
- Question ❶ b): *What do you notice about the numbers you can see?*

IN FOCUS This gives children the chance to start forming an understanding of partitioning, in the context of solving additions with missing numbers.

PRACTICAL TIPS Model clearly how to take your time when thinking about puzzles or problem solving. It might help for children to practise simpler puzzles, such as:

11 = 10 + square

20 = heart + 10

ANSWERS

Question ❶ a): Triangle = 5, Circle = 50.

Question ❶ b): Star = 28

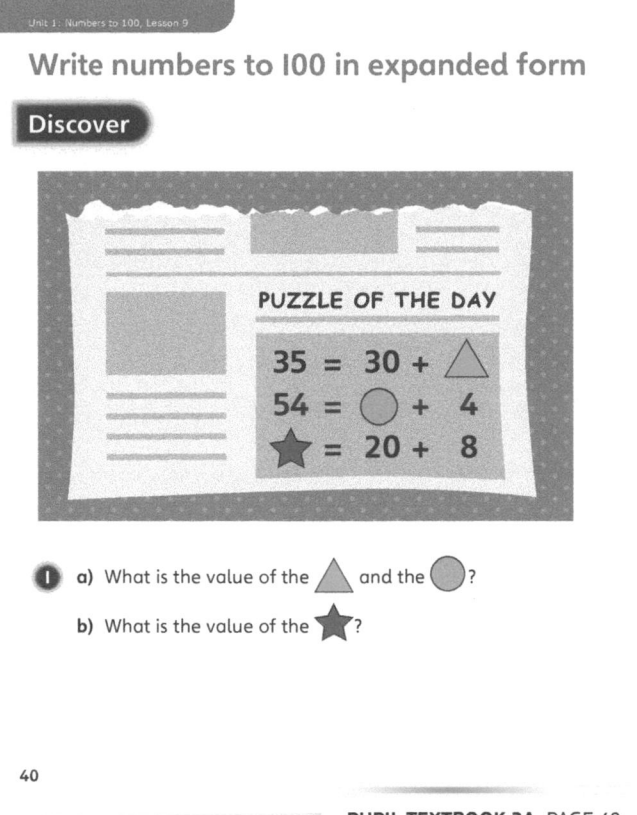

PUPIL TEXTBOOK 2A PAGE 40

Share

WAYS OF WORKING Whole class teacher led

ASK

- Question ❶ a): *Can you see how the base 10 equipment helps with solving this puzzle?*
- Question ❶ b): *Do you notice that additions can be used to show the 10s and 1s?*

IN FOCUS In question ❶ a), children are starting to use simple addition equations to show partitions into 10s and 1s. By question ❶ b) they should be relating the base 10 equipment very clearly to the numbers and signs in the equations.

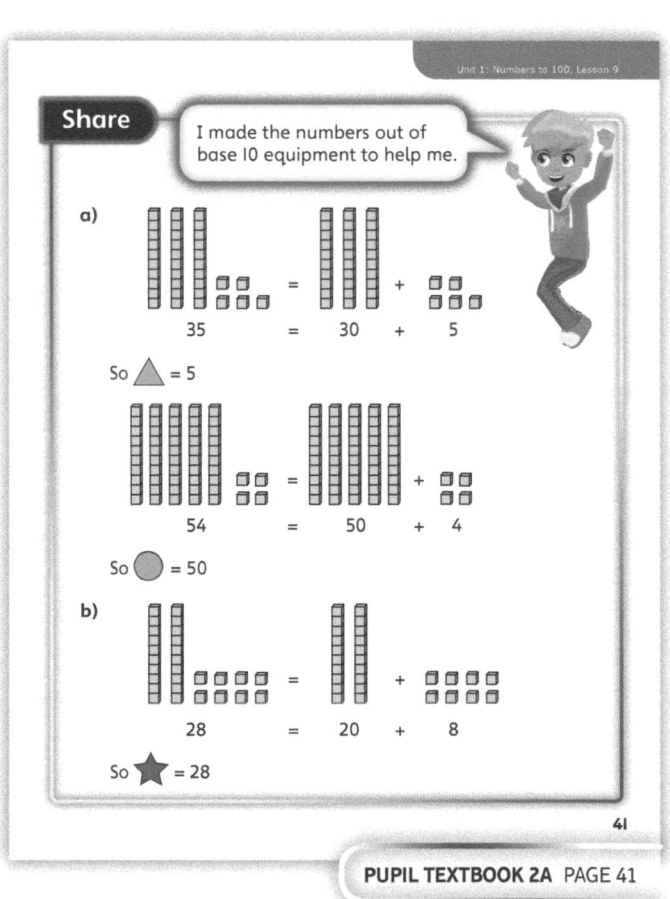

PUPIL TEXTBOOK 2A PAGE 41

Think together

WAYS OF WORKING Whole class teacher led (I do, We do, You do)

ASK

- Question **1**: *What is the whole? What are the parts? Can you see the 10s and the 1s that make a whole 2-digit number? What is missing; the 10s, the 1s or the whole?*
- Question **2**: *What is missing; the 10s and 1s, or the whole?*
- Question **3**: *Does this remind you of any learning from a previous lesson?*

IN FOCUS In question **1**, children recognise the missing element from a number sentence involving 10s and 1s and a whole. This is scaffolded by base 10 equipment. In question **2**, children are completing number sentences involving 10s and 1s. Question **3** links this learning to more flexible partitioning.

STRENGTHEN Model very clearly how to write number sentences alongside using corresponding base 10 equipment. Practise with some 2-digit numbers to build confidence.

DEEPEN Challenge children to explore flexible partitioning of a number more fully, as in question **3**. Ask: *Can you explain what happens to one part when the other part changes?*

ASSESSMENT CHECKPOINT Assess whether children can explain the value of each part of their number sentence, and relate it to base 10 equipment representations.

ANSWERS

Question **1** a): 15

Question **1** b): 3

Question **1** c): 20

Question **2** a): 40 + 6

Question **2** b): 50 + 3

Question **2** c): 61

Question **2** d): 73

Question **3**: 52 = 50 + **2**

52 = 40 + **12**

52 = 30 + **22**

52 = 20 + **32**

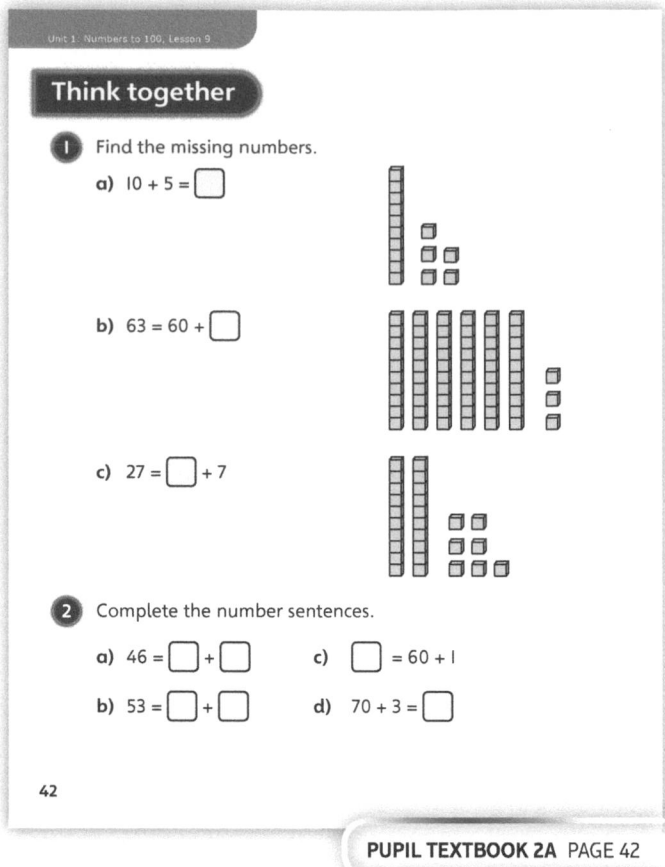

PUPIL TEXTBOOK 2A PAGE 42

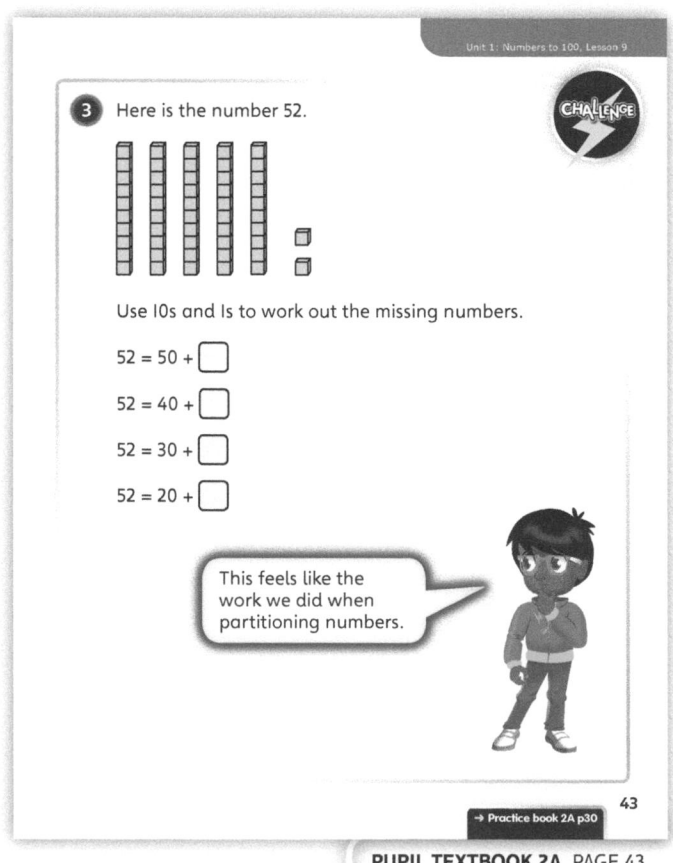

PUPIL TEXTBOOK 2A PAGE 43

Practice

WAYS OF WORKING Independent thinking

IN FOCUS Question ❶ has missing number sentences scaffolded by base 10 equipment. In question ❷, children match place value additions to 2-digit numbers. In question ❸, children complete place value additions with missing 1s. Question ❹ has place value additions with missing 10s and question ❺ has place value additions with missing 10s and 1s. Question ❻ addresses potential misconceptions around order of 10s and 1s in an addition. Question ❼ introduces more flexible partitioning.

STRENGTHEN Present number sentences alongside base 10 equipment, to reinforce the link between partitioning and the addition calculations.

DEEPEN Challenge children to find more than five different partitioning additions for the numbers 94, 83, 61 and 59.

ASSESSMENT CHECKPOINT Assess whether children can complete an addition number sentence for a given 2-digit number.

ANSWERS Answers for the **Practice** part of the lesson can be found in the *Power Maths* online subscription.

Reflect

WAYS OF WORKING Pair work

IN FOCUS The **Reflect** part of the lesson is an open question for children to demonstrate an awareness of how to partition and write an addition number sentence for a given 2-digit number.

ASSESSMENT CHECKPOINT Can children show the 10s and 1s parts of a number and reflect this in their number sentences?

ANSWERS Answers for the **Reflect** part of the lesson can be found in the *Power Maths* online subscription.

After the lesson ⏸

- Ensure that children understand how these additions are linked with all the place value learning from this unit.

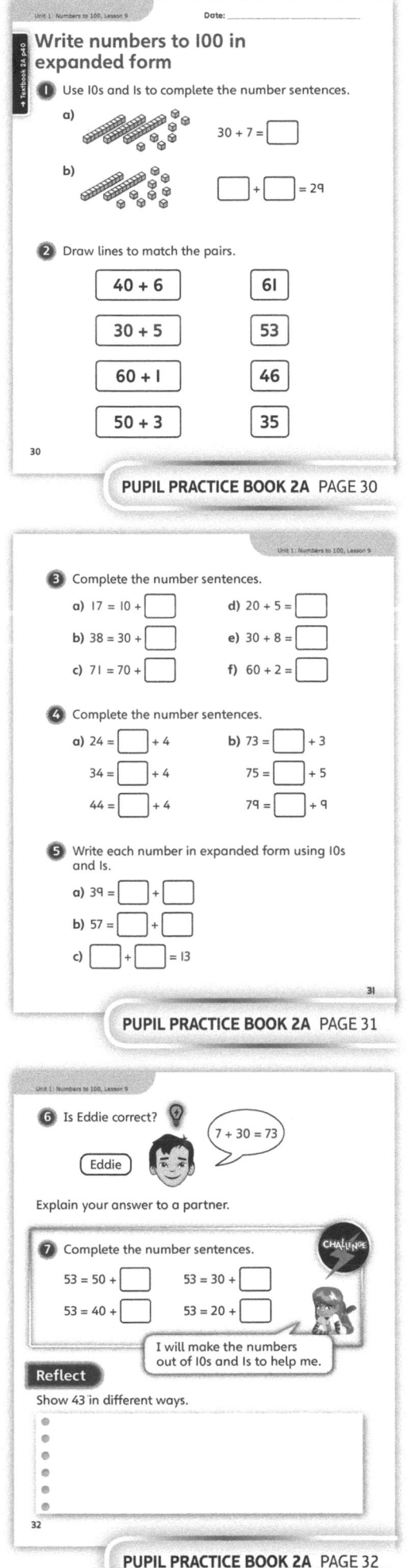

10s on a number line to 100

Learning focus

In this lesson, children develop a deeper understanding of number lines, including number lines that do not show every number, only multiples of 10.

Before you teach ⏸

- Can children find multiples of 10 on a 100 square?
- Can they count in 10s to 100?

NATIONAL CURRICULUM LINKS

Year 1 Number – number and place value

Identify, represent and estimate numbers using different representations, including the number line.

ASSESSING MASTERY

Children can identify a multiple of 10 shown on a 0 to 100 number line marked in 10s.

COMMON MISCONCEPTIONS

Children may not have a strong enough number knowledge or the flexibility needed to interpret number lines. For example, children using 'finger-spacing' may think that every number line needs to count in 1s, with a finger gap between each one. Ask:
- *Will this number line fit on the page? How else could you draw it?*

STRENGTHENING UNDERSTANDING

Give opportunities for children to rehearse counting up in 10s on a number line.

GOING DEEPER

Encourage children to practise drawing their own number lines with increasing accuracy, using pencil and ruler skills.

KEY LANGUAGE

In lesson: number line, tens (10s)

Other language to be used by the teacher: compare

STRUCTURES AND REPRESENTATIONS

Number line

RESOURCES

Mandatory: rulers, number lines

 In the eTextbook of this lesson, you will find interactive links to a selection of teaching tools.

Quick recap

Challenge children to draw a number line with 0 at one end and 10 at another end, and then to fill in all the numbers from 1 to 9 in the correct places.

Discover

10s on a number line to 100

Discover

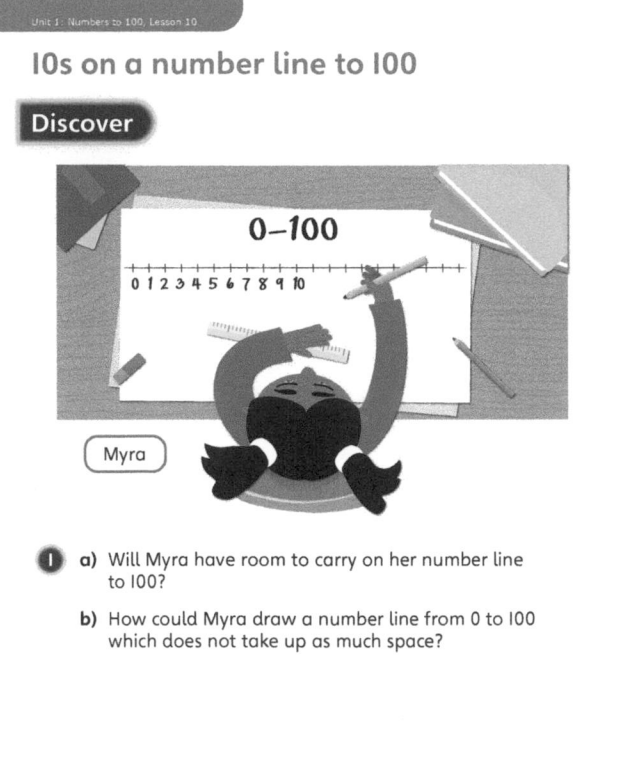

WAYS OF WORKING Pair work

ASK

- Question ❶ a): *If you draw a number line to 100 and write each number by counting in 1s, can you fit it on one page?*
- Question ❶ b): *Can you think of a way to show a number line counting in 10s?*

IN FOCUS This gives children a chance to explore the idea that a number line may not always show every single number.

PRACTICAL TIPS Draw number lines on the board or on the playground to explore how long the line will become if you try to show every single number.

ANSWERS

Question ❶ a): No, as Myra's number line goes up in 1s.

Question ❶ b): Her number line could go up in 10s instead.

❶ a) Will Myra have room to carry on her number line to 100?

  b) How could Myra draw a number line from 0 to 100 which does not take up as much space?

44

PUPIL TEXTBOOK 2A PAGE 44

Share

WAYS OF WORKING Whole class teacher led

ASK

- Question ❶ a): *Will they run out of room? What could they do to solve this problem?*
- Question ❶ b): *Can you see how to count in 10s on this line?*

IN FOCUS Question ❶ a) demonstrates the practical problem of showing every number on a number line if you count in 1s up to 100. Question ❶ b) introduces the practical adaptation of showing only the 10s up to 100.

Share

a) Myra's number line goes up in 1s.

I can see that Myra cannot fit her number line up to 100 on the card.

b) Some number lines go up in 10s instead of 1s.

We know where the other numbers go, but do not always need to label them all.

45

PUPIL TEXTBOOK 2A PAGE 45

Think together

Whole class teacher led (I do, We do, You do)

ASK

- Question ❶: *Where does the counting start? Can you count in 10s?*
- Question ❷: *Can you predict which numbers are nearer to 100?*
- Question ❸: *What is the same and what is different about each number line?*

IN FOCUS In question ❶, children are identifying a 10 shown on a partially labelled number line. In question ❷, they are locating a given multiple of 10 on a number line. Question ❸ requires children to recognise the link between counting in 1s and counting in 10s.

STRENGTHEN Give opportunities for children to rehearse how to point and count in 10s along a number line.

DEEPEN Ask children to explore variety in number lines. Challenge them to draw vertical number lines as well as horizontal number lines.

ASSESSMENT CHECKPOINT Assess whether children can identify a multiple of 10 indicated on a number line that is marked in 10s.

ANSWERS

Question ❶: A = 20, B = 40, C = 60

Question ❷:

Question ❸ a): The top number line counts in 1s and the bottom number counts in 10s.

Question ❸ b): Same: it is marked in 10s from 0 to 100.

Different: It is vertical, the other number lines are horizontal.

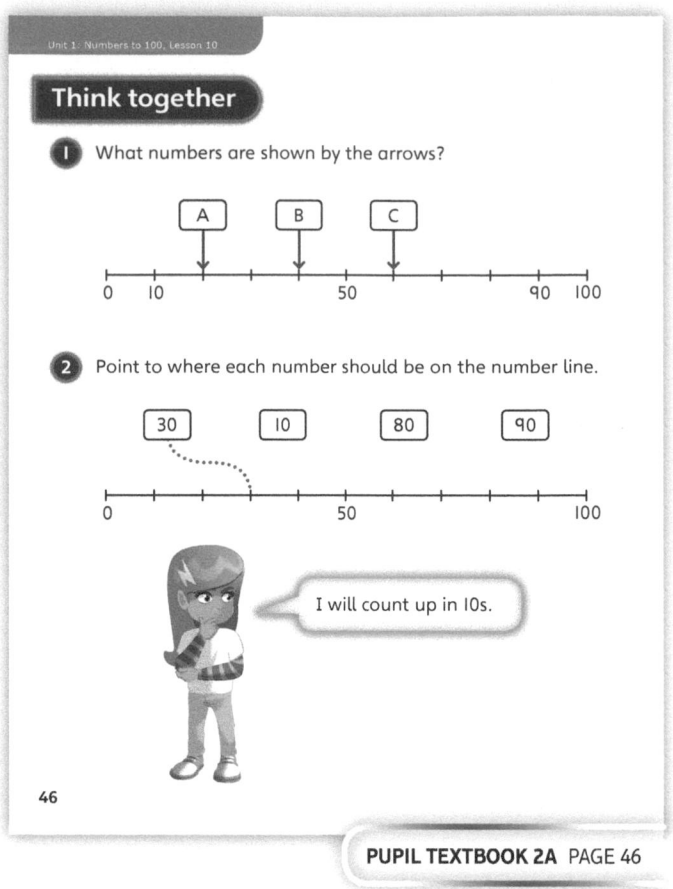

PUPIL TEXTBOOK 2A PAGE 46

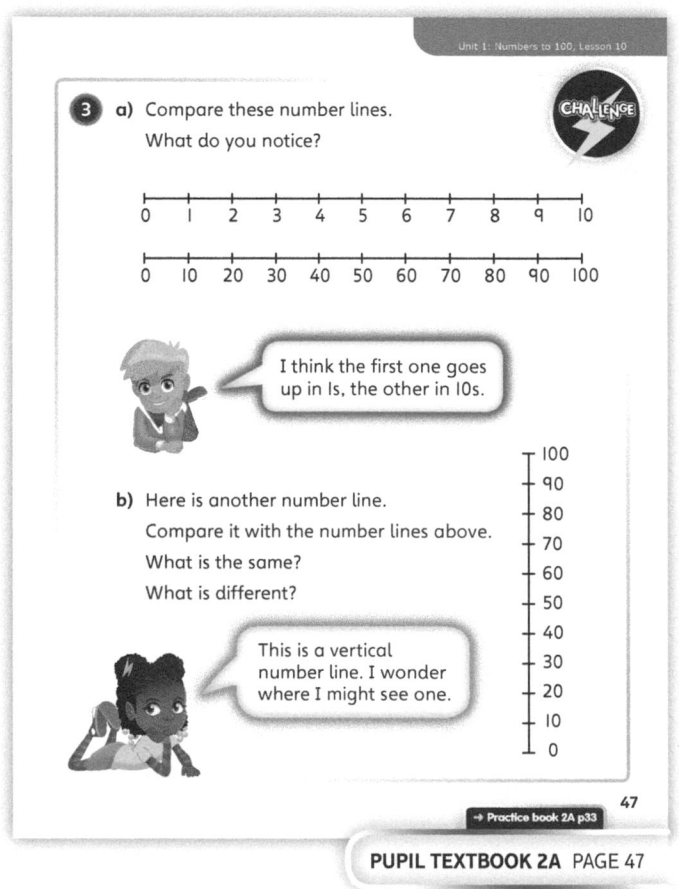

PUPIL TEXTBOOK 2A PAGE 47

Practice

Independent thinking

IN FOCUS In question ① children are identifying a multiple of 10 indicated on a number line, and in question ②, they are locating a multiple of 10 on the number line. For question ③ children need to count in 10s along a number line in order to complete it. Question ④ links base 10 equipment to the number line representation and question ⑤ requires children to work flexibly with variations of number lines.

STRENGTHEN Repeat question ③ with children counting and pointing at the line as they go along.

DEEPEN Use question ⑤ to explore possible variations of number lines. Ask: *What other variations of number lines can you think of?*

ASSESSMENT CHECKPOINT Use question ④ to assess whether children are linking place value understanding with number line work.

ANSWERS Answers for the **Practice** part of the lesson can be found in the *Power Maths* online subscription.

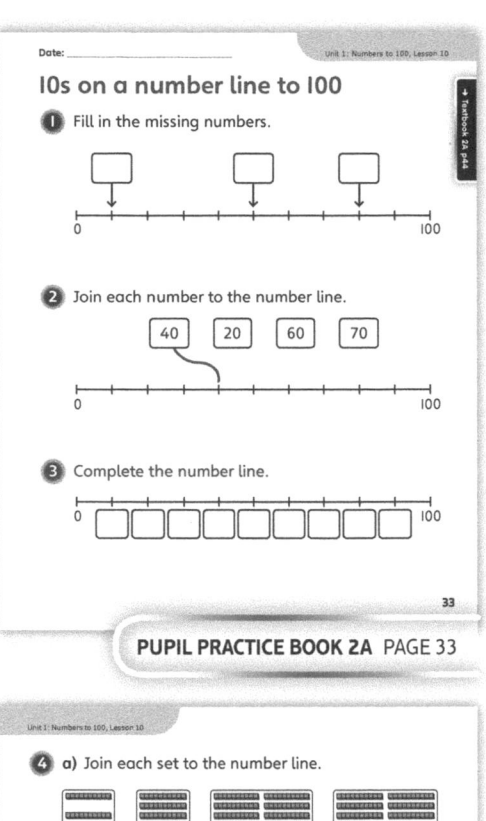

PUPIL PRACTICE BOOK 2A PAGE 33

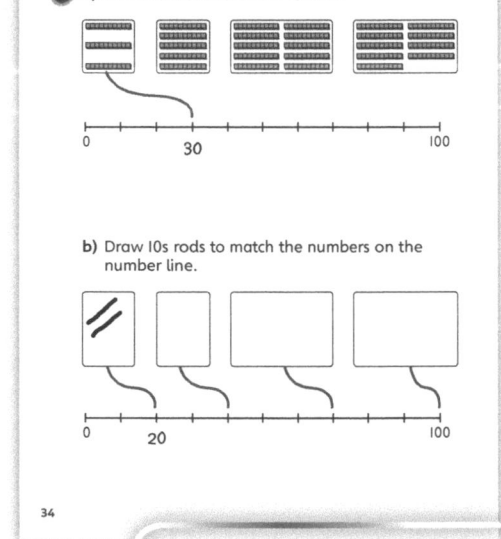

PUPIL PRACTICE BOOK 2A PAGE 34

Reflect

Independent thinking

IN FOCUS The **Reflect** part of the lesson gives children an opportunity to show a deeper understanding of the structure of a 0 to 100 number line.

ASSESSMENT CHECKPOINT Assess whether children are showing an awareness of how to show only the 10s on their number line.

ANSWERS Answers for the **Reflect** part of the lesson can be found in the *Power Maths* online subscription.

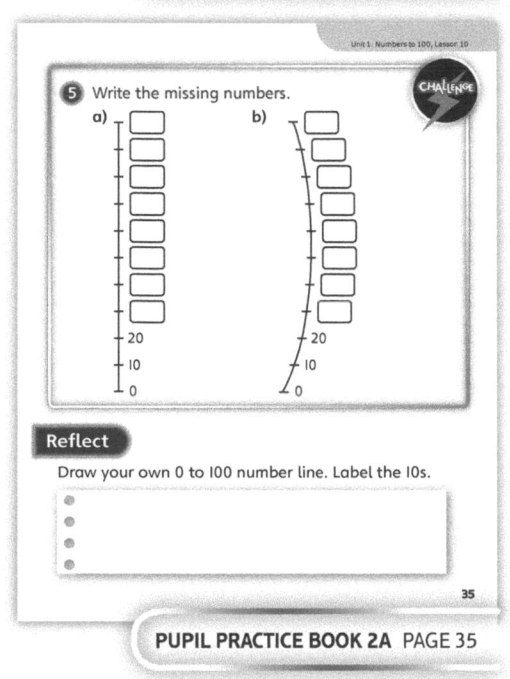

PUPIL PRACTICE BOOK 2A PAGE 35

After the lesson ⏸

- Ask children to count up in 10s on a partially labelled 0 to 100 number line.

10s and 1s on a number line to 100

Learning focus

In this lesson, children develop a deeper understanding of number lines, including number lines that do not start on 0, but start instead on a multiple of 10.

Before you teach

- Can children count up and back in 1s, within 100?
- Can they count up in 1s from any given number?
- Can they count up in 1s from any multiple of 10?

NATIONAL CURRICULUM LINKS

Year 1 Number – number and place value

Identify, represent and estimate numbers using different representations, including the number line.

Recognise the place value of each digit in a two-digit number (tens, ones).

ASSESSING MASTERY

Children can identify a 2-digit number shown between two multiples of 10 on a number line that is marked in 1s.

COMMON MISCONCEPTIONS

Children may need time to explore how number lines can start from different numbers, as they develop further flexibility in their thinking about number lines. Ask:
- *Does this number line start at 0? What number does it start from?*

STRENGTHENING UNDERSTANDING

Show a number line from 0 to 20. Cover up the part that shows 0 to 10. Discuss how the number line could start at any number, not always 0.

GOING DEEPER

Learn how to count back from a number line, as well as forwards.

KEY LANGUAGE

In lesson: tens (10s), ones (1s), number line

Other language to be used by the teacher: mark

STRUCTURES AND REPRESENTATIONS

Number line

RESOURCES

Optional: number lines

 In the eTextbook of this lesson, you will find interactive links to a selection of teaching tools.

Quick recap 🔁

Challenge children to draw a 0 to 100 number line with 10s marked. Ask children to point to different numbers on the number line as you call them out.

Discover

WAYS OF WORKING Pair work

ASK

- Question ① a): *What numbers can you see on each of these number lines?*
- Question ① b): *What information does each number line give you?*

IN FOCUS This requires children to compare two number lines, one with only the multiples of 10 labelled and one where all the multiples of 1 are also shown, but are not labelled. Children consider how they can use the given marks and labels to locate a given 2-digit number.

PRACTICAL TIPS Provide enlarged number lines that start at one multiple of 10 and end at the next multiple of 10, with the 1s in between marked but not labelled with numbers. Encourage children to count up along the line, pointing at the mark as they say each number.

ANSWERS

Question ① a): Asha's number line goes up in 10s. Filip's goes up in 10s and 1s. They both go from 0 to 100 and they are both horizontal.

b): The 4th mark after 30 shows 34.

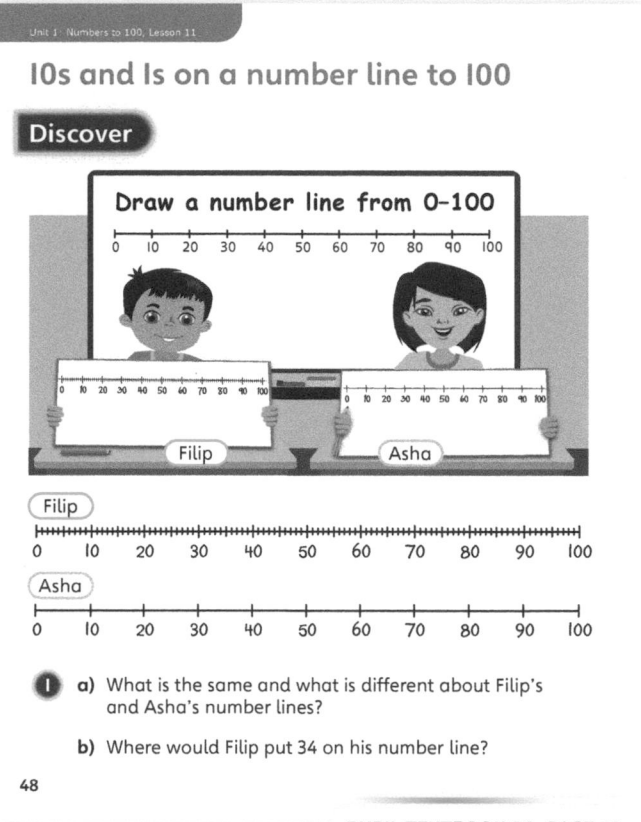

10s and 1s on a number line to 100

Discover

Draw a number line from 0–100

Filip

Asha

① a) What is the same and what is different about Filip's and Asha's number lines?

b) Where would Filip put 34 on his number line?

48

PUPIL TEXTBOOK 2A PAGE 48

Share

WAYS OF WORKING Whole class teacher led

ASK

- Question ① a): Whose number line do you prefer? Why?
- Question ① b): How did Filip know where to put the number? Where would Asha put the number on her number line?

IN FOCUS Question ① a) draws children's attention to features of number lines, including the marks that can be used to show 10s and 1s. In question ① b), children use these features to locate a given 2-digit number.

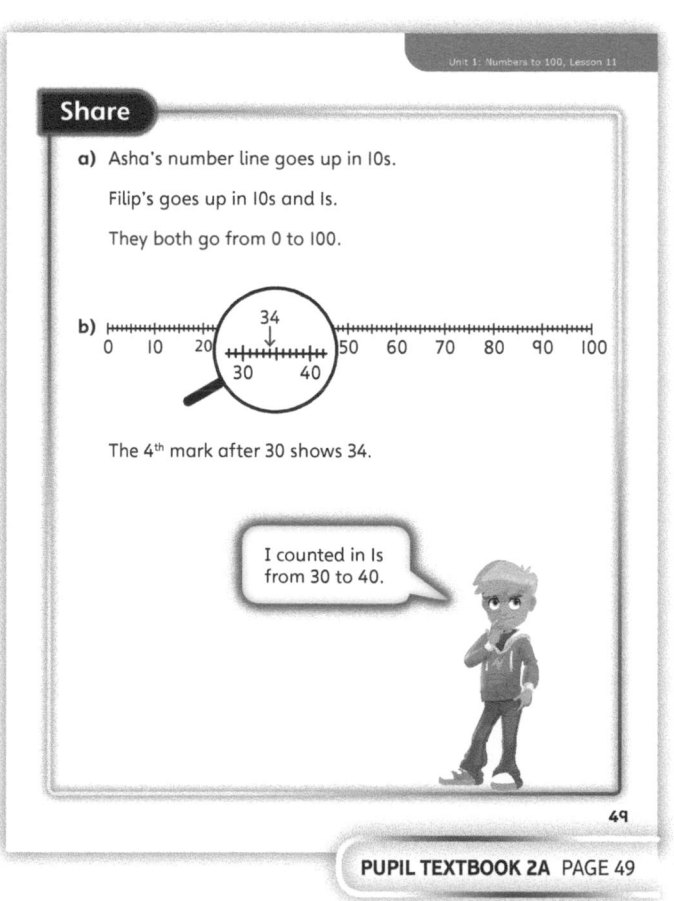

Share

a) Asha's number line goes up in 10s.

Filip's goes up in 10s and 1s.

They both go from 0 to 100.

b)

The 4th mark after 30 shows 34.

I counted in 1s from 30 to 40.

49

PUPIL TEXTBOOK 2A PAGE 49

Think together

Whole class teacher led (I do, We do, You do)

ASK

Question ❶: *What is the starting number? What is the end number? Where should you start counting?*
Question ❷: *Can you solve this by counting from 50? Is there another way?*
Question ❸: *Can you see the little marks? Can you try to be as accurate as possible?*

IN FOCUS In question ❶ children are counting in 1s from a multiple of 10. In question ❷ they will need to count up or count back from multiples of 10 to identify a number. Children should be encouraged to make reasonable efforts at precision when locating numbers on the line in question ❸.

STRENGTHEN Give children opportunities to practise counting up in 1s from any given multiple of 10.

DEEPEN Ask children to explore how to count back from multiples of 10, and discuss when this is a more efficient way to identify a number.

ASSESSMENT CHECKPOINT Assess whether children can identify a number indicated between two multiples of 10.

ANSWERS

Question ❶: 73, 74, 75, 76, 77, 78, 79

Question ❷: 59

Question ❸: Children identify where 42, 65 and 89 are on the number line.

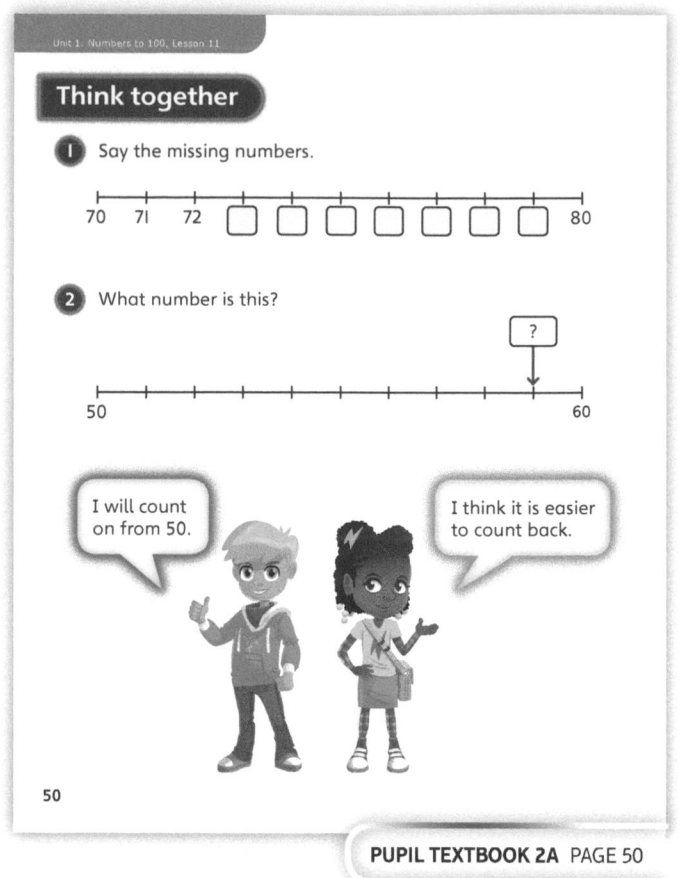

PUPIL TEXTBOOK 2A PAGE 50

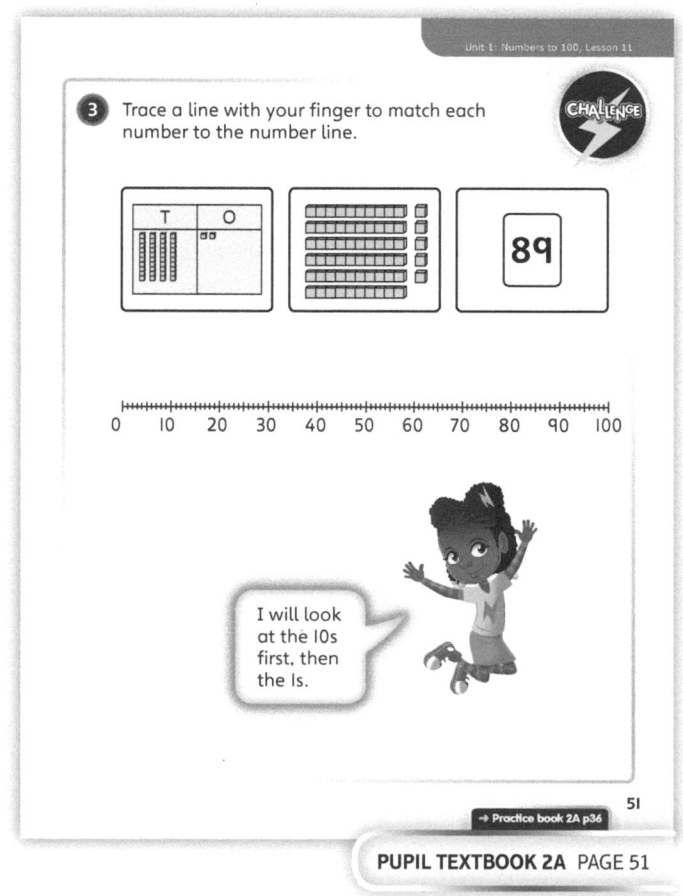

PUPIL TEXTBOOK 2A PAGE 51

Practice

WAYS OF WORKING Independent thinking

IN FOCUS In question ❶ children identify a number from its position on the number line. Question ❷ requires children to count in 1s from a multiple of 10 to the next multiple of 10 and to show this on a number line. In question ❸ children are locating given numbers on a number line between multiples of 10. Question ❹ links the learning to place value. Question ❺ requires children to work with precision and efficiency when labelling numbers between multiples of 10 on number lines showing 0 to 100.

STRENGTHEN Provide variations of the number lines in question ❷ for children to rehearse how to show the numbers between multiples of 10 on a number line.

DEEPEN Use the number lines from question ❺ for children to explore efficient counting strategies, including counting back from a given multiple of 10.

ASSESSMENT CHECKPOINT Use question ❹ to assess whether children can show a deep understanding of the location of 2-digit numbers on a number line and how this is linked to place value.

ANSWERS Answers for the **Practice** part of the lesson can be found in the *Power Maths* online subscription.

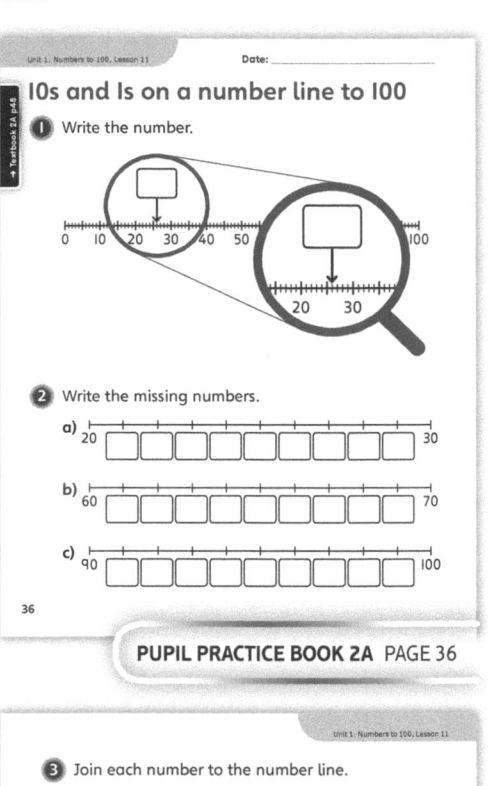

PUPIL PRACTICE BOOK 2A PAGE 36

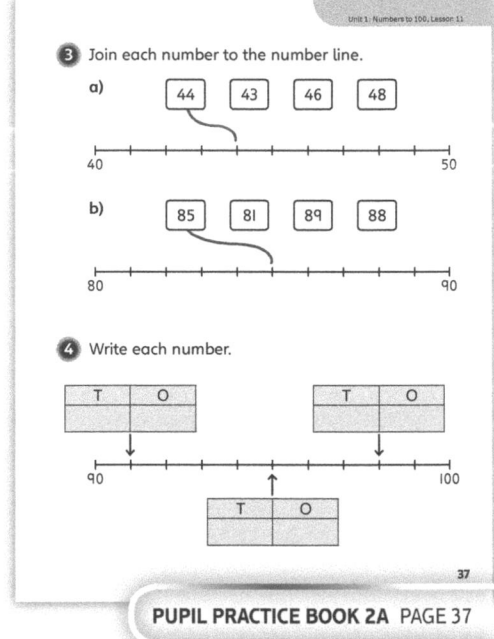

PUPIL PRACTICE BOOK 2A PAGE 37

WAYS OF WORKING Independent thinking

IN FOCUS The **Reflect** part of the lesson gives children the opportunity to show an understanding of the structure of number lines between two multiples of 10.

ASSESSMENT CHECKPOINT Assess whether the number lines that children have drawn show the intervals accurately.

ANSWERS Answers for the **Reflect** part of the lesson can be found in the *Power Maths* online subscription.

After the lesson ⏸

- Check that children can identify a given 2-digit number shown on a number line.

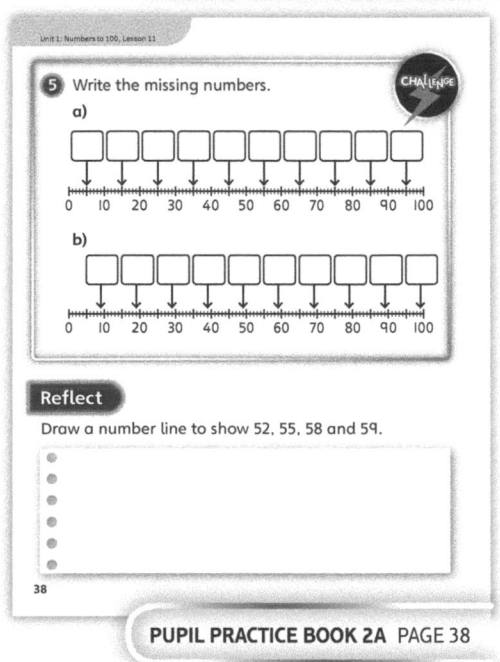

PUPIL PRACTICE BOOK 2A PAGE 38

Estimate numbers on a number line

Learning focus

In this lesson, children further develop their understanding of number lines for representing numbers within the range 0 to 100.

Before you teach ⏸

- Have children had the opportunity to rehearse counting in 10s on a number line?
- Can children identify multiples of 10 on an unlabelled number line?
- Can they locate 2-digit numbers on a number line between two multiples of 10?

NATIONAL CURRICULUM LINKS

Year 1 Number – number and place value

Identify, represent and estimate numbers using different representations, including the number line.

ASSESSING MASTERY

Children can make reasonable estimates for the location of a 1- or 2-digit number on a partially marked number line.

COMMON MISCONCEPTIONS

Children may find it challenging to work with estimates because they think of it as merely a guess, or of less value than the exact answer. Ask:
- *What is your estimate? What is the answer? How did your estimate help you?*

STRENGTHENING UNDERSTANDING

Model how to make and justify an estimate, and also how to 'change your mind' and make an improvement after further thinking or discussion.

GOING DEEPER

If children are ready, ask them to work with 'half-way numbers' between two multiples of 10.

KEY LANGUAGE

In lesson: estimate, guess, number line

Other language to be used by the teacher: half-way

STRUCTURES AND REPRESENTATIONS

Number line

RESOURCES

Mandatory: number line

Optional: number cards, sticky notes

 In the eTextbook of this lesson, you will find interactive links to a selection of teaching tools.

Quick recap 🔁

Play an estimation game. Ask children to estimate how many steps it will take you to walk from one end of the classroom to the other. Record the estimates and then take the steps to check.

Discover

Unit 1: Numbers to 100, Lesson 12

Estimate numbers on a number line

WAYS OF WORKING Pair work

ASK

- Question ❶: *What do you notice about the number line?*
- Question ❶ a): *Could you use it to count on in 10s from 0 to 100?*
- Question ❶ b): *Where do you think the number 82 might go? Why?*

IN FOCUS This first gives children the chance to rehearse how to identify multiples of 10 on the number line. They then start to make estimates and approximations about where numbers will be located.

PRACTICAL TIPS Use a large version of the number line, for example on the whiteboard. Let children practise counting up in 1s as you slowly move a pointer from 0 smoothly to 10, then 20, and so on along the line.

ANSWERS

Question ❶ a): Yes.

Question ❶ b): 82 lies between 80 and 90 on the number line, close to 80.

Discover

The arrow is pointing to the number 20.

I am going to draw an arrow to number 82.

Tim

❶ a) Is the teacher correct?

b) Where will Tim's arrow go?

52

PUPIL TEXTBOOK 2A PAGE 52

Share

WAYS OF WORKING Whole class teacher led

ASK

- Question ❶ a): *Where would be a good place to start writing the numbers from? Do you need to write all of the numbers?*
- Question ❶ b): *Can you know exactly where to put the number? What other numbers will 82 be near to? Will it be nearer to 0 or nearer to 100?*

IN FOCUS In question ❶ a), children are rehearsing identifying multiples of 10 on a number line. In question ❶ b) they begin to think about how to make a reasonable estimation for the location of a 2-digit number on a 0 to 100 number line, with the multiples of 10 labelled.

Share

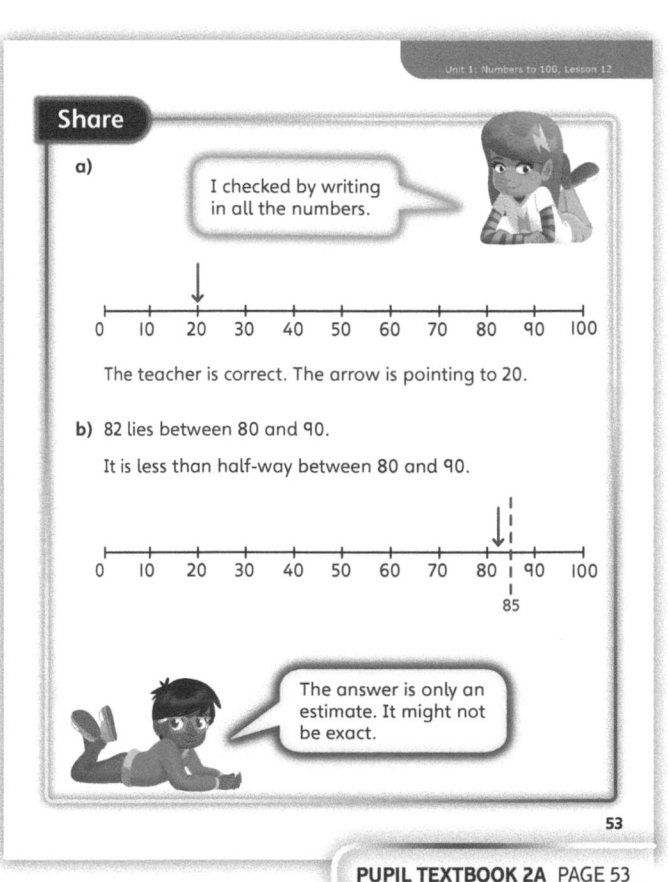

a)

I checked by writing in all the numbers.

The teacher is correct. The arrow is pointing to 20.

b) 82 lies between 80 and 90.

It is less than half-way between 80 and 90.

85

The answer is only an estimate. It might not be exact.

53

PUPIL TEXTBOOK 2A PAGE 53

Think together

Think together

WAYS OF WORKING Whole class teacher led (I do, We do, You do)

ASK

- Question ❶: *Can you say any of the numbers exactly? Do you need to estimate any?*
- Question ❷: *What is the same and what is different about each number line?*
- Question ❸: *Do you think of these as a big number or a small number?*

IN FOCUS In question ❶ children identify the location of numbers precisely or estimate where needed. In question ❷ they need to adapt their thinking for different styles of number line. For question ❸, children need to harness their understanding of the relative size or magnitude of each number.

STRENGTHEN Work together with a practical number line and numbers written on sticky notes, so that estimates can be reconsidered and moved, rather than being seen as 'wrong' answers.

DEEPEN Prompt children to work with number lines presented as different lengths and different orientations, and to think about how their estimates need to adapt each time the number line is different.

ASSESSMENT CHECKPOINT Assess whether children can make sensible estimates about the location of numbers on a variety of number lines. Question ❷ in particular will offer insight into the key learning for this lesson.

ANSWERS

Question ❶ a): 10, 50 and 85

Question ❶ b): Close to 100.

Question ❷:

Question ❸:

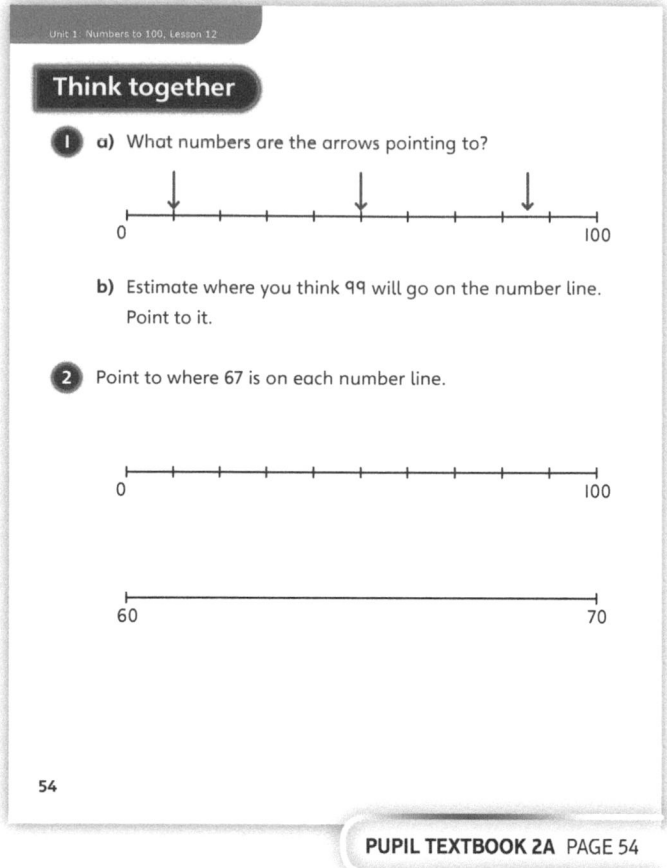

PUPIL TEXTBOOK 2A PAGE 54

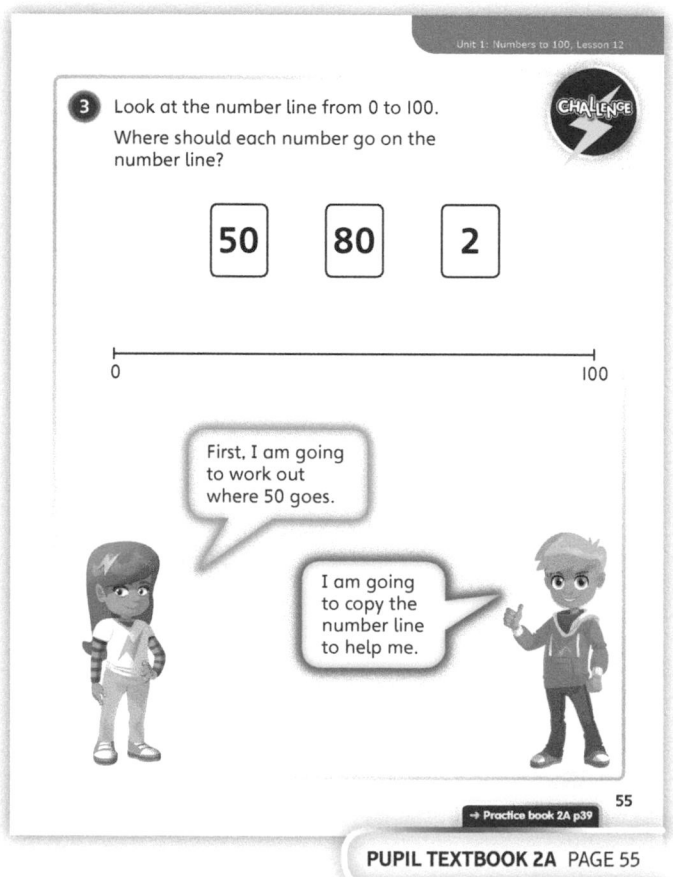

PUPIL TEXTBOOK 2A PAGE 55

Practice

WAYS OF WORKING Independent thinking

IN FOCUS For question ①, children are identifying or estimating the location of numbers as appropriate, on numbers lines with intervals marked. In question ②, the intervals are not marked. In question ③, children are making an estimate based on how close a number is to a multiple of 10. For questions ④, ⑤ and ⑥, children are locating numbers on a variety of number lines with reasonable precision.

STRENGTHEN Let children practise locating numbers on sticky notes with large number lines.

DEEPEN Challenge children to investigate number lines of different lengths, perhaps even as long as the playground.

ASSESSMENT CHECKPOINT Assess whether children can use the intervals on numbers lines to help them locate numbers. Question ③ will demonstrate a good level of understanding for this lesson.

ANSWERS Answers for the **Practice** part of the lesson can be found in the *Power Maths* online subscription.

Reflect

WAYS OF WORKING Pair work

IN FOCUS The **Reflect** part of the lesson requires children to make estimations about the location of given numbers on a blank 0 to 100 number line.

ASSESSMENT CHECKPOINT Assess whether children can justify their thinking based on their understanding of numbers to 100.

ANSWERS Answers for the **Reflect** part of the lesson can be found in the *Power Maths* online subscription.

After the lesson ⏸

- Ask children to guess the location of numbers on a blank number line.

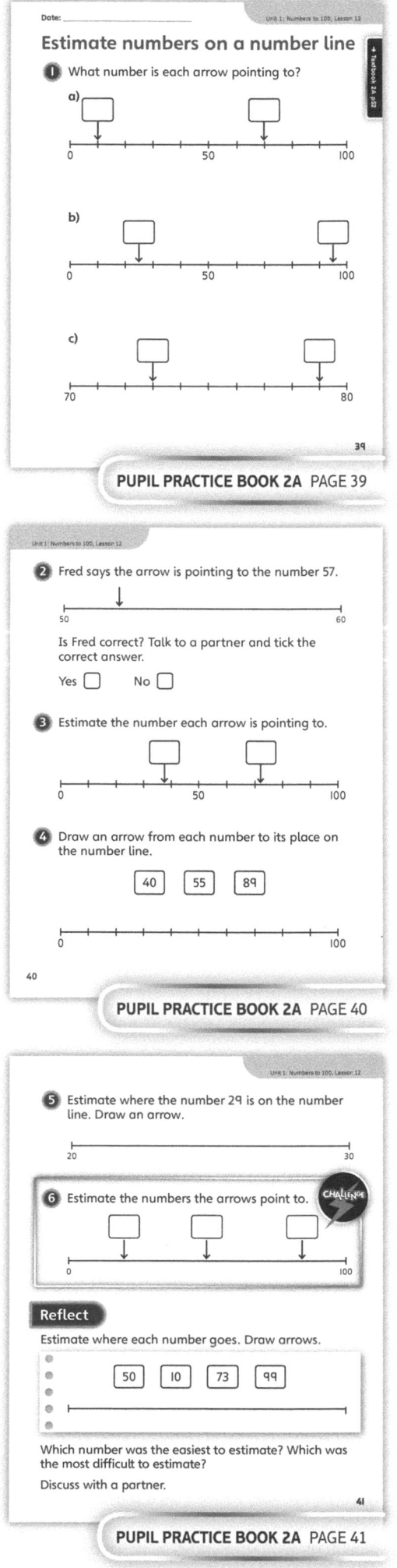

PUPIL PRACTICE BOOK 2A PAGE 39

PUPIL PRACTICE BOOK 2A PAGE 40

PUPIL PRACTICE BOOK 2A PAGE 41

Compare numbers ❶

Learning focus

In this lesson, children will develop their understanding of comparing numbers. Children will start to use their understanding of place value to aid them in their comparisons.

Before you teach ⏸

- Are children confident with the language of more and fewer?
- Will children be confident with comparing numbers clearly in different representations?

NATIONAL CURRICULUM LINKS

Year 1 Number – number and place value

Compare and order numbers from 0 up to 100; use <, > and = signs.

Identify, represent and estimate numbers using different representations, including the number line.

ASSESSING MASTERY

Children can prove that one group is greater than another, by matching up and comparing concrete and pictorial representations of numbers.

COMMON MISCONCEPTIONS

Children may not compare the numbers in an efficient way. Ask:
- *Is there a quicker way to make the comparison? Could you show the comparison in another way?*

STRENGTHENING UNDERSTANDING

Introduce the lesson through role playing a scene of two dolls sharing sweets. Share the sweets in different ways, both equally and unequally. Ask who has more and who has fewer. If the sweets are shared unequally, ask children how they could change the amounts so it is fair.

GOING DEEPER

Deepen understanding of different contexts by offering short word problems. For example: *Tim has 63p and Milly has 70p. Who has the most money to spend at the shops?* and *Taylor needs 49 cm of ribbon to wrap a present. He found 45 cm of ribbon. Does he have enough?* Encourage children to represent these problems with the resources they have used previously.

KEY LANGUAGE

In lesson: fewer, more, less, greater, tens (10s), ones (1s), compare

Other language to be used by the teacher: larger, bigger, smaller, fewer, equal

STRUCTURES AND REPRESENTATIONS

Multilink cubes, base 10 equipment

RESOURCES

Mandatory: multilink cubes, base 10 equipment

Optional: counters

 In the eTextbook of this lesson, you will find interactive links to a selection of teaching tools.

Quick recap 🔁

Ask children to think of different special numbers: a small number, a big number, a huge number and a tiny number. Talk about different contexts for the numbers they suggest. For example, 100 years old is very old for a human, but 20 years old is very old for a pet dog or cat.

Discover

Pair work

ASK

- Question ① a): *How have the children organised their cookies? Why do you think they did that?*
- Question ① b): *Who do you predict has fewer? How did you make your prediction?*
- Question ① b): *Is it fair at the moment?*
- Question ① b): *Can you write Matt's number of cookies as a written number?*

IN FOCUS The picture reinforces the previous lessons by grouping the cookies in 10s.

PRACTICAL TIPS Provide children with multilink cubes in two colours to make the two numbers. Encourage them to show the numbers as groups of 10 and some 1s.

ANSWERS

Question ① a): Matt has 43 and Anna has 50.

Question ① b): Matt has the fewest cookies.

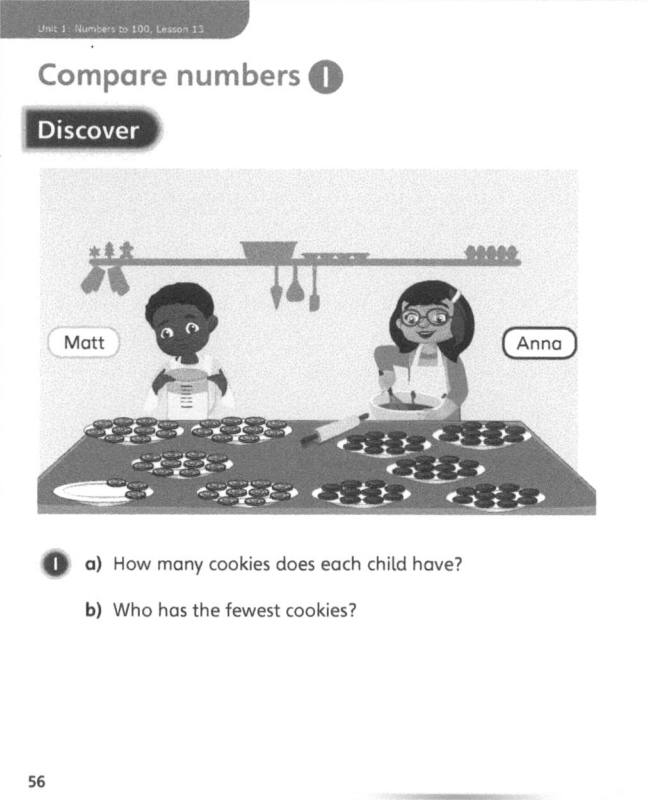

Compare numbers ①

Discover

① a) How many cookies does each child have?

b) Who has the fewest cookies?

56

PUPIL TEXTBOOK 2A PAGE 56

Share

WAYS OF WORKING Whole class teacher led

ASK

- Question ① a): *Should we compare the 10s or 1s first?*
- Question ① b): *Why is it important to compare the last row?*
- Question ① b): *Did anyone start by comparing the last row? Would this work all the time?*

IN FOCUS Questions ① a) and ① b) remind children of the importance of arranging base 10 equipment systematically in 10s and 1s. Question ① b) involves comparing 3 ones and a 10 (or 10 ones) in the last row.

Share

a) and b)

I counted the 10s and then the 1s.

Matt / Anna

Matt has fewer cookies than Anna.

Anna has more cookies than Matt.

I used base 10 equipment to represent the cookies.

57

PUPIL TEXTBOOK 2A PAGE 57

Think together

WAYS OF WORKING Whole class teacher led (I do, We do, You do)

ASK

- Question **1**: *What should you do first?*
- Question **1** a): *Which number is bigger?*
- Question **2**: *Are the cubes organised in a way that makes it easy to compare?*

IN FOCUS Questions **1** and **2** scaffold children in their ability to compare numbers, gradually reducing the amount of pre-completed elements. The 10s and 1s are arranged differently in questions **1** and **2**, giving children the opportunity to work out which they think is clearer.

STRENGTHEN Encourage children to make concrete representations of the numbers using cubes or other countable objects. Remind them to arrange the objects so that they can easily see which is the bigger number.

DEEPEN Question **3** presents numbers arranged in different ways. Ask children whether the objects are arranged so they can count them quickly, and whether they could change the arrangements to make their counting more efficient.

ASSESSMENT CHECKPOINT Questions **1** and **2** assess whether children can compare representations of two numbers. Children can recognise, and demonstrate through their comparisons, that they need to compare the 10s first, then the 1s.

ANSWERS

Question **1** a): 30 is less than 43.

Question **1** b): 43 is equal to 43.

Question **2**: 30 is greater than 24.
Ros has more cubes.

Question **3** a): Mo has 70 straws and Jan has 40 straws.

Question **3** b): Children point to 40 and 70 on the number line.

Question **3** c): Jan has fewer straws. Mo has more straws.

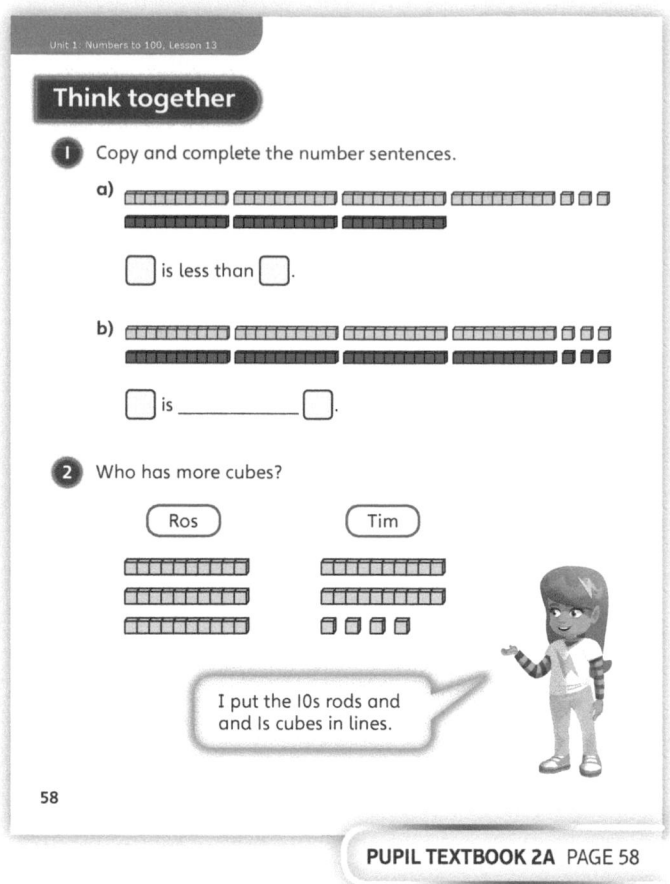

PUPIL TEXTBOOK 2A PAGE 58

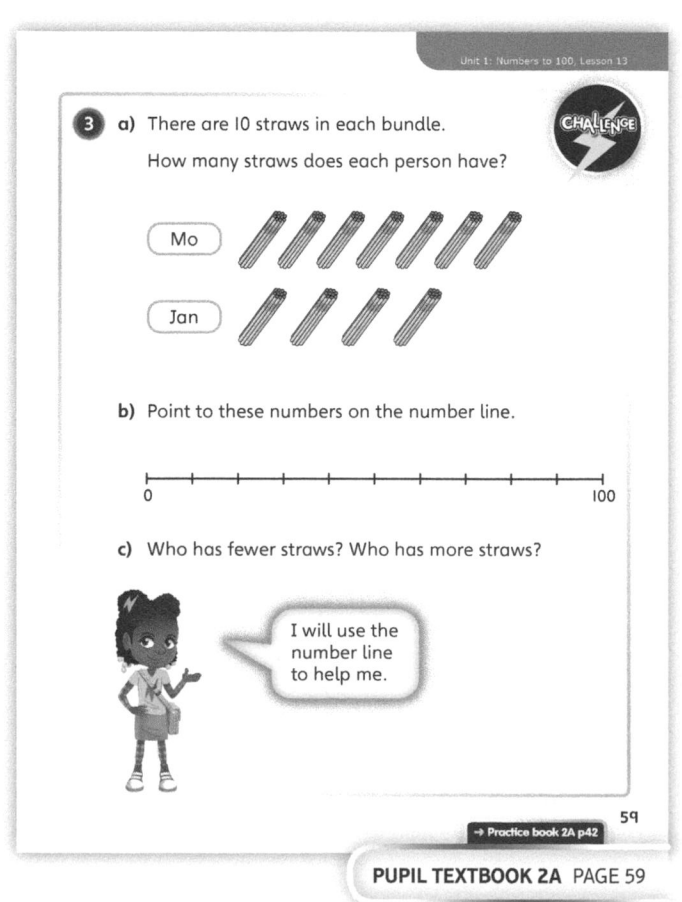

PUPIL TEXTBOOK 2A PAGE 59

Practice

WAYS OF WORKING Independent thinking

IN FOCUS Questions ❶ and ❷ give children the opportunity to practise comparing with different representations of numbers. Question ❶ scaffolds their solutions while question ❷ introduces a comparison sentence. Question ❸ reminds children of different representations of numbers they have studied.

STRENGTHEN In question ❸, which uses the language of 'more than' and 'less than', encourage children to represent the numbers in concrete ways. Ask: *How can you make the numbers easier to compare? Can you represent all the numbers in the same way?*

DEEPEN In question ❹, use the number line to deepen children's ability to compare numbers. Ask: *How does the number line help us see which is the greater number?* In question ❺, discuss how using regular arrangements of cubes makes it easier to count and compare them accurately. Ask: *How many 10s and 1s does Anya have? How could the cubes be laid out to help with making the comparison?*

ASSESSMENT CHECKPOINT Questions ❶, ❷ and ❸ should help you to determine whether children can compare numbers using multiple representations, and recognise that it is important and more efficient to compare the 10s first, then the 1s.

ANSWERS Answers for the **Practice** part of the lesson appear in the *Power Maths* online subscription.

Reflect

WAYS OF WORKING Pair work

IN FOCUS The **Reflect** part of the lesson requires children to discuss and define keywords for comparing numbers with their partner.

ASSESSMENT CHECKPOINT Assess whether children can suggest meaningful and correct examples using the words fewer, less, more and greater.

ANSWERS Answers for the **Reflect** part of the lesson appear in the *Power Maths* online subscription.

After the lesson ⏸

- Were children confident using the vocabulary of comparison?
- Ideas of 'more' and 'fewer' occur in many situations – including games, role play (for example, shopping activities) and classroom jobs like organising resources or putting pupils into groups.
- What opportunities can you identify to reinforce and apply this lesson's learning?

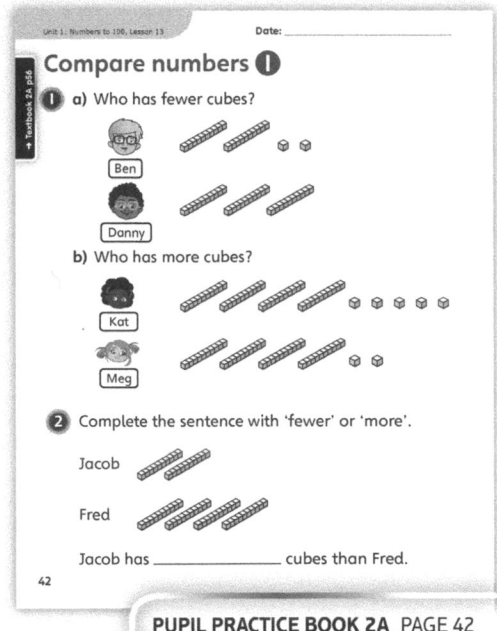

PUPIL PRACTICE BOOK 2A PAGE 42

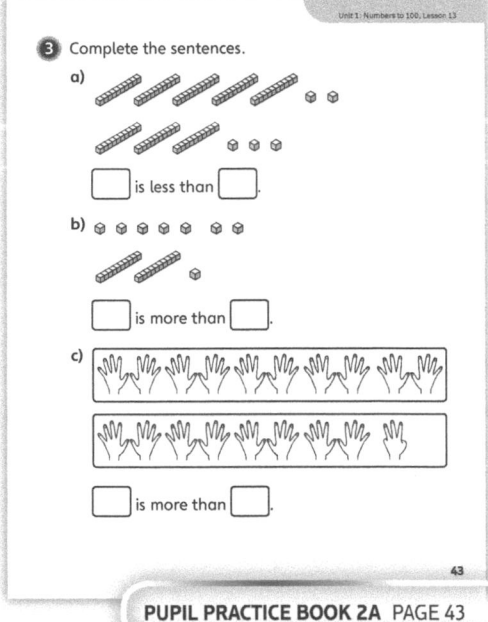

PUPIL PRACTICE BOOK 2A PAGE 43

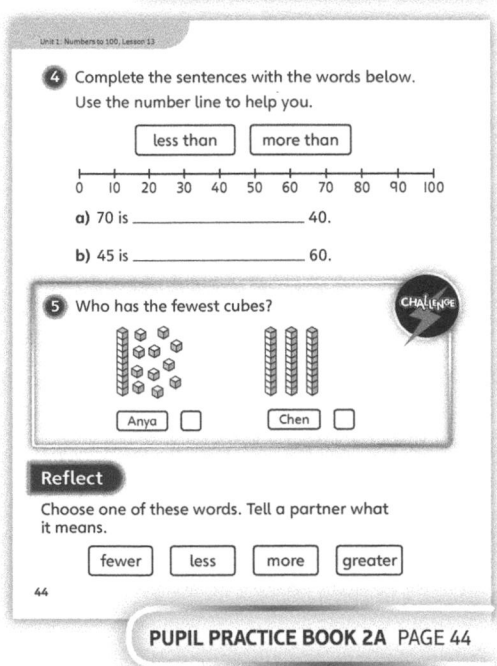

PUPIL PRACTICE BOOK 2A PAGE 44

Compare numbers ❷

Learning focus

In this lesson, children will continue developing their ability to compare numbers, using more abstract representations.

Before you teach ⏸

- Are children ready to move into the more abstract representations?
- How could you use this lesson to further develop children's problem-solving abilities?

NATIONAL CURRICULUM LINKS

Year 1 Number – number and place value

Compare and order numbers from 0 up to 100; use <, > and = signs.

ASSESSING MASTERY

Children can confidently compare numbers using place value to help and can recognise and explain why it is important to compare the larger part of two numbers first. Children are able to use more abstract representations to represent the numbers they are comparing.

COMMON MISCONCEPTIONS

Children may find comparing numbers with a 0 in the 1s column confusing. Reinforce that they need to compare the 10s column first. Ask:

- *How can you represent the numbers? What will you do first? The 10s numbers are the same. What will you do now?*

STRENGTHENING UNDERSTANDING

Ensure that children who are unsure about comparing concrete and pictorial representations of numbers are given the opportunity to practise and develop this skill. Provide a large printed place value grid in which children can arrange the numbers in a concrete way, such as with cubes, before completing the more abstract grid. Possible approaches include comparing cube towers grouped in 10s and 1s, and comparing groups of objects through classroom jobs or role play, for example, pencils on tables or sharing sweets between two cuddly toys.

GOING DEEPER

Give children a 2-digit number (for example, 86). Ask how they could make the number smaller or larger by only changing the 1s. Then ask how they could make the number smaller or larger by only changing the 10s. This problem could give children the opportunity to work with problems that have either more than one solution or no solution at all.

KEY LANGUAGE

In lesson: more, less, greater, fewer, tens (10s), ones (1s), place value, number line, compare, equal, <, >

Other language to be used by the teacher: larger, bigger, smaller

STRUCTURES AND REPRESENTATIONS

Base 10, place value grid

RESOURCES

Optional: multilink cubes, large printed place value grid

 In the eTextbook of this lesson, you will find interactive links to a selection of teaching tools.

Quick recap

Rehearse the meaning of the < and > signs and practise using them in number statements with numbers between 0 and 10.

Discover

Pair work

ASK

- Question **1**: *What could you compare in this picture? Which tree has the fewest leaves? Are there more leaves on the trees or on the ground?*
- Question **1** a): *How can you tell who has more leaves?*
- Question **1** b): *How many more leaves should Asif pick up so he has the same number as Beth?*
- Question **1** b): *How many different ways could you show the numbers?*

IN FOCUS Questions **1** a) and **1** b) both require children to consider how to compare numbers without seeing pictorial representations of the numbers.

PRACTICAL TIPS This lesson is broadly abstract, however some children will be more confident comparing numbers by thinking about the order when counting along a number line, and some will compare numbers by thinking about the 10s digits and then the 1s digits. Provide number lines and base 10 equipment for those who need it.

ANSWERS

Question **1** a): Beth has more leaves.

Question **1** b): 43 < 57.

PUPIL TEXTBOOK 2A PAGE 60

Share

WAYS OF WORKING Whole class teacher led

ASK

- Question **1** a): *How does the place value grid help show the value of the number?*
- Question **1** a): *What column do you compare first?*
- Question **1** b): *How do you know which sign to use to show the comparison?*

IN FOCUS Questions **1** a) and **1** b) ask children to compare numbers with the more abstract representations and to use comparison signs, developing the skill of comparing numbers.

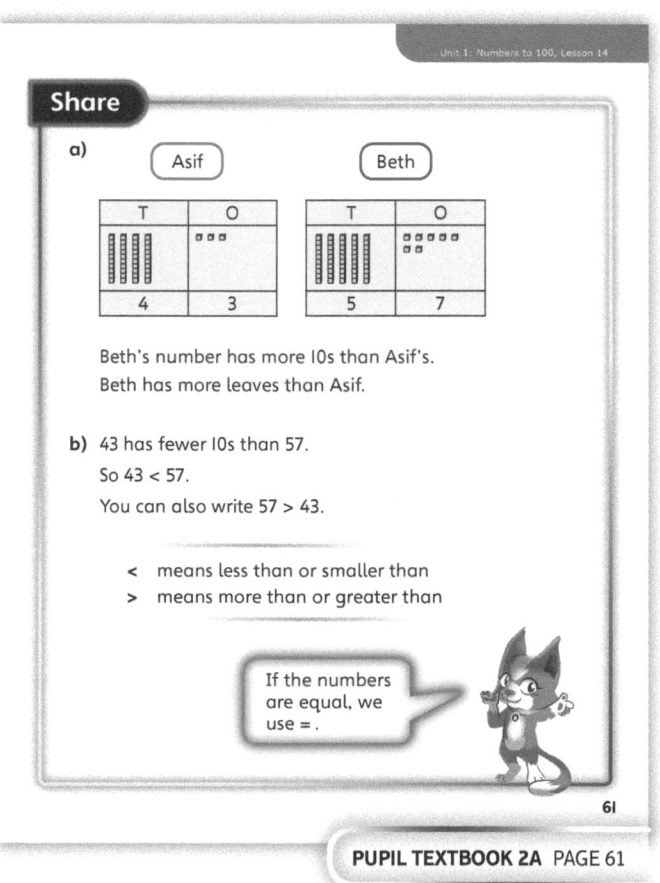

PUPIL TEXTBOOK 2A PAGE 61

Think together

Whole class teacher led (I do, We do, You do)

ASK

- Question ❶: *What should you do first?*
- Question ❶: *What sign will you use to show the comparison?*
- Question ❶: *Is there another way to write the comparison?*
- Question ❶: *How does the place value grid show the value of the number? How will you use it to compare the numbers?*
- Question ❸: *What column will you look at first when comparing the numbers?*

IN FOCUS Questions ❶ and ❷ scaffold children's move from comparing numbers with a different amount of 10s to numbers with the same amount of 10s.

STRENGTHEN Some children may be confused by the fact that both numbers in question ❷ have the same number of 10s. Discuss that when numbers have the same amount of 10s, the smaller number is always the one that has fewer 1s. Encourage them to count the cubes in the place value grid to check.

DEEPEN Use Astrid and Ash's comments to discuss whether all the comparisons can be solved by looking only at the 10s. Ask: *Can all the comparisons be solved by only looking at the 10s? Do you ever need to look at the 1s? Can you find an example where you cannot compare only using the 10s?*

ASSESSMENT CHECKPOINT Use question ❸ to assess whether children can compare numbers using their understanding of place value grids, comparing the 10s and then the 1s, if necessary.

ANSWERS

Question ❶ a): 75

Question ❶ b): 54 < 75

Question ❷: 62

Question ❸ a): 64 > 26

Question ❸ b): 57 < 70

Question ❸ c): 57 > 54

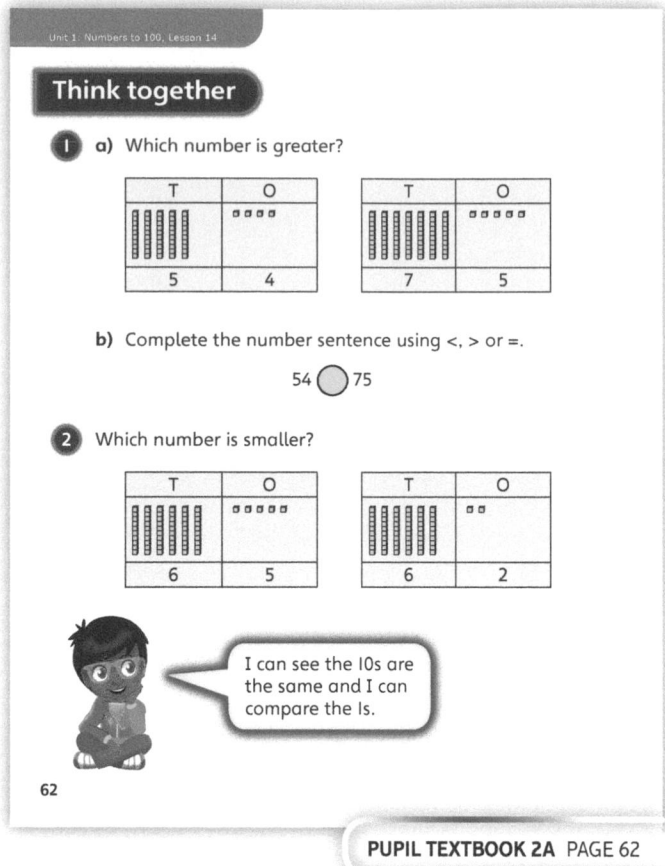

PUPIL TEXTBOOK 2A PAGE 62

PUPIL TEXTBOOK 2A PAGE 63

Practice

WAYS OF WORKING Independent thinking

IN FOCUS Questions ❶ and ❷ require the children to compare numbers using place value grids, with question ❸ moving to comparing numbers without the support of using place value grids. Question ❹ ensures that children are confident expressing <, > and = in words.

THINK DIFFERENTLY Question ❺ introduces missing number statements where there may be more than one possible answer. Ask: *How many different answers can you find?*

STRENGTHEN In question ❹, when using numbers written in ways that require some problem solving (for example, '4 tens and 8 ones'), support children by asking: *What is this number? Can you show me the number using base 10 equipment or in a place value grid?*

DEEPEN In question ❻ ask: *Can you explain how you solved the question? Do you think Flo's method is the best way of solving this? Can you create a similar number puzzle?*

ASSESSMENT CHECKPOINT Use questions ❸ and ❹ to assess whether children can compare 2-digit numbers represented in an abstract form.

ANSWERS Answers for the **Practice** part of the lesson appear in the *Power Maths* online subscription.

Reflect

WAYS OF WORKING Pair work

IN FOCUS The **Reflect** part of the lesson requires children to apply and explain their knowledge about comparing 2-digit numbers. Ask children to share their reasoning with their partner, justifying their opinions.

ASSESSMENT CHECKPOINT Assess whether children can fluently explain their reasoning when comparing two numbers, using the appropriate concepts and vocabulary.

ANSWERS Answers for the **Reflect** part of the lesson appear in the *Power Maths* online subscription.

After the lesson ⏸

• Were children confident comparing numbers in abstract form?
• Were children able to recognise when there could be more than one correct solution?

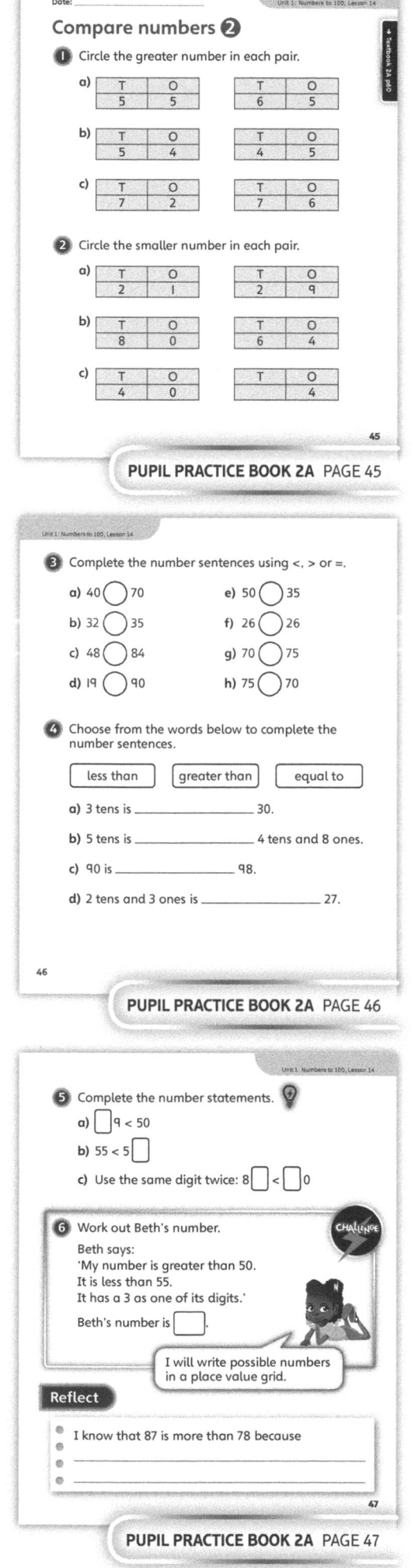

PUPIL PRACTICE BOOK 2A PAGE 45

PUPIL PRACTICE BOOK 2A PAGE 46

PUPIL PRACTICE BOOK 2A PAGE 47

Order numbers

Learning focus

In this lesson, children draw on their skills in comparing and place value, and use them to find effective and efficient ways to order three or more 1- and 2-digit numbers.

Before you teach

- Can children discuss how to compare two 2-digit numbers?
- Have they had the opportunity to practise using the < and > signs?
- Can you give children the chance to rehearse the key vocabulary of more, less, greater, bigger, smaller?

NATIONAL CURRICULUM LINKS

Year 1 Number – number and place value

Compare and order numbers from 0 up to 100; use <, > and = signs.

ASSESSING MASTERY

Children can use their understanding of place value to justify the order of a set of numbers.

COMMON MISCONCEPTIONS

Children may not apply reasoning when ordering. For example, they may not recognise that if *a* is less than *b*, and *b* is less than *c*, then *a* must also be less than *c*. Ask:
- *Which number is the biggest? Which number is the smallest? How do you know?*

STRENGTHENING UNDERSTANDING

Give children a 0 to 100 number line, and prompt them to use their number line skills when thinking about ordering numbers.

GOING DEEPER

Challenge children to explore the place value of the numbers more fully, using comparison of the 10s digits first, and then the 1s.

KEY LANGUAGE

In lesson: greater, greatest, tallest, most, less, least, shortest

Other language to be used by the teacher: sort

STRUCTURES AND REPRESENTATIONS

Number line, place value grid

RESOURCES

Optional: number lines, number cards, place value grids

 In the eTextbook of this lesson, you will find interactive links to a selection of teaching tools.

Quick recap

Give a drawing challenge to prompt discussion about comparing and ordering. Ask children to draw a small version, a medium version and a large version of an object on a chosen theme, for example an animal, vehicle or house.

Discover

WAYS OF WORKING Pair work

ASK

- Question 1: *What do you think the table shows?*
- Question 1 b): *How tall is Dan's plant?*
- Question 1 b): *How tall is Eva's plant?*
- Question 1 b): *How tall is Felix's plant?*

IN FOCUS The picture introduces the concept of ordering more than one number. It may engage children to repeat this activity with plants in the school garden.

PRACTICAL TIPS Use a real pot plant, or a picture or model, to demonstrate how the children have used cubes to measure the height of a plant.

ANSWERS

Question 1 a): Eva has the tallest sunflower (plant C).

Question 1 b): Dan has the shortest sunflower (plant B).

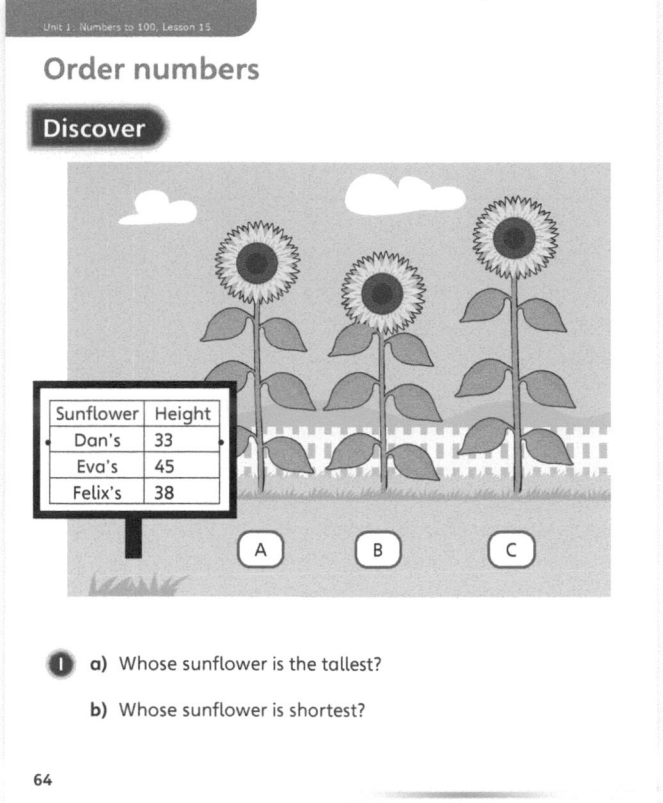

Order numbers

Discover

Sunflower	Height
Dan's	33
Eva's	45
Felix's	38

1 a) Whose sunflower is the tallest?

b) Whose sunflower is shortest?

64

PUPIL TEXTBOOK 2A PAGE 64

Share

WAYS OF WORKING Whole class teacher led

ASK

- Question 1 a): *How do the place value grids relate to the measurements? What does each digit mean? Why is the 10s digit the best digit to compare?*
- Question 1 b): *Look at the number line. How can this help us think about the order?*

IN FOCUS In question 1 a), children are introduced to efficient methods for comparing a set, through place value understanding. In question 1 b) they are connecting the ordinality and cardinality of numbers, through the representation of the number line, and recognising that the smallest number is the shortest flower.

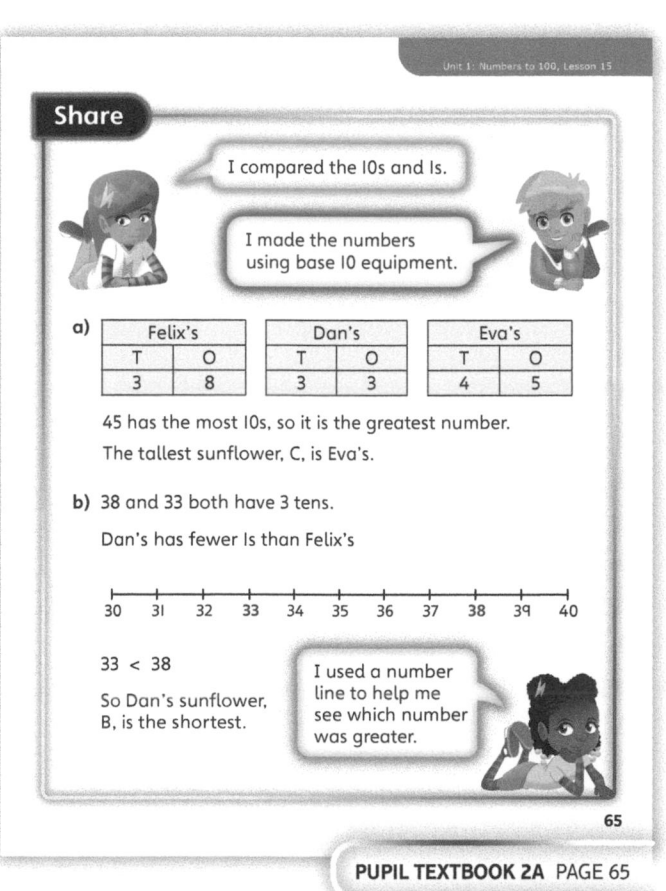

Share

I compared the 10s and 1s.

I made the numbers using base 10 equipment.

a)

Felix's		Dan's		Eva's	
T	O	T	O	T	O
3	8	3	3	4	5

45 has the most 10s, so it is the greatest number.
The tallest sunflower, C, is Eva's.

b) 38 and 33 both have 3 tens.

Dan's has fewer 1s than Felix's

30 31 32 33 34 35 36 37 38 39 40

33 < 38

So Dan's sunflower, B, is the shortest.

I used a number line to help me see which number was greater.

65

PUPIL TEXTBOOK 2A PAGE 65

Think together

WAYS OF WORKING **WAYS OF WORKING** Whole class teacher led (I do, We do, You do)

ASK

- Question **1**: *Which digits will you compare first?*
- Question **2**: *Is there only one answer? What is the same and what is different about the numbers you have been given?*
- Question **3**: *Are some guesses better than others? What numbers are crossed out? Why?*

IN FOCUS In question **1**, children are using place value knowledge to compare and order numbers. In question **2**, they are finding possible numbers that fall within a given range. For question **3**, children are required to use reasoning to develop a trial and improve method for finding a mystery number.

STRENGTHEN Provide a 0 to 100 number line to support children's thinking throughout.

DEEPEN Prompt discussions about a good strategy for question **3**, for example looking for a number near the middle of the interval, to rule out more options.

ASSESSMENT CHECKPOINT Use question **2** to assess whether children can demonstrate their reasoning and use of place value skills.

ANSWERS

Question **1**: 67 > 63 > 31; 31 < 63 < 67

Question **2**: Any number between 31 and 39.

Any number between 56 and 64.

Numbers between 1 and 99. The number in the first box must be larger than the number in the second box.

Question **3**: Any number between 11 and 79 inclusive.

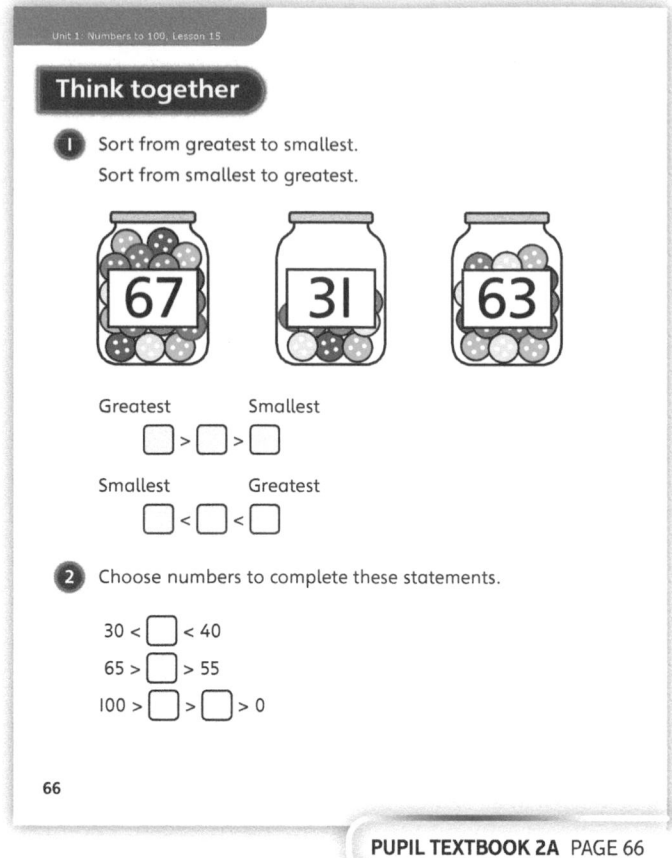

PUPIL TEXTBOOK 2A PAGE 66

PUPIL TEXTBOOK 2A PAGE 67

Practice

WAYS OF WORKING Independent thinking

IN FOCUS In questions ❶ and ❷ children are comparing using place value grids to find first the smallest number and then the greatest number in a set. In question ❸ children are sorting a set of numbers from smallest to greatest, and in question ❹ from greatest to smallest. For question ❺ children need to find a number between two numbers, in a given range. In question ❻ children are solving ordering puzzles using digit cards.

STRENGTHEN Provide 0 to 100 number lines throughout to support children's comparison skills.

DEEPEN In pairs or small groups, children can make and solve their own versions of question ❸.

ASSESSMENT CHECKPOINT Use questions ❸ and ❹ to assess the key learning, whether children can order sets of three 1- or 2-digit numbers using their knowledge of place value.

ANSWERS Answers for the **Practice** part of the lesson can be found in the *Power Maths* online subscription.

Reflect

WAYS OF WORKING Pair work

IN FOCUS The **Reflect** part of the lesson requires children to use their place value skills to explain how they compare and order three 2-digit numbers.

ASSESSMENT CHECKPOINT Assess whether children can explain their reasoning to justify the order of the numbers based on place value.

ANSWERS Answers for the **Reflect** part of the lesson can be found in the *Power Maths* online subscription.

After the lesson ⏸

- Check that children can order a set of any three 2-digit numbers that have been chosen at random.

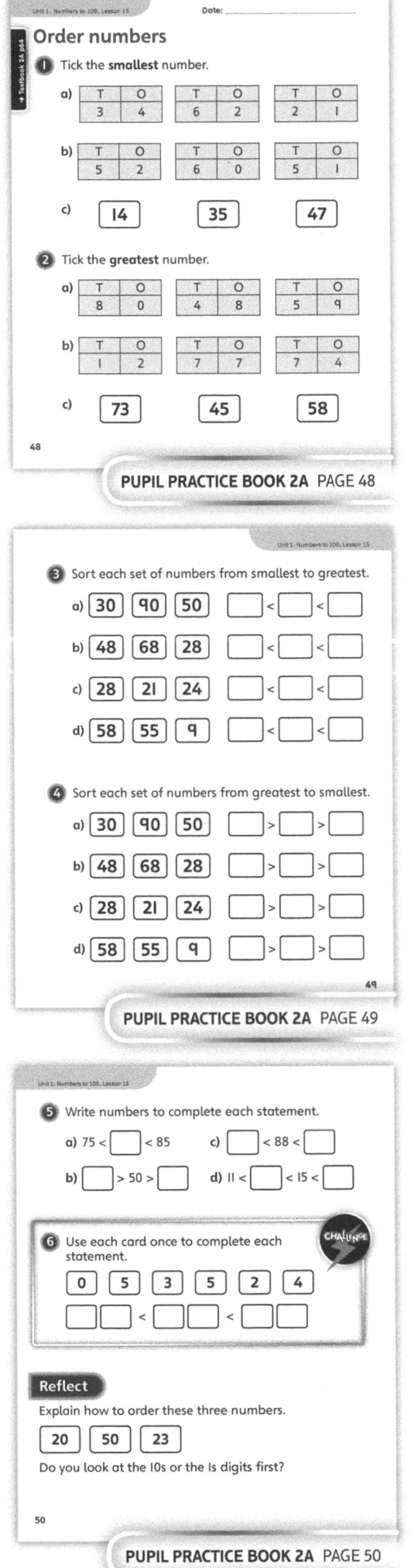

PUPIL PRACTICE BOOK 2A PAGE 48

PUPIL PRACTICE BOOK 2A PAGE 49

PUPIL PRACTICE BOOK 2A PAGE 50

Count in 2s, 5s and 10s

Learning focus

In this lesson, children will learn to count forwards and backwards in 2s, 5s and 10s.

Before you teach ⏸

- Are the children confident at spotting patterns around them?
- How could you use this to develop their reasoning skills in this lesson?

NATIONAL CURRICULUM LINKS

Year 1 Number – number and place value

Count in steps of 2, 3, and 5 from 0, and in tens from any number, forward and backward.

ASSESSING MASTERY

Children can count reliably forwards and backwards in steps of 2, 5 and 10. They can recognise patterns within their counting, using their knowledge of place value, and can show the patterns using different representations.

COMMON MISCONCEPTIONS

Children may have trouble counting backwards. Using visual representations along a number line, ask:
- *What number am I pointing at? Show me where to point if I counted backwards 2 or 5 or 10. How much do I have now? How was that different to counting forwards?*

STRENGTHENING UNDERSTANDING

This concept could be approached before the lesson through role-playing a visit to the shops. For example, if bananas come in bunches of 2, ask children how many bananas in 1 bunch, 2 bunches, 3 bunches, and so on; if oranges are sold in bags of 5, ask how many oranges there are in 3 bags.

GOING DEEPER

Ask children to investigate which numbers are not counted in the 2s, 5s and 10s when counting from 0. For example, 3 and 9 will not be counted in any of them. Ask: *Which numbers will never be counted in any of the counts? Can you explain why?*

KEY LANGUAGE

In lesson: altogether, count, pair, twos (2s), fives (5s), tens (10s), less, more

Other language to be used by the teacher: steps, forwards, backwards, increase, decrease, bigger, smaller, pattern

STRUCTURES AND REPRESENTATIONS

Number line, ten frame, number track

RESOURCES

Mandatory: number lines, multilink cubes

Optional: base 10 equipment, counters

 In the eTextbook of this lesson, you will find interactive links to a selection of teaching tools.

Quick recap 🔁

Sing 2, 4, 6, 8 counting songs together to practise counting in 2s.

Discover

Unit 1: Numbers to 100, Lesson 16

Count in 2s, 5s and 10s

Discover

WAYS OF WORKING Pair work

ASK

- Question **1** a): *What do you notice about the children? Can you see them in pairs?*
- Question **1** b): *Is it always best to count in 1s? Are there any other ways to count?*

IN FOCUS Children are first counting to 12 in 2s and then counting to 120 in 5s.

PRACTICAL TIPS Use practical equipment such as cubes or other countable items to represent the schoolchildren. Ask children to move them and sort them into pairs before counting in 2s.

ANSWERS

Question **1** a): 2, 4, 6, 8, 10, 12

Question **1** b): 10, 20, 30, 40, 50, 60

It is easier to count you all if you line up in pairs.

SCHOOL

1 **a)** Count the children onto the bus in pairs.

    b) Count the fingers and thumbs of 6 of the children.

68

PUPIL TEXTBOOK 2A PAGE 68

Share

WAYS OF WORKING Whole class teacher led

ASK

Question **1** a): *If you start counting in 2s, how will you know when to stop counting?*

Question **1** b): *Do you know the counting in 5s pattern? What rhythm can you hear?*

IN FOCUS In question **1** a), children are counting in 2s and demonstrating that they know when the count is complete. In question **1** b), they are doing the same, but this time with counting in 5s.

Share

a)

2   4   6   8   10   12

I counted them all in 2s.

b)

0 5 10 15 20 25 30 35 40 45 50 55 60

I counted in 5s.

I could have counted in 10s.

0 10 20 30 40 50 60

69

PUPIL TEXTBOOK 2A PAGE 69

Think together

WAYS OF WORKING Whole class teacher led (I do, We do, You do)

ASK

- Question **1**: *How do you know whether to count in 1s, 2s, 5s or 10s? What clues are there?*
- Question **2**: *What number starts the count? Are you counting up or down? How can you tell?*
- Question **3**: *What counting patterns will help you? Which of these counting patterns is easiest for you? Why?*

IN FOCUS In question **1**, children need to recognise and continue counting patterns in 2s, 5s and 10s, starting at 0. In question **2**, they are counting up or back in different steps from a given starting number. In question **3**, they are linking counting patterns on and back to 'less' and 'more'.

STRENGTHEN Give children opportunities to practise counting within 20, using objects to count. Ensure that they point at each object and count in time.

DEEPEN In question **3**, ask: *Why will Flo's idea help you? Could you use anything else to help you check your ideas? Do your answers match the patterns you saw earlier in the lesson?* Be aware that question **3** d) requires children to cross into 3-digit numbers.

ASSESSMENT CHECKPOINT Assess whether children can arrange a group of 20 items in order to accurately count them in 2s, 5s and 10s.

ANSWERS

Question **1** a): 25, 30, 35, 40, 45, 50

Question **1** b): 30, 40, 50

Question **1** c): 8, 10, 12, 14, 16, 18, 20, 22, 24, 26, 28, 30, 32, 34, 36, 38, 40, 42, 44, 46, 48, 50

Question **2** a): 34, 36, 38, 40, 42, 44, 46

Question **2** b): 45, 50, 55, 60, 65, 70

Question **2** c): 100, 90, 80, 70, 60, 50

Question **3** a): 68, 72

Question **3** b): 76, 80

Question **3** c): 65, 75

Question **3** d): 80, 100

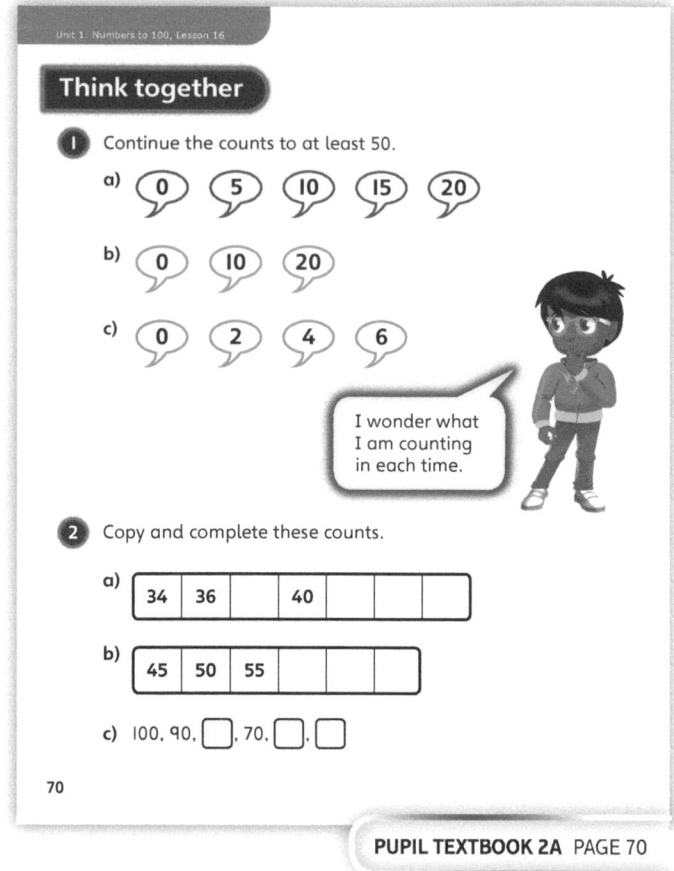

PUPIL TEXTBOOK 2A PAGE 70

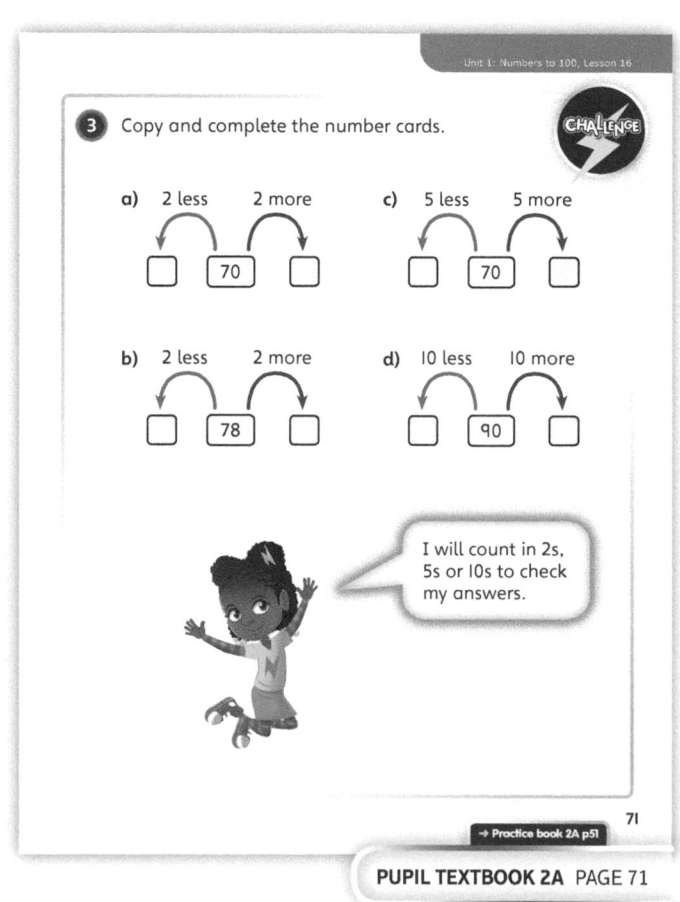

PUPIL TEXTBOOK 2A PAGE 71

Practice

WAYS OF WORKING Independent thinking

IN FOCUS In question ❶ children are counting in 2s, 5s or 10s from 0 and in question ❷ they complete counts from given starting points on a number track or number line. The final part of question ❷ shows a descending count. Question ❸ requires children to count representations of physical items. In question ❺ children recognise which numbers will appear in two different counting patterns.

STRENGTHEN For question ❹, offer children a number line to 100. Ask which numbers are greater or less than 20 or 50. Help children to recognise a pattern in the numbers Joe counts.

DEEPEN Give children practice of counting up and down in 2s from any given number and encourage them to find the patterns on a 100 square for different counts.

THINK DIFFERENTLY Question ❹ requires children to recognise numbers that are not in a given counting pattern.

ASSESSMENT CHECKPOINT Use question ❷ to assess whether children can accurately complete a variety of counts in 2s, 5s or 10s.

ANSWERS Answers for the **Practice** part of the lesson can be found in the *Power Maths* online subscription.

Reflect

WAYS OF WORKING Pair work

IN FOCUS Children are exploring how far they can continue the 2s, 5s and 10s counts.

ASSESSMENT CHECKPOINT Assess whether children can continue each of the counts, and whether they identify patterns that will help them to count accurately.

ANSWERS Answers for the **Reflect** part of the lesson can be found in the *Power Maths* online subscription.

After the lesson ⏸

- Give opportunities to count a number of real items (for example, PE equipment or other children) in 2s, 5s and 10s.

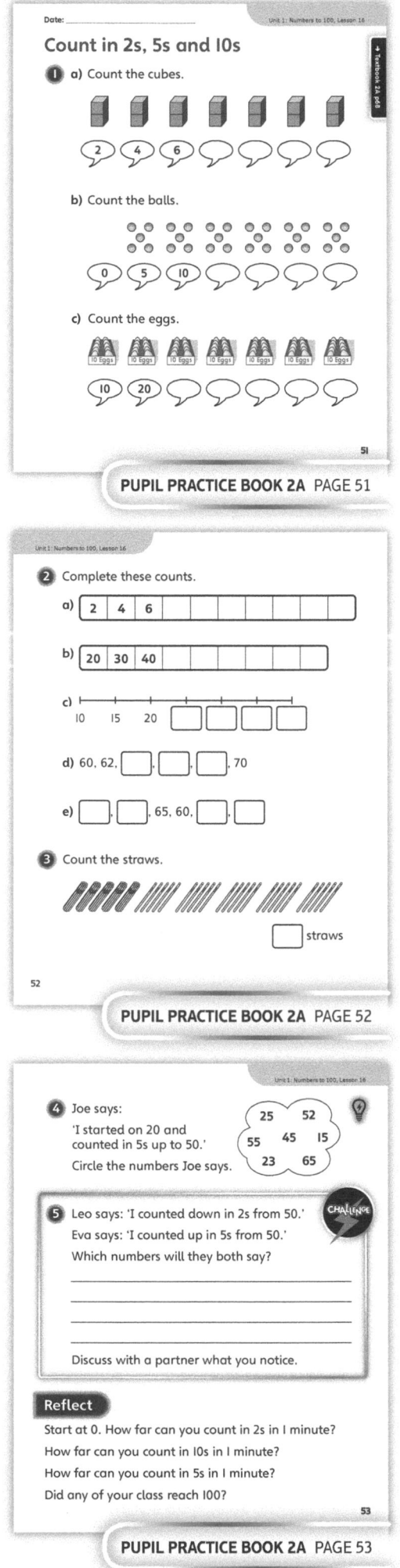

PUPIL PRACTICE BOOK 2A PAGE 51

PUPIL PRACTICE BOOK 2A PAGE 52

PUPIL PRACTICE BOOK 2A PAGE 53

Count in 3s

Learning focus

In this lesson, children will learn to count forwards and backwards in 3s.

Before you teach 🅿

- Could you provide a counting song to help develop children's confidence and fluency?

NATIONAL CURRICULUM LINKS

Year 1 Number – number and place value

Count in steps of 2, 3 and 5 from 0, and in tens from any number, forwards and backwards.

ASSESSING MASTERY

Children can count reliably forwards and backwards in steps of 3. They can recognise patterns within their counting and can show the patterns using different representations.

COMMON MISCONCEPTIONS

Children may think that every number ending in '3' is a multiple of 3. Using a number line and visual representations of the number, ask:

- *Can you count up one 3 from 0? What number would you be at if you counted on another 3? What do you notice about this number? What about the next one? Does every number have to end in a 3 when counting in 3s?*

STRENGTHENING UNDERSTANDING

To introduce this concept, children could be put into teams of three during a PE lesson. Alternatively, children could be asked to leave and enter a room in groups of three. After each group has entered or left, ask how many children are in the room and how many have left.

GOING DEEPER

Ask children to investigate what happens when they start counting in 3s at a number other than 0. Before they start, ask: *What do you think might change when counting from a different number? What might stay the same? Can you show me your reasoning with a picture or with equipment?*

KEY LANGUAGE

In lesson: count, threes (3s)

Other language to be used by the teacher: steps, forwards, backwards, increase, decrease, bigger, smaller, check

STRUCTURES AND REPRESENTATIONS

Number line, number track, 100 square

RESOURCES

Mandatory: number line, dice, sticks

Optional: multilink cubes, counters, collections of countable objects

 In the eTextbook of this lesson, you will find interactive links to a selection of teaching tools.

Quick recap

Ask children to make groups of 3 items, for example counters or cubes, or just 3 fingers on each hand. Play dice games where you keep going until you roll a 3.

Discover

WAYS OF WORKING Pair work

ASK

- Question ❶: *What kind of pattern did Andy make? How would you describe it?*
- Question ❶ a): *How many sticks did he use for the top triangle?*
- Question ❶ a): *How many sticks did he use for the second row of triangles? How many sticks did he use for the third row of triangles?*
- Question ❶ b): *What do you notice?*
- Question ❶ b): *Can you make the pattern Andy has made? How many sticks do you need in total?*
- Question ❶ b): *Can you spot any other patterns?*

IN FOCUS Andy's pattern is made from triangles, each using three sticks. This introduces children to the idea of counting in 3s to find the total number of sticks made from 6 triangles.

PRACTICAL TIPS Provide sticks for children to make Andy's pattern, counting in 3s as they make each triangle.

ANSWERS

Question ❶ a): Andy used 18 sticks.

Question ❶ b): Andy needs 12 more sticks.

Counts in 3s

Discover

I made this pattern using sticks.

❶ a) How many sticks did Andy use?

b) Andy wants to add another row of triangles at the bottom.
How many more sticks does he need?

72

PUPIL TEXTBOOK 2A PAGE 72

Share

WAYS OF WORKING Whole class teacher led

ASK

- Question ❶: *How have the triangles been organised to make the counting easier?*
- Question ❶: *Does it show 3 more clearly?*
- Question ❶ b): *How else could you show 3?*

IN FOCUS Both question ❶ a) and ❶ b) make clear how to count in 3s and introduce the patterns that are evident in this method of counting.

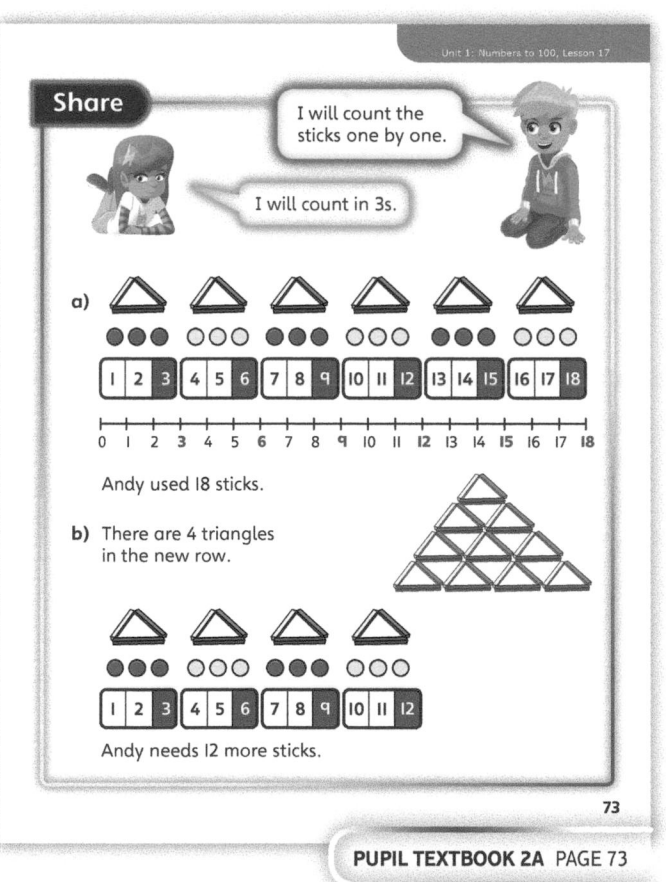

Share

I will count the sticks one by one.

I will count in 3s.

a)

Andy used 18 sticks.

b) There are 4 triangles in the new row.

Andy needs 12 more sticks.

73

PUPIL TEXTBOOK 2A PAGE 73

Think together

WAYS OF WORKING Whole class teacher led (I do, We do, You do)

ASK

- Question ❶: *Why are some boxes blank?*
- Question ❶: *How could you use the pictures to help?*
- Question ❶: *What is the same about questions ❶ a) and b)? What is different?*
- Question ❷: *What is different about the counting this time?*
- Question ❷: *How will you know where to begin counting?*
- Question ❸: *How can you prove your ideas?*

IN FOCUS Questions in this section offer children different ways of counting in 3s from different starting points.

STRENGTHEN For question ❷, provide children with coloured cubes to build the castles. Each time they add another set of three cubes, ask how many cubes they had to start with and the size of each group. Encourage them to arrange the groups so that they are easier to count.

DEEPEN Use question ❶ to begin practising counting back. For example, what if the birds flew away, 3 at a time? For question ❸, ask children whether they can predict what the next common number will be, justifying their answer. Ask: *What patterns are in the numbers Jake and Zara both write?*

ASSESSMENT CHECKPOINT Questions ❶ and ❷ will help you to decide whether children can count up in 3s and recognise the patterns within their counting.

ANSWERS

Question ❶ a): The missing numbers are: 3, 6, 9, 12. There are 12 trees.

Question ❶ b): The missing numbers are: 3, 6, 9, 12, 15. There are 15 birds.

Question ❷: The missing numbers are: 15, 18, 21, 24. Steve used 24 blocks altogether.

Question ❸: Jake's counting:
2, 4, 6, 8, 10, 12,14
Zara's counting:
3, 6, 9, 12, 15, 18, 21
They will both write 6 and 12. Common numbers beyond these will be all the multiples of 6.

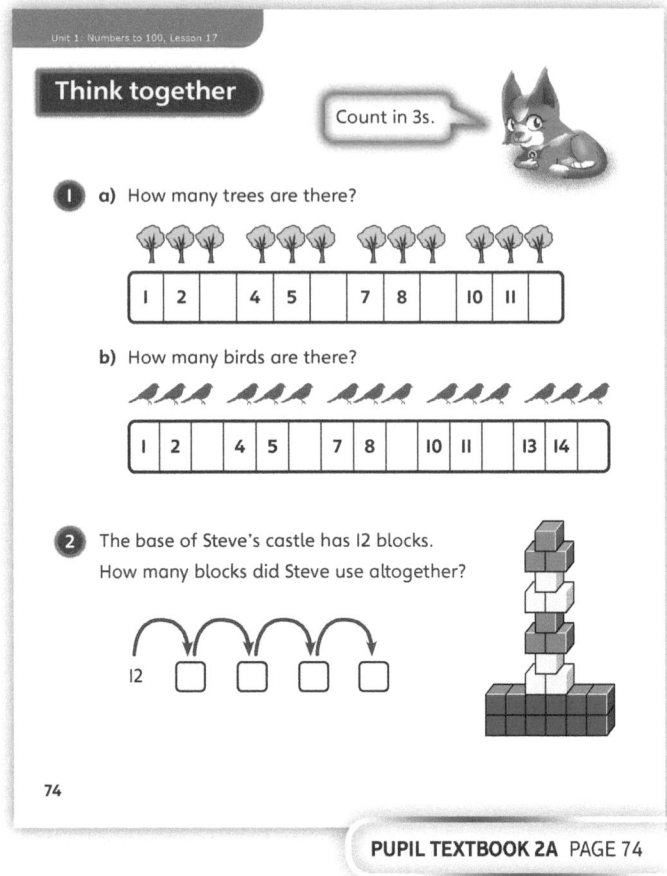

PUPIL TEXTBOOK 2A PAGE 74

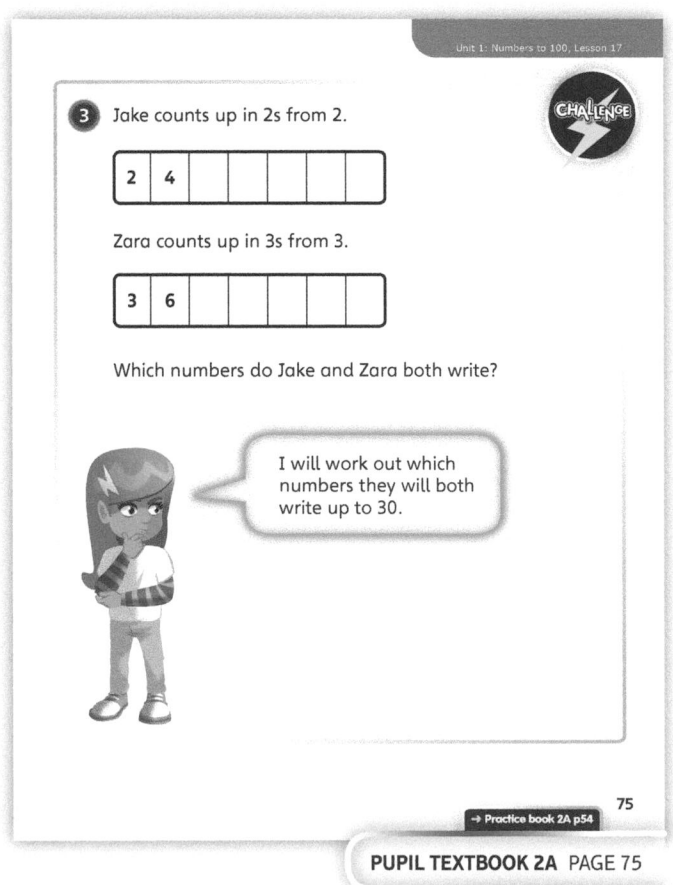

PUPIL TEXTBOOK 2A PAGE 75

Practice

WAYS OF WORKING Independent thinking

IN FOCUS Questions **1**, **2** and **4** give children opportunity to count in 3s using different representations.

THINK DIFFERENTLY Question **5** provides an interesting opportunity to recognise patterns and to consider how they change in the different number grids.

STRENGTHEN Provide a large number line or number track. For each question, encourage children to represent the count using equipment arranged on one of the number tracks. Ask: *Can you show me how you can use a number line or track to help you?*

DEEPEN Question **6** offers a good opportunity for problem-solving, reasoning and justification. For parts b) and d), ask children to explain how they find the middle number. Ask how they can check their answer.

ASSESSMENT CHECKPOINT Use question **3** to assess whether children can count in 3s without concrete or pictorial representation. Question **5** should help you to decide whether children can recognise the different patterns evident in different ways of counting. Question **6** assesses whether children can count backwards in 3s as well as forwards.

ANSWERS Answers for the **Practice** part of the lesson can be found in the *Power Maths* online subscription.

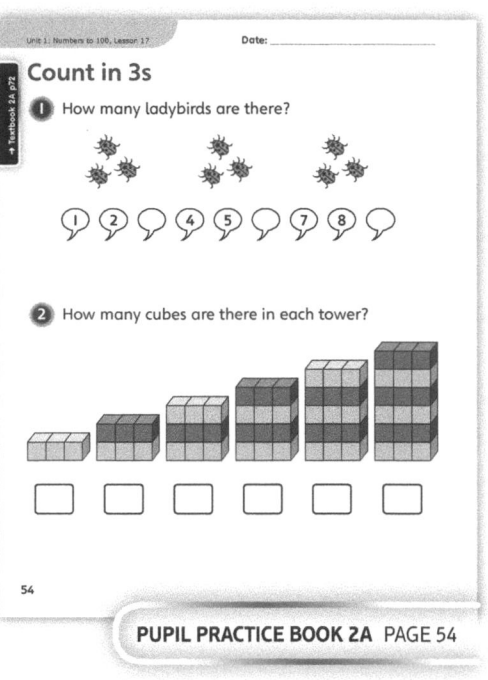

PUPIL PRACTICE BOOK 2A PAGE 54

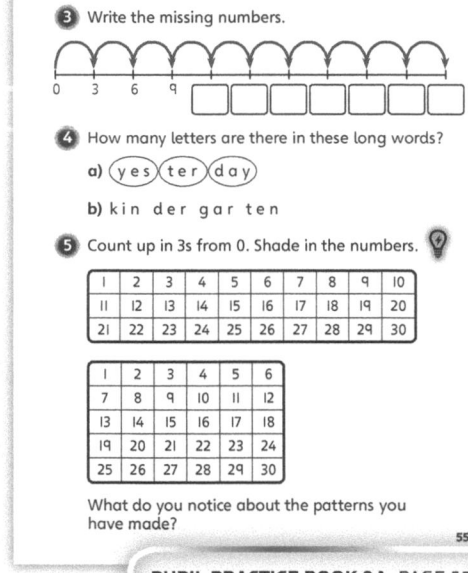

PUPIL PRACTICE BOOK 3A PAGE 55

Reflect

WAYS OF WORKING Pair work

IN FOCUS The **Reflect** part of the lesson requires children to identify which numbers in the cloud Jodie would say, by counting from 0 in 3s. When children have had some time to devise their solution and develop their reasoning and justification, ask them to share with their partner.

ASSESSMENT CHECKPOINT Assess whether children can count reliably in 3s and prove their ideas with confident use of the representations used up until this point.

ANSWERS Answers for the **Reflect** part of the lesson can be found in the *Power Maths* online subscription.

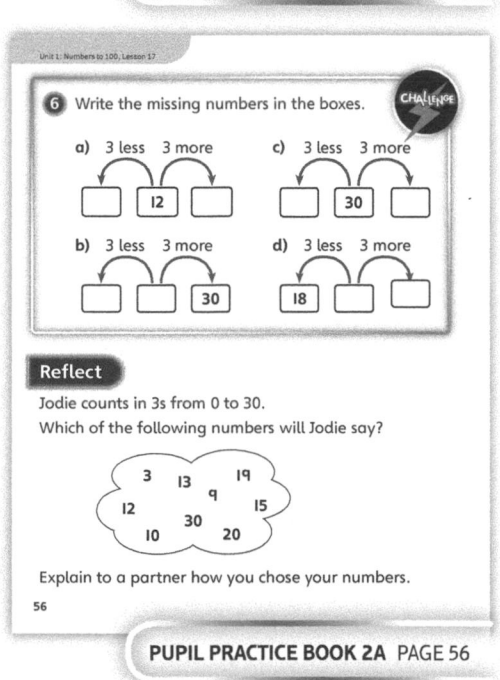

PUPIL PRACTICE BOOK 2A PAGE 56

After the lesson ⏸

- Were children as confident counting in 3s as in 2s, 5s and 10s?
- Were children able to identify patterns when counting in 3s?

End of unit check

> Don't forget the unit assessment grid in your *Power Maths* online subscription.

WAYS OF WORKING Group work – adult led

IN FOCUS Questions ❶–❹ all focus on children's understanding of place value and the different representations of numbers they have met across the course of the unit.

Question ❺ assesses understanding of 10s on a number line.

Think!

WAYS OF WORKING Pair work

IN FOCUS This question has been chosen to assess children's understanding of place value, counting in 10s and 1s, and use and understanding of the different representations they have worked with.

Focus children on proving their ideas. Ask:
- *How could you show these numbers in a way that makes comparing them clearer?*
- *Could you prove your ideas using resources?*
- *Could you prove it with a picture?*

ANSWERS AND COMMENTARY Children who demonstrate mastery of this concept will be able to recognise the multiple representations of 93 and explain how they are different to the representation of 39. They will recognise and explain the differences between the numbers, potentially using concrete, pictorial and abstract evidence, and will be able to explain how their evidence supports their ideas. They will recognise and explain the value of 10 and 1 and relate this to each number, using their ability to compare numbers to support their reasoning. Children may also choose to use a place value grid to support their thinking.

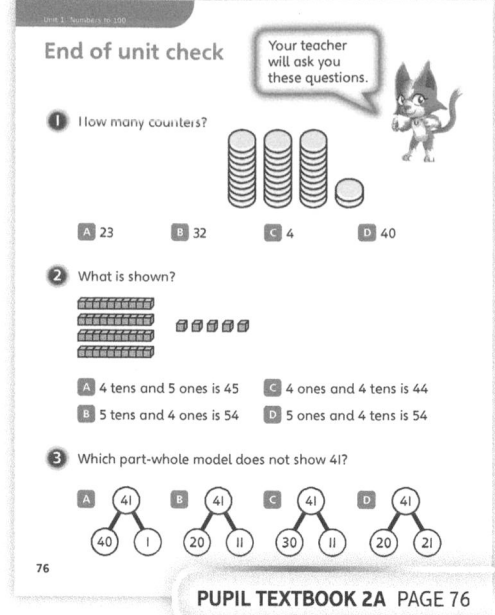

PUPIL TEXTBOOK 2A PAGE 76

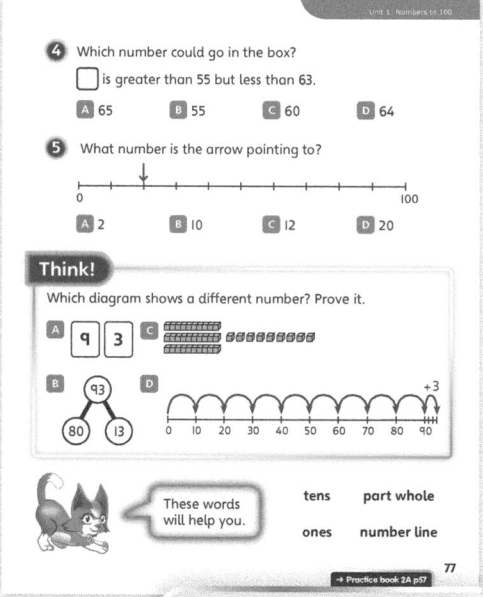

PUPIL TEXTBOOK 2A PAGE 77

Q	A	WRONG ANSWERS AND MISCONCEPTIONS	STRENGTHENING UNDERSTANDING
1	B	A may suggest children have transposed the 10s and 1s. C may suggest they have counted the columns of 10 as 1. D may suggest that children have continued counting in 10s when counting the 1s.	Place value: Give children opportunities to partition 2-digit numbers in different ways. This may be in the context of:
2	A	B and D may suggest children may have transposed the 10s and 1s. C and D may suggest that children are muddling 10s and 1s. C and B may suggest children have miscounted.	• dividing items between two people • investigating how many ways a number can be split into two groups. Children could be encouraged to put the two amounts into a part-whole model.
3	B	C and D may suggest children are unsure about other ways to partition a number beyond its 10s and 1s.	Counting:
4	C	A, B and D may suggest children's understanding of place value and ordering numbers needs reinforcing.	To help children retain the number sequences, it would be beneficial to find, or make up, songs or chants they can sing.
5	D	Answers of A, B or C suggest children have not fully understood how to recognise counting 10s on a number line.	

My journal

WAYS OF WORKING Independent thinking

ANSWERS AND COMMENTARY Children may record:

I can prove that 39 is a different number because it has 3 tens and 9 ones. All the other numbers have 9 tens and 3 ones.

If children are struggling to compare the numbers, ask:

• *Which representations do you recognise?*
• *What is the same about the representations and what is different?*
• *Can you tell me how they work?*
• *Could you show the number in a different way?*
• *Could you use that method for all the numbers shown?*
• *What do you notice about the representations you have made? What is the same? What is different?*

If children are encouraged to make the number using a representation they are comfortable with, they should find it easier to make the comparisons. When children have explained their observations in words, ask:

• *How could you record that on the page?*
• *What words will you need to use?*
• *Can you use any from the word bank?*
• *Can you say the sentence first?*

Power check

WAYS OF WORKING Independent thinking

ASK

• *What steps could you count in before this unit?*
• *What steps can you count in now?*
• *How confident are you about finding the 10s and 1s in a number?*
• *How confident do you feel about ordering two 2-digit numbers? How about three?*
• *Do you think you could use a place value grid on your own?*

Power play

WAYS OF WORKING Pair work

IN FOCUS Use this **Power play** to see if children can work together to follow a route through the maze counting in 2s and 5s. Children should recognise that the 100 they land on when counting in 5s should be closer than the one they land on when counting in 2s.

ANSWERS AND COMMENTARY Look closely at the pattern the children have followed. If they have successfully followed both then this suggests that they are confident in counting in steps of 2 or 5. If they have made a mistake on either of them this could suggest that more practice with that counting pattern is needed.

After the unit ⏸

• Counting in steps of 2, 3, 5 and 10 can be applied to many different real-life contexts (for example, 2p, 5p and 10p when shopping, through games and grouping objects or teams). How could you get the children to apply their learning from this unit in other areas of the curriculum, especially counting in 3s?
• Did the unit assessment show any misconceptions that the class still has? How will you support and develop this area of learning?

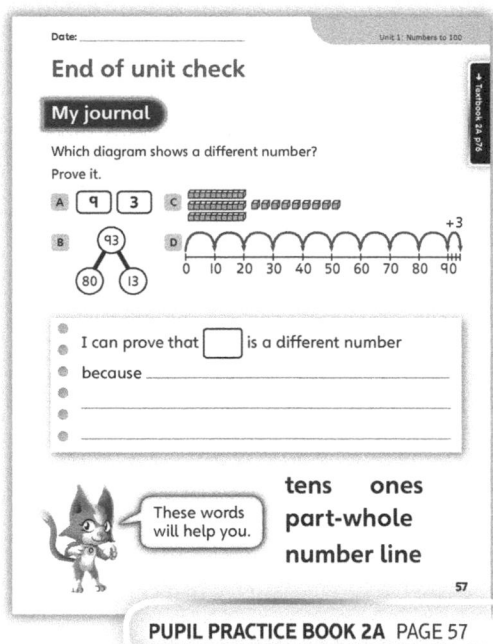

PUPIL PRACTICE BOOK 2A PAGE 57

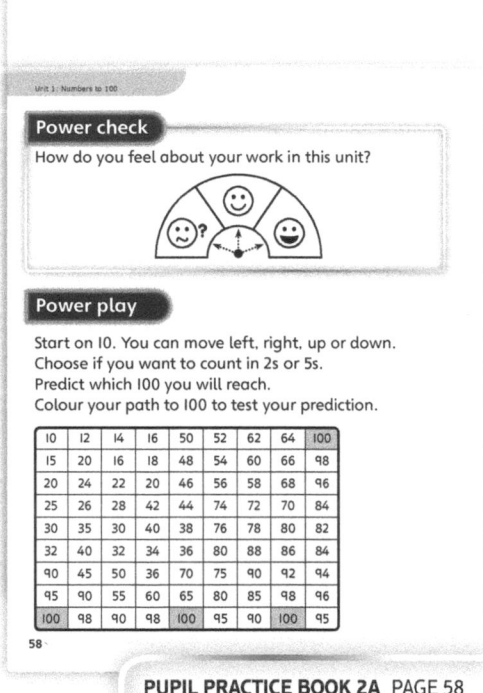

PUPIL PRACTICE BOOK 2A PAGE 58

Strengthen and **Deepen** activities for this unit can be found in the *Power Maths* online subscription.

Unit 2
Addition and subtraction ❶

Mastery Expert tip! 'It is important for this unit that children understand the importance of learning key number facts. Children should be encouraged to spot number bonds within 10 throughout the unit, all around them in the school environment and in other subject lessons.'

Don't forget to watch the Unit 2 video!

WHY THIS UNIT IS IMPORTANT

In this unit, children will build upon the number bonds to 10 that they will have learned in Year 1.

Children are introduced to writing fact families of equations, and to relating addition and subtraction operations. As a result, children learn to use the inverse of one operation to check calculations using the other operation. Children will also be introduced to the concept of 'make 10' to aid mental calculations.

The key learning is being able to work confidently with numbers as 10s and 1s, and to understand the counting patterns, as well as the application of number bonds, to deal with 10s and 1s in efficient ways.

WHERE THIS UNIT FITS

→ Unit 1: Numbers to 100
→ **Unit 2: Addition and subtraction (1)**
→ Unit 3: Addition and subtraction (2)

This unit builds on the previous unit and applies children's place value understanding to addition and subtraction problems. A good understanding of place value and of counting patterns within 10, 20 and 100 is vital for approaching the addition and subtraction techniques with confidence.

Before they start this unit, it is expected that children:
• know how to partition 2-digit numbers into 10s and 1s
• understand the value of each digit in a 2-digit number
• know and apply number bonds within 10.

ASSESSING MASTERY

Children who have mastered this unit will be able to relate each number in a calculation to what it represents within a context. Children will be able to use a variety of manipulatives to represent addition and subtraction and use these alongside efficient mental methods or methods using jottings. Children will also become fluent at recalling and applying their number bonds within 10 to addition and subtraction equations.

COMMON MISCONCEPTIONS	STRENGTHENING UNDERSTANDING	GOING DEEPER
Children may think the commutative property of addition problems can be applied to subtractions when it cannot.	Allow children to work in pairs to complete problems where one manipulates resources and the other records using the column method.	Use subtraction to check addition calculations and vice versa.
Children may not understand the importance of using known facts within calculations and instead use inefficient strategies such as counting on in 1s using their fingers.	Providing children with number facts that are useful to complete calculations will increase the likelihood of children understanding the importance of memorisation.	Children should represent the same calculation in as many different ways as possible using different manipulatives to do so.

Unit 2: Addition and subtraction ❶

UNIT STARTER PAGES

Use these pages to introduce the unit focus to children as a whole class. You can use the characters to explore different ways of working.

STRUCTURES AND REPRESENTATIONS

Part-whole model: This model helps children understand that two or more parts combine to make a whole. It will also help children understand how addition and subtraction are linked and can be used to calculate an unknown part or whole.

Number line: This model helps children visualise the order of numbers. It helps children to count on and back from a number. Number lines are used to show jumps of different amounts to help children understand the 'make 10' strategy.

10s and 1s: Use place value equipment and other representations such as ten frames to reinforce understanding and application of the partitions of a 2-digit number.

100 square: This model shows how numbers link to each other and how numbers change when 10 is added or subtracted to or from a number. This model is especially useful to help children make links to the column method.

1	2	3	4	5	6	7	8	9	10
11	12	13	14	15	16	17	18	19	20
21	22	23	24	25	26	27	28	29	30
31	32	33	34	35	36	37	38	39	40
41	42	43	44	45	46	47	48	49	50
51	52	53	54	55	56	57	58	59	60
61	62	63	64	65	66	67	68	69	70
71	72	73	74	75	76	77	78	79	80
81	82	83	84	85	86	87	88	89	90
91	92	93	94	95	96	97	98	99	100

KEY LANGUAGE

There is some key language that children will need to know as a part of the learning in this unit:

➜ part, whole and part-whole

➜ add, added, plus, total, altogether, sum (+), calculation

➜ count, count on, count back, left

➜ subtract, take away, difference, minus (−)

➜ 1s, 10s, 10 more, 10 less, place value, column, 1-digit number, 2-digit number

➜ number sentence, number bonds, known fact, fact family, multiples

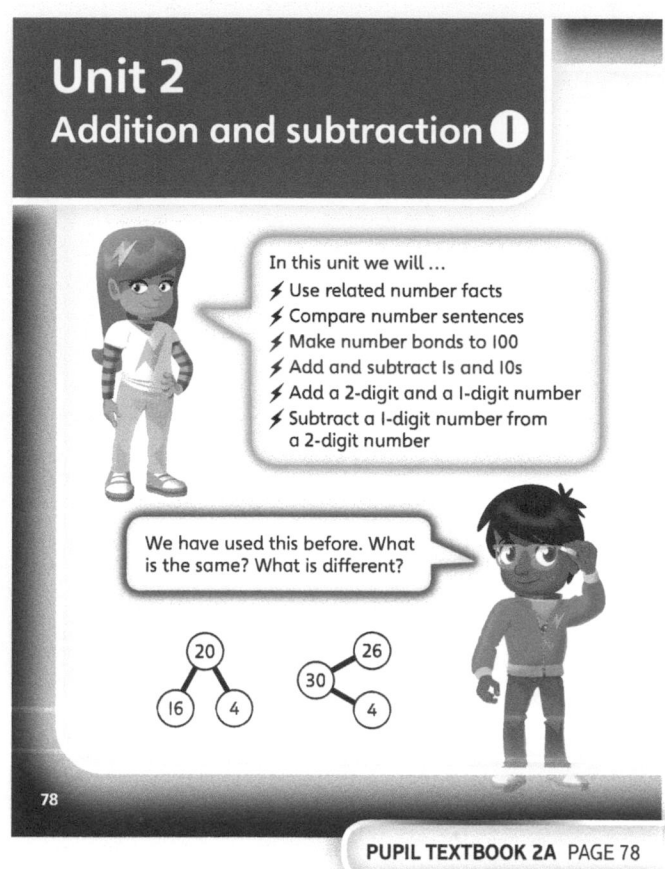

PUPIL TEXTBOOK 2A PAGE 78

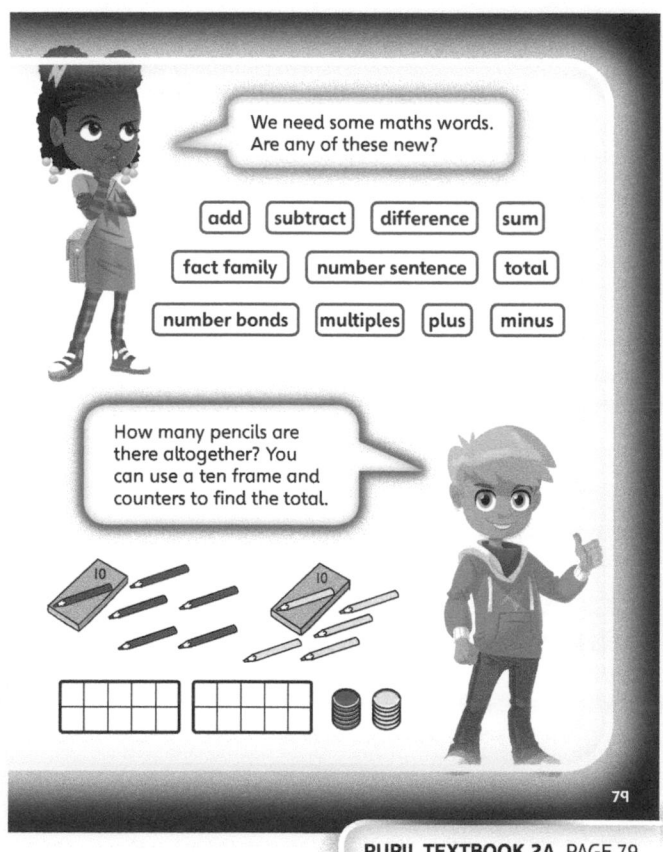

PUPIL TEXTBOOK 2A PAGE 79

Fact families

Learning focus

In this lesson, children will focus on bonds within 20, using the part-whole diagram to help them see these visually. The focus is on finding fact families and recording known facts in different ways within addition and subtraction calculations.

Before you teach

- Based on teaching of the part-whole model in Year 1, are there any additional misconceptions that need to be addressed?
- How will you support children to understand the differences between addition and subtraction when recording numbers in calculation scaffolds?

NATIONAL CURRICULUM LINKS

Number – addition and subtraction

Recall and use addition and subtraction facts to 20 fluently, and derive and use related facts up to 100.

ASSESSING MASTERY

Children can write down fact families from a part-whole model and identify what each number within a calculation represents.

COMMON MISCONCEPTIONS

Children may put the numbers the wrong way in the scaffold for addition or subtraction calculations. Similarly, they may believe subtraction is commutative. Ask:
- *Which two numbers are the parts? Which number is the whole? If you subtract the whole from a part will you be left with a part?*

Children may incorrectly interpret the part-whole model and as a result may not understand what the number sentence is telling them. Ask:
- *What does each number represent? Where is that number found in the part-whole model or pictorial representation?*

STRENGTHENING UNDERSTANDING

Reinforce learning by encouraging children to make the representations shown using concrete manipulatives. Children can then move the parts and wholes onto part-whole models and into subtraction and addition calculation scaffolds.

GOING DEEPER

Encourage children to record all possible number sentences from a given part-whole model. This could include situations where all numbers are given and those involving an unknown quantity which children can represent using a '?' in their number sentences.

KEY LANGUAGE

In lesson: addition, subtraction, +, –, =, fact family, number sentence, in total, number bond, altogether

Other language to be used by the teacher: calculation, unknown, equals, equivalent

STRUCTURES AND REPRESENTATIONS

Part-whole model, addition and subtraction calculation scaffold

RESOURCES

Mandatory: cubes or counters, blank part-whole model, blank addition and subtraction calculation scaffolds

Optional: base 10 equipment, digit cards, physical resources to make the parts and wholes represented in questions

 In the eTextbook of this lesson, you will find interactive links to a selection of teaching tools.

Quick recap

As a class, make a list of all of the number bonds that equal 5.

Then work together to find these missing numbers:

$2 + ? = 5$ $5 - ? = 2$

$5 - 1 = ?$ $4 + ? = 5$

Discover

Unit 2: Addition and subtraction (1), Lesson 1

Fact families

Discover

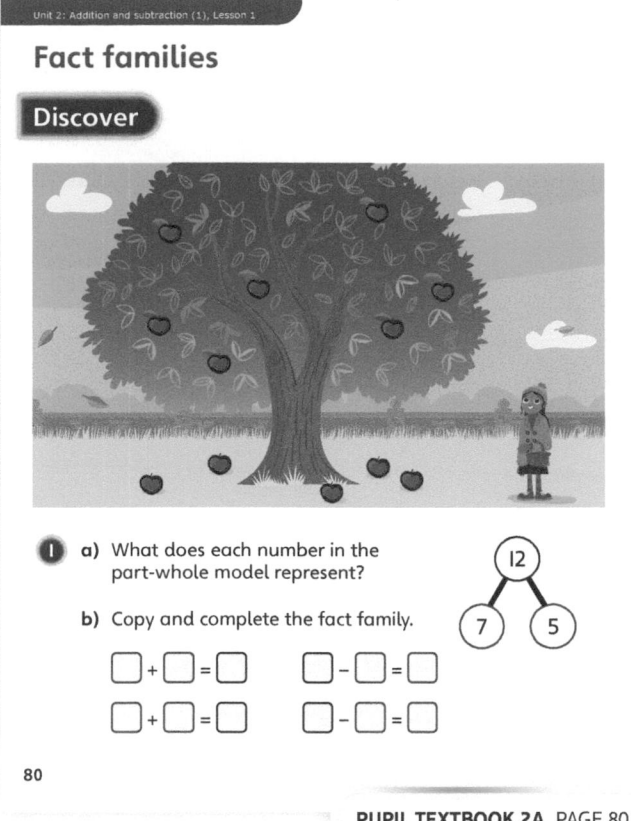

WAYS OF WORKING Pair work

ASK

- Question **1** a): *Which numbers are the parts? What do they represent?*
- Question **1** a): *Does it matter if the 7 is in the left or right part?*
- Question **1** b): *Does the order that the numbers are placed in the addition and subtraction sentences matter?*

IN FOCUS Question **1** presents children with a real-life problem to help them understand what the different parts and the whole represent in a part-whole model.

PRACTICAL TIPS Provide cubes for children to make the parts and then combine them to make the whole, showing this as 1 ten and 2 ones.

ANSWERS

Question **1** a): The number 7 represents apples in the tree.
The number 5 represents apples on the ground.
The number 12 represents apples in total.

Question **1** b): 7 + 5 = 12, 5 + 7 = 12
These number sentences tell you how many apples there are altogether.
12 – 5 = 7
This number sentence tells you how many apples are in the tree.
12 – 7 = 5
This number sentence tells you how many apples are on the ground.

1 a) What does each number in the part-whole model represent?

b) Copy and complete the fact family.

☐ + ☐ = ☐   ☐ – ☐ = ☐

☐ + ☐ = ☐   ☐ – ☐ = ☐

80

PUPIL TEXTBOOK 2A PAGE 80

Share

WAYS OF WORKING Whole class teacher led

ASK

- Question **1** a): *Which part of the part-whole model represents the number of apples in the tree?*
- Question **1** a): *Which part of the part-whole model represents the number of apples on the ground?*
- Question **1** b): *What are you finding out when you subtract?*

IN FOCUS Question **1** b) requires children to identify what each number within the addition and subtraction sentences represents. Encourage them to explain in full sentences what is being calculated each time.

DEEPEN Challenge children to explain what is the same and what is different about the addition and subtraction sentences. Ask them to write the same number sentences with the equals sign at the beginning (for example, 12 = 7 + 5).

Share

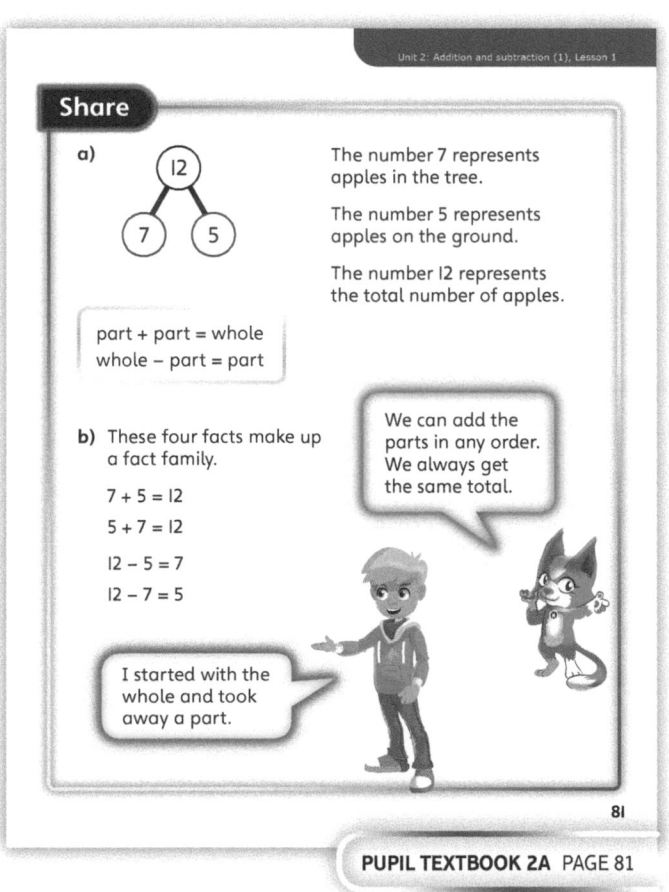

a)

The number 7 represents apples in the tree.

The number 5 represents apples on the ground.

The number 12 represents the total number of apples.

part + part = whole
whole – part = part

b) These four facts make up a fact family.

7 + 5 = 12

5 + 7 = 12

12 – 5 = 7

12 – 7 = 5

We can add the parts in any order. We always get the same total.

I started with the whole and took away a part.

81

PUPIL TEXTBOOK 2A PAGE 81

Think together

WAYS OF WORKING Whole class teacher led (I do, We do, You do)

ASK

• Question ❶: *What does each number within the part-whole model represent?*
• Question ❶: *What does each number within each number sentence represent?*
• Question ❷ b): *Is the position of the whole within subtraction sentences important?*

IN FOCUS Question ❶ requires children to complete a fact family by identifying what each number within the part-whole model represents and what the result of each number sentence shows within the context of the question.

STRENGTHEN To strengthen understanding of the equals sign and the order of numbers within subtraction sentences, provide children with separate pieces of paper, with parts of a number sentence written on each piece, to move around. For example, write 8 – 5 on one piece of paper, = on another, and 3 on a third piece of paper.

Alternatively, provide children with number cards and blank addition and subtraction sentence scaffolds to help them understand that the three numbers do not change; only their location changes.

DEEPEN Question ❸ requires children to use a provided number sentence and to draw on prior knowledge of number bonds to 10 to work out the other facts in that family. Ask children to draw the corresponding part-whole models.

ASSESSMENT CHECKPOINT Assess how children are calculating the number of objects in each part. Are they counting in 1s or are they able to use their understanding of fact families to find the answers efficiently?

ANSWERS

Question ❶: 3 + 5 = 8   8 – 3 = 5
                5 + 3 + 8   8 – 5 = 3

Question ❷ a): 10 + 6 = 16 is correct.

Question ❷ b): 16 – 10 = 6 is correct.

Question ❸ a):

Question ❸ b): 3 + 4 = 7   7 – 4 = 3
                4 + 3 = 7   7 – 3 = 4
                7 = 4 + 3   4 = 7 – 3
                7 = 3 + 4   3 = 7 – 4

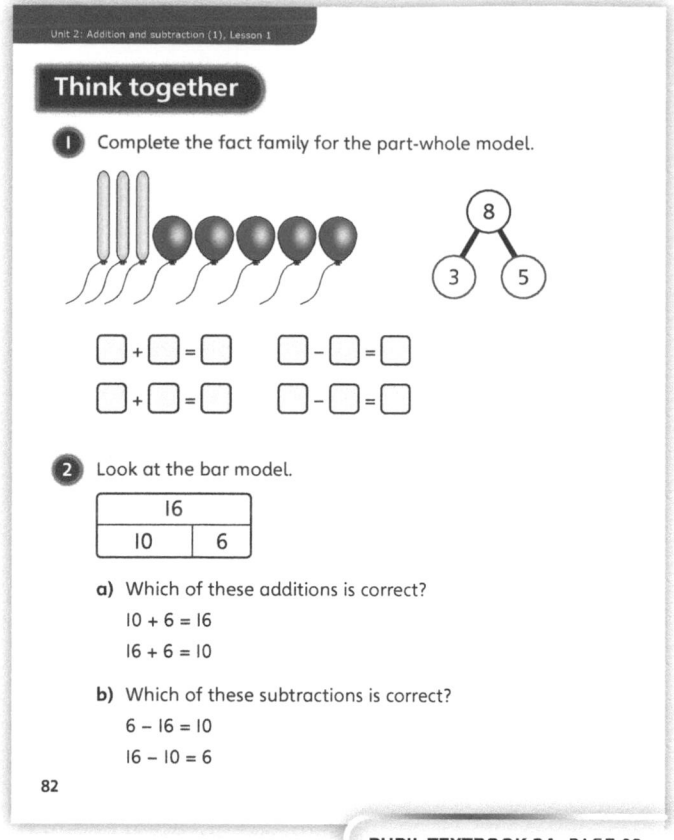

Think together

❶ Complete the fact family for the part-whole model.

☐ + ☐ = ☐   ☐ – ☐ = ☐
☐ + ☐ = ☐   ☐ – ☐ = ☐

❷ Look at the bar model.

16	
10	6

a) Which of these additions is correct?
   10 + 6 = 16
   16 + 6 = 10

b) Which of these subtractions is correct?
   6 – 16 = 10
   16 – 10 = 6

82

PUPIL TEXTBOOK 2A PAGE 82

❸ Meg writes down a number bond.

CHALLENGE

3 + 4 = 7

a) Complete the part-whole model for the bond.

b) Write down the fact family for the part-whole model.

I can work out other facts using one I am given.

I think I can write down 8 facts. I remember doing these last year.

83

→ Practice book 2A p59

PUPIL TEXTBOOK 2A PAGE 83

Practice

WAYS OF WORKING Independent thinking

IN FOCUS Throughout these questions, children are exposed to different part-whole models and number sentences exploring fact families. In question ②, the bar model is also used to show parts and the whole.

STRENGTHEN Provide children with scaffolds to help them determine the location of the different digits within calculations; for example, part − part = whole and whole − part = part.

DEEPEN Question ⑤ illustrates what happens when the two parts are the same. There will only be two calculations in the fact family as there are no commutative facts. Challenge children to give other examples that demonstrate why this is the case.

ASSESSMENT CHECKPOINT Assess whether children can justify the number facts they find in questions ④ and ⑤. Can they state whether each number within a calculation is a part or a whole and how they know they are correct?

ANSWERS Answers for the **Practice** part of the lesson can be found in the *Power Maths* online subscription.

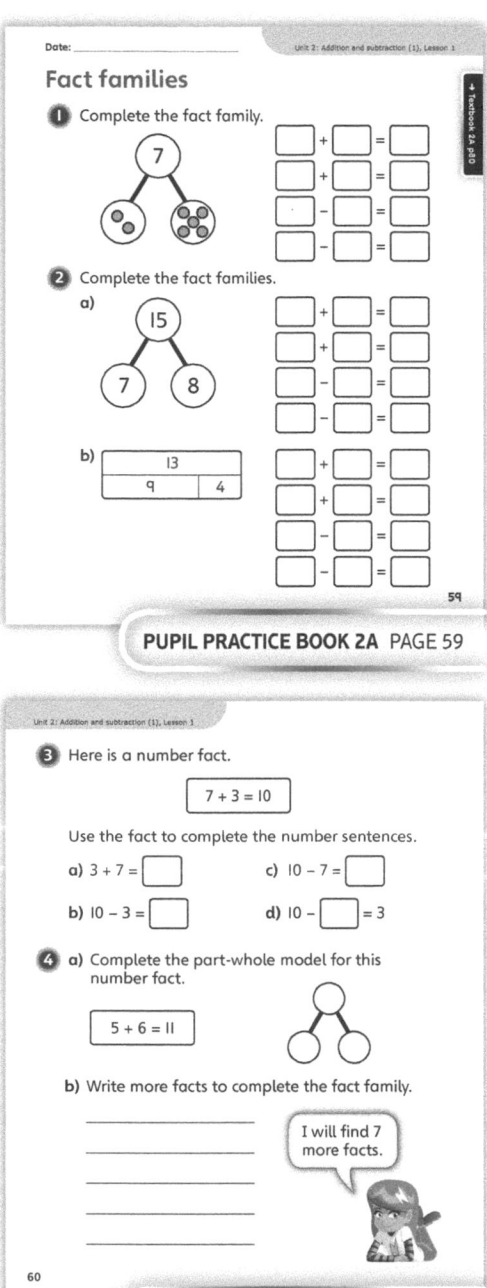

PUPIL PRACTICE BOOK 2A PAGE 59

PUPIL PRACTICE BOOK 2A PAGE 60

Reflect

WAYS OF WORKING Independent thinking

IN FOCUS By the **Reflect** part of the lesson, children should be able to write four number sentences (two addition and two subtraction) with the equals sign at the end of the calculation.

ASSESSMENT CHECKPOINT Assess whether children can verbalise how they know they have found all the possible number sentences. Can they use the language of parts and whole to describe their sentences (for example, can they read 9 + 8 = 17 as part + part = whole)?

ANSWERS Answers for the **Reflect** part of the lesson can be found in the *Power Maths* online subscription.

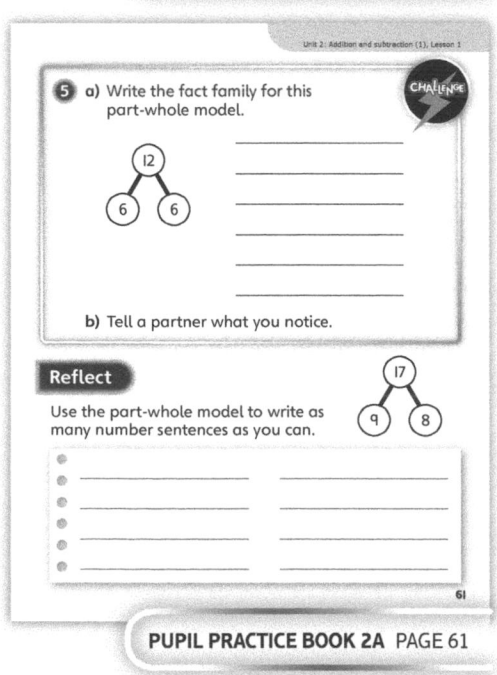

After the lesson

- Are children confident recording four number sentences for each part-whole model?
- Do children understand the importance of the location of the parts and the wholes within addition and subtraction calculations?

PUPIL PRACTICE BOOK 2A PAGE 61

Learn number bonds

Learning focus

In this lesson, children explore strategies for learning number bonds and consider which facts they need to learn off by heart.

Before you teach

- Play some games for children to practise showing doubles up to 5 + 5 on fingers?
- Play a 'What comes next?' number game. Say a 2-digit number and ask children to say the next number.

NATIONAL CURRICULUM LINKS

Number – addition and subtraction

Recall and use addition and subtraction facts to 20 fluently, and derive and use related facts up to 100.

ASSESSING MASTERY

Children can discuss the number bond facts that they already know well, and those which they intend to learn more fully.

COMMON MISCONCEPTIONS

Some children may feel that there is an overwhelming number of facts to learn and find this an obstacle to learning. Help them to explore how each fact can help them to work out others. Ask:
- *If you know that fact, what else can you say? Do you need to work this out or is it something you already know?*

STRENGTHENING UNDERSTANDING

Use this lesson as an opportunity to show children how to make tasks more manageable by picking just a few facts to master at a time.

GOING DEEPER

Develop metacognition through discussion of learning and memory. Ask children to describe and explain their own strategies, or what it 'feels like' to know something really well.

KEY LANGUAGE

In lesson: number bond, fact

STRUCTURES AND REPRESENTATIONS

Number bond addition grid

RESOURCES

Optional: number track

 In the eTextbook of this lesson, you will find interactive links to a selection of teaching tools.

Quick recap 🔁

Play 'Say the next number' together. Say a number between 0 and 9 and ask children to call out the next number.

Then play 'Say the previous number'. Say a number between 1 and 10 and ask children to call out the previous number.

Discover

WAYS OF WORKING Pair work

ASK

- Question ➊: *What do you notice? What do you wonder?*
- Question ➊: *Can you see any patterns?*

IN FOCUS This lets children explore how they might help someone who is feeling confused about number bonds and facts. The addition grid lets them work systematically and look for patterns.

PRACTICAL TIPS Spend time exploring the grid together before beginning the lesson. Ask children if they can see any patterns and allow open questioning and investigation of the numbers in the grid.

ANSWERS

Question ➊ a): Gita is working out the facts about adding zero. When you add zero the number does not change.

Question ➊ b): Zac is adding 1. When you add 1, count to the next number.

Share

WAYS OF WORKING Whole class teacher led

ASK

- Question ➊ a): *What is the same and what is different about each of Gita's additions?*
- Question ➊ b): *What do you notice about Zac's facts?*

IN FOCUS Question ➊ a) demonstrates the effect of adding zero to any number. In Question ➊ b) children will see the link between adding one and counting on to the next number.

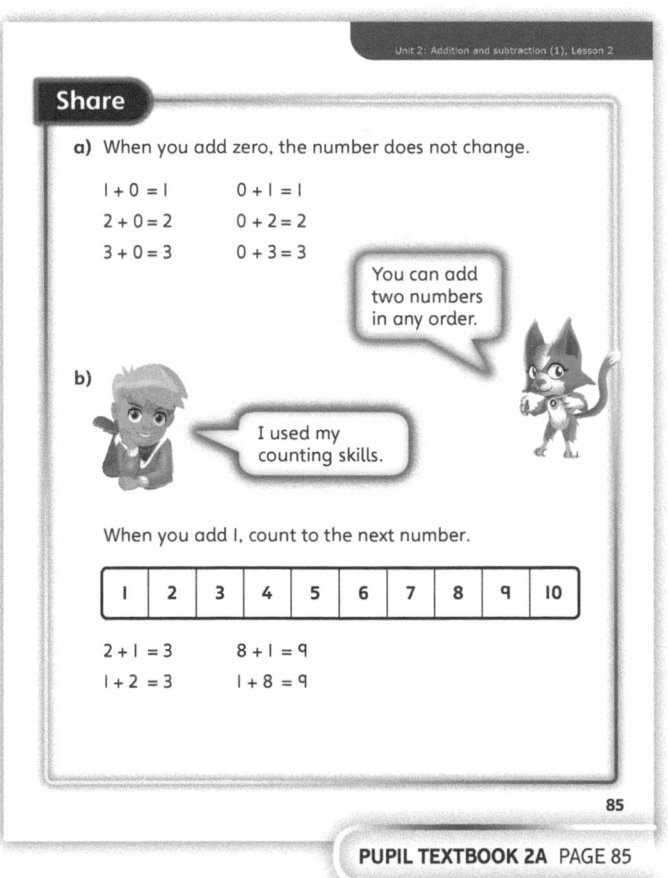

Think together

Whole class teacher led (I do, We do, You do)

ASK

- Question **1**: *Can you see any patterns?*
- Question **2**: *How can you work out subtractions if you know additions?*
- Question **3**: *Which facts do you know well? Which would you like to learn next?*

IN FOCUS In question **1** children are using the addition grid to find number bonds to 10. Question **2** requires children to work with doubles and the related subtraction facts. In question **3**, children are using metacognitive strategies to build an awareness of their own learning journey.

STRENGTHEN Use an addition grid like the one in the **Discover** task. Ask children to first colour or cover the facts that they know well. Then choose a set of two or three facts they would like to learn next.

DEEPEN Challenge children to investigate patterns that they can find in the grid. Can they explain why any of these patterns occur?

ASSESSMENT CHECKPOINT Use question **1** to develop a sense of which facts children should focus on learning next.

ANSWERS

Question **1**: 10 + 0 = 10
9 + 1 = 10
8 + 2 = 10
7 + 3 = 10
6 + 4 = 10
5 + 5 = 10
4 + 6 = 10
3 + 7 = 10
2 + 8 = 10
1 + 9 = 10
0 + 10 = 10

Question **2**: 2 + 2 = **4**   4 − 2 = **2**
3 + 3 = **6**   6 − 3 = **3**
4 + 4 = **8**   8 − 4 = **4**

Question **3** a): There will be a variety of answers here depending on children's own explanations.

Question **3** b): Children's recall of facts by heart will vary.

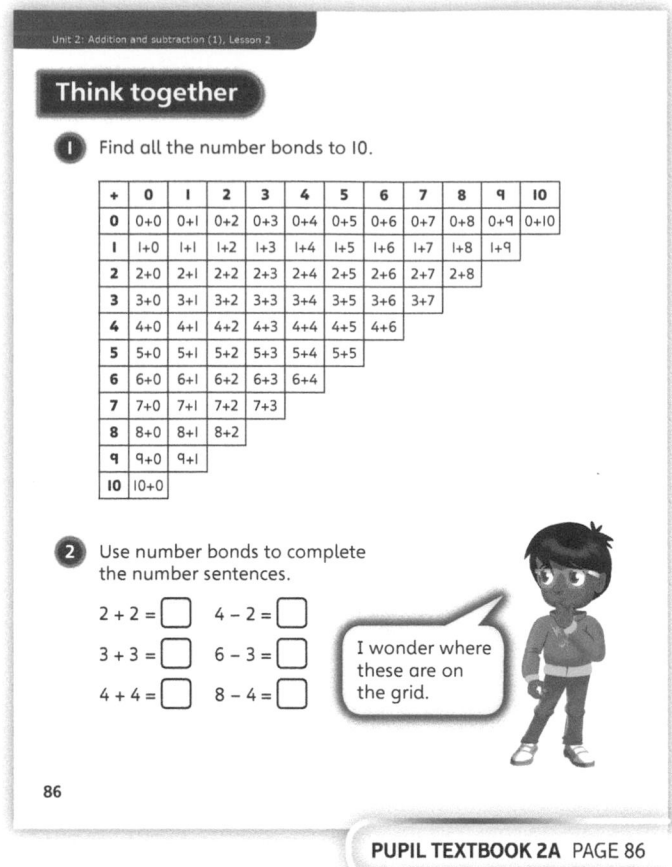

PUPIL TEXTBOOK 2A PAGE 86

PUPIL TEXTBOOK 2A PAGE 87

Practice

WAYS OF WORKING Independent thinking

IN FOCUS In question **1**, children explore what happens when adding 0 and question **2** requires them to complete number sentences by adding and subtracting 1. Question **3** is about number bonds to 10 as both addition and subtraction facts. The number sentences in question **4** feature doubles and near doubles. Question **5** has a variety of missing number calculations and also presents number facts in the part-whole model.

STRENGTHEN Support children in learning strategies for remembering and recalling these facts. For example, one more and one less can be learned as a strategy of counting on to the next number or back to the previous number.

DEEPEN Challenge children to investigate patterns of odd and even numbers that they can find on the number bond addition grid.

ASSESSMENT CHECKPOINT Use Question **2** to assess whether children can work with number facts that involve adding and subtracting 1. Can they recall the facts, or do they need to count on or back each time?

ANSWERS Answers for the **Practice** part of the lesson can be found in the *Power Maths* online subscription.

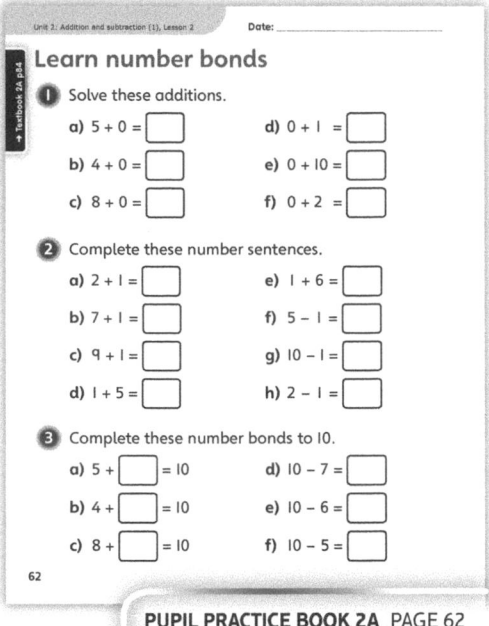

PUPIL PRACTICE BOOK 2A PAGE 62

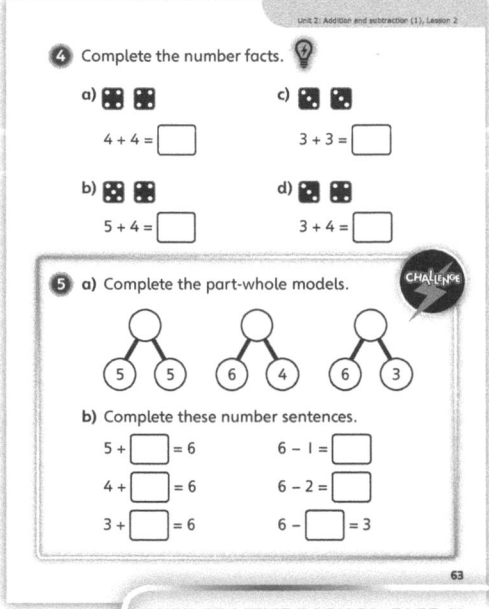

PUPIL PRACTICE BOOK 2A PAGE 63

Reflect

WAYS OF WORKING Independent thinking

IN FOCUS This extended **Reflect** activity will allow children to consider their learning journey about number bonds and facts. What do they know well now? What will they learn next?

ASSESSMENT CHECKPOINT Self-assessment is key in this activity. How are children feeling about the experience of learning number bonds and facts? Children can take this chance to choose one or two facts that they would like to 'over-learn' together.

ANSWERS Answers for the **Reflect** part of the lesson can be found in the *Power Maths* online subscription.

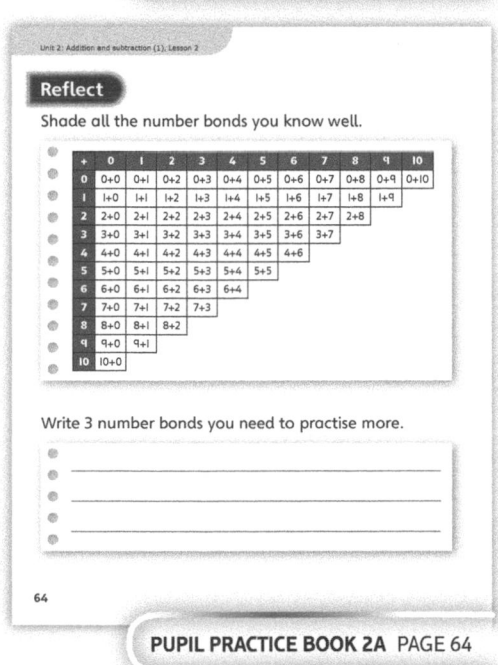

After the lesson

- Choose one or two bonds to 'over-learn' together, with children taking some responsibility for helping each other.
- How will you all support each other as a class to learn children's chosen facts?

PUPIL PRACTICE BOOK 2A PAGE 64

Add and subtract two multiples of 10

Learning focus

In this lesson, children use known facts with 1s to determine other facts with corresponding multiples of 10.

Before you teach

- Can children think of number facts that they could use independently?
- How will you encourage children to make links with known facts?

NATIONAL CURRICULUM LINKS

Number – addition and subtraction

Recall and use addition and subtraction facts to 20 fluently, and derive and use related facts up to 100.

ASSESSING MASTERY

Children can use number bonds within 10 to determine related facts with multiples of 10, rather than calculating these facts as a result of using addition or subtraction in an inefficient way. Children can identify why each helps with the related calculation.

COMMON MISCONCEPTIONS

Children may not see the purpose of using known facts and may find alternative ways to calculate the answer. Ask:
- *What fact did you use to help with this calculation?*

STRENGTHENING UNDERSTANDING

Ask children to make the numbers using concrete manipulatives alongside the abstract calculations. They should explain the parts that they have made and describe how the known fact relates to the unknown fact (for example, 2 ones + 3 ones = 5 ones so 2 tens + 3 tens = 5 tens).

GOING DEEPER

Encourage children to describe more than one fact that they could use to help them find the solution to a new problem. This will highlight their flexibility and fluency of number facts.

KEY LANGUAGE

In lesson: facts, number sentence, signs, digits, ones (1s), tens (10s)

Other language to be used by the teacher: number bonds, addition, subtraction, link, relate

STRUCTURES AND REPRESENTATIONS

Part-whole model, bar model

RESOURCES

Mandatory: base 10 equipment

Optional: blank number sentence scaffold

 In the eTextbook of this lesson, you will find interactive links to a selection of teaching tools.

Quick recap ↻

Ask children to count on to 100 in 10s. Then ask them to count down from 100 in 10s. Look together for all the 10s on a 100 square.

Discover

WAYS OF WORKING Pair work

ASK

- Question ❶ a): *What do you know about 2 + 3?*
- Question ❶: *What other facts can you work out from this?*

IN FOCUS The numbers in question ❶ allow children to make simple links between the number of pencils Milo and Mr Abbot each have.

PRACTICAL TIPS Encourage children to make links by asking them to draw both sets of pencils or make both sets of numbers with resources.

ANSWERS

Question ❶ a): Milo has 5 pencils.

Question ❶ b): Mr Abbot has 50 pencils.

Add and subtract two multiples of 10

Discover

❶ **a)** How many pencils does Milo have?

b) How many pencils does Mr Abbot have?

88

PUPIL TEXTBOOK 2A PAGE 88

Share

WAYS OF WORKING Whole class teacher led

ASK

- Question ❶ b): *What is the same and what is different about the number of pencils Milo and Mr Abbot each have?*
- Question ❶ b): *How many parts do Milo and Mr Abbot each have?*

IN FOCUS When tackling question ❶ b), children should be able to make links between the number of pencils that Milo and Mr Abbot have: they should be able to see that the number of 1s that Milo has is the same as the number of 10s that Mr Abbot has.

Share

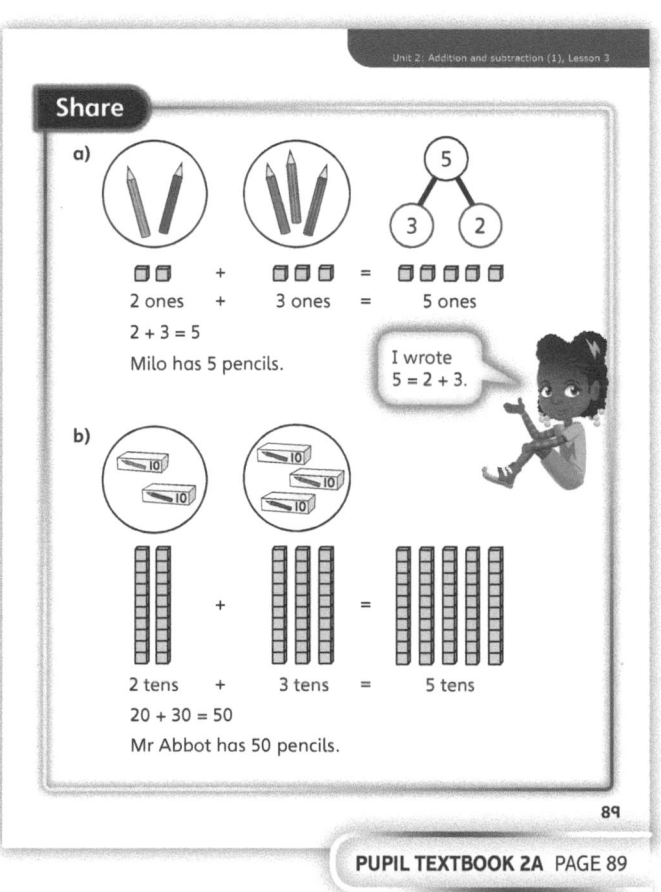

a)
2 ones + 3 ones = 5 ones
2 + 3 = 5
Milo has 5 pencils.

I wrote 5 = 2 + 3.

b)
2 tens + 3 tens = 5 tens
20 + 30 = 50
Mr Abbot has 50 pencils.

89

PUPIL TEXTBOOK 2A PAGE 89

Think together

WAYS OF WORKING Whole class teacher led (I do, We do, You do)

ASK

• Question **1**: *How does 4 + 3 = 7 help you to know that 40 + 30 = 70?*
• Question **2**: *How does the part-whole model help you with the second calculation?*
• Question **3**: *Whose strategy is more efficient, Astrid's or Flo's?*

IN FOCUS Question **1** is similar to the **Discover** question, while question **2** removes the pictorial representations and includes only the model. Encourage children to verbalise how the known fact helps them with the unknown fact. Question **3** builds on work from the previous lesson as well as requiring children to use a known fact to help with the calculations.

STRENGTHEN Making the known and unknown facts using base 10 equipment will strengthen children's understanding of how the facts are linked. Encourage them to explain the facts (for example, 5 ones + 1 one = 6 ones, so 5 tens + 1 ten = 6 tens).

DEEPEN Give children a part-whole model for 50, 20 and 30. Ask them to write the fact family that matches the part-whole model. Can they find all the additions and subtractions?

ASSESSMENT CHECKPOINT Children should be able to explain the links between the calculations in questions **1** to **3**. Use their explanations to assess whether children are using known facts or if they are simply calculating each answer.

ANSWERS

Question **1** a): 4 + 3 = 7
　　　　　There are **7** pencils.
　　　　　4 + 3 = 7
　　　　　There are **7** apples.
　　　　　4 ones + 3 ones = **7** ones
　　　　　4 tens + 3 tens = **7** tens

Question **1** b): 4 + 3 = **7**
　　　　　40 + 30 = **70**

Question **2**: 2 + 6 = **8**
　　　　　20 + 60 = **80**

Question **3**: 50 + 10 = **60**
　　　　　10 + **50** = 60
　　　　　60 − 50 = **10**
　　　　　10 = **60** − 50

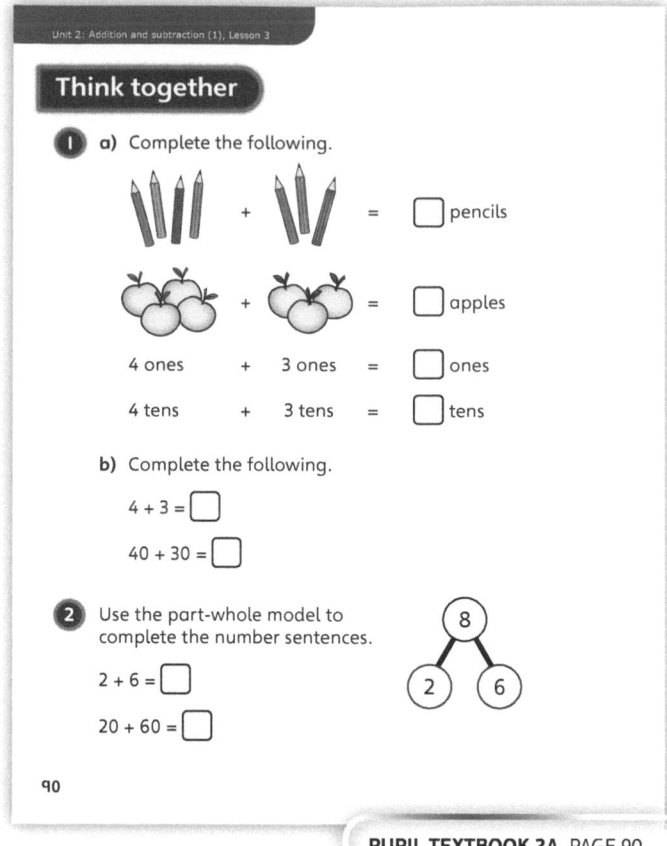

PUPIL TEXTBOOK 2A PAGE 90

PUPIL TEXTBOOK 2A PAGE 91

Practice

Independent thinking

IN FOCUS All the **Practice** questions require children to make links to known facts in order to calculate unknown facts. In question ❶ the known and unknown facts are displayed alongside each other, but this support is gradually withdrawn throughout the later questions, to test children's understanding further.

STRENGTHEN If children are finding the concept difficult, they should continue to use concrete resources alongside the abstract calculations they are required to complete. Encourage them to describe the parts and the wholes they are making and to explain what is the same and what is different about the known and unknown facts.

DEEPEN Question ❹ challenges children to make links between number bonds to 10 and how to add and subtract two multiples of 10. They should use the part-whole models in part a) to complete the number sentences in part b).

ASSESSMENT CHECKPOINT Assess whether children can justify how the known fact helps them calculate the unknowns. Ask them to explain the link in full sentences.

ANSWERS Answers for the **Practice** part of the lesson can be found in the *Power Maths* online subscription.

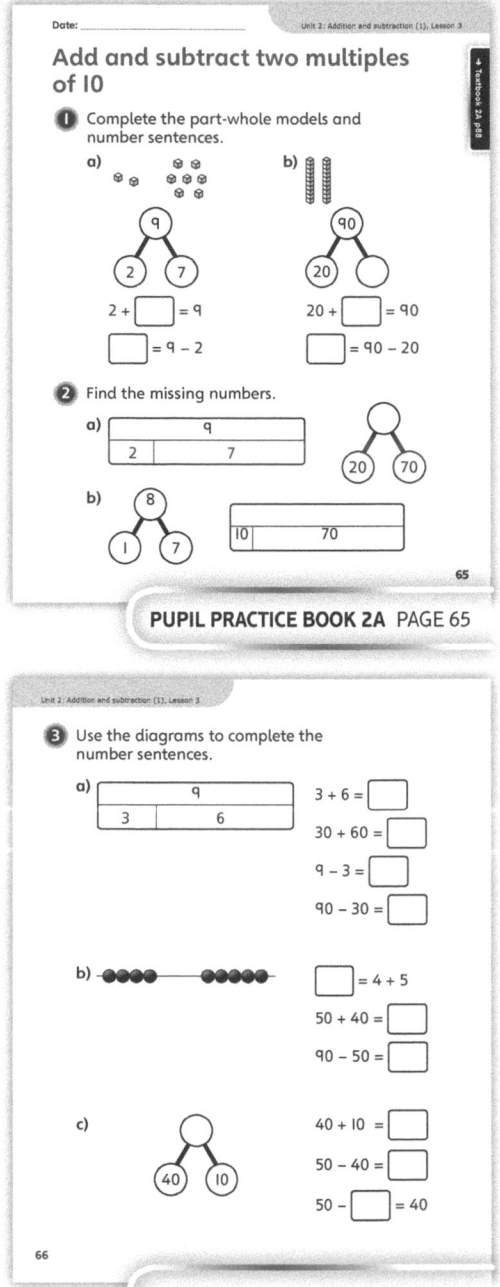

PUPIL PRACTICE BOOK 2A PAGE 65

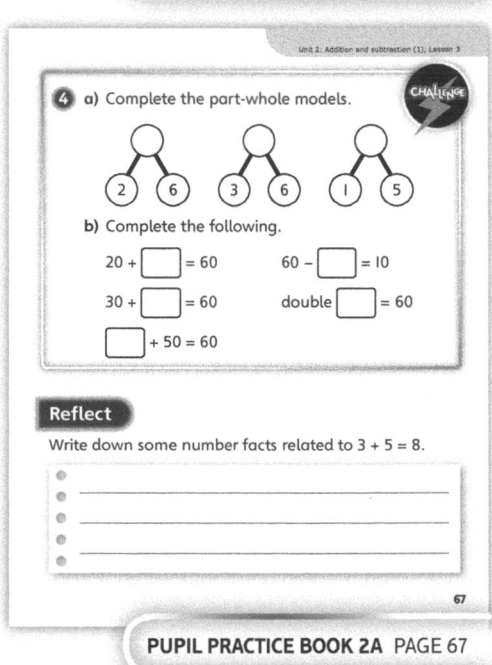

PUPIL PRACTICE BOOK 2A PAGE 66

Reflect

Independent thinking

IN FOCUS The **Reflect** part of the lesson allows children to show what they have learnt in this and previous lessons. Encourage them to include number facts involving 10s numbers. This will show whether they fully understand how to derive facts about multiples of 10 from calculations with 1s.

ASSESSMENT CHECKPOINT Assess whether children can explain how they know their number facts are correct, based on the number fact they have been given.

ANSWERS Answers for the **Reflect** part of the lesson can be found in the *Power Maths* online subscription.

After the lesson

- What were the most common misconceptions that arose during the lesson?
- How will you address these misconceptions before the next lesson?

Complements to 100 (tens)

Learning focus

In this lesson, children use their knowledge of number bonds to learn complements of 100; 10 + 90, 20 + 80, 30 + 70, and so on.

Before you teach

- Can children count in 10s to 100?
- Have children had the opportunity to explore 10s place value equipment?
- How could children practise number bonds to 10?

NATIONAL CURRICULUM LINKS

Number – addition and subtraction

Recall and use addition and subtraction facts to 20 fluently, and derive and use related facts up to 100.

ASSESSING MASTERY

Children can add multiples of 10 to make 100.

COMMON MISCONCEPTIONS

Some children may not automatically make the link between adding 1s and adding 10s. Ask:
- *Does this remind you of anything you have seen before? How many 1s are in 10? How many 10s are in 1 hundred? How could that help you?*

STRENGTHENING UNDERSTANDING

Spend some time exploring the fact that 100 is made up of 10 tens. Make sure children also recall that 10 is made up of 10 ones.

GOING DEEPER

Work with children to link the learning in this lesson with subtraction. For each addition that they encounter, ask: *How could you rewrite that as a subtraction? How do you know the answer?*

KEY LANGUAGE

In lesson: tens (10s), one hundred (100), bonds

Other language to be used by the teacher: altogether

STRUCTURES AND REPRESENTATIONS

Rekenrek

RESOURCES

Mandatory: place value equipment

Optional: Rekenrek, bead string

 In the eTextbook of this lesson, you will find interactive links to a selection of teaching tools.

Quick recap

As a class, make a list of all of the number bonds that equal 10. Then work together to find these missing numbers:

2 + ? = 10   10 − ? = 2

10 − 1 = ?   4 + ? = 10

Discover

WAYS OF WORKING Pair work

ASK

- Question ❶: What do you notice about this picture?
- Question ❶ a): What is the same and what is different in each row?
- Question ❶ a): What do you think is the reason for having two different colours of beads?

IN FOCUS This begins by supporting children in recognising 100 as 10 tens. They then move on to explore the number composition of 100. It is shown with a piece of equipment called a 'Rekenrek', which is a little like an abacus.

PRACTICAL TIPS If you do not have access to a rekenrek, then you could use other apparatus such as bead strings or a 10 by 10 grid.

ANSWERS

Question ❶ a): There are 100 beads in total.

Question ❶ b): 60 + **40** = 100

Complements to 100 (tens)

Discover

❶ a) How many beads are there altogether?

b) Complete the number sentence.

$60 + \boxed{} = 100$

92

PUPIL TEXTBOOK 2A PAGE 92

Share

WAYS OF WORKING Whole class teacher led

ASK

- *Question ❶ a): How many beads are in each row? How many rows are there?*
- *Question ❶ b): Can you see 6 tens? Can you see 4 tens? How many 10s is 6 tens plus 4 tens?*

IN FOCUS In question ❶ a) children are using the Rekenrek to explore the composition of the number 100 and familiarise themselves with the fact that there are 10 tens in 1 hundred. They then use this to add 10s to make 100 in question ❶ b).

Share

a) There are 10 beads in a row.
There are 10 rows of beads.

I know that this is called a rekenrek.

There are 100 beads in total.

b) There are 60 beads to the left.
There are 40 beads to the right.
There are 100 beads altogether.

Two numbers that go together to make 100 are called **complements** to 100.

60 + 40 = 100

93

PUPIL TEXTBOOK 2A PAGE 93

Think together

ASK

- Question ❶: *Can you see 10s here?*
- Question ❷: *How many 10s make 100?*
- Question ❸: *What number bonds could help you?*

IN FOCUS In questions ❶ and ❷, children are exploring multiples of 10 complements to 100, with the Rekenrek or the 100 square to help them visualise this. Question ❸ encourages a systematic approach to learning.

STRENGTHEN Ask children to use fingers to count aloud in 10s. Link this to known number bonds.

DEEPEN How many subtractions can children find that are linked to multiples of 10 number bonds to 100? Do they see a pattern?

ASSESSMENT CHECKPOINT Question ❸ will allow you to assess whether children can find all the multiples of 10 number bonds to 100. Look out for children working in a systematic way.

ANSWERS

Question ❶ a): $50 + 50 = 100$

Question ❶ b): Children make their own number bonds to 100.

Question ❷: $70 + \textbf{30} = 100$

Question ❸ $10 + 90 = 100$
$20 + 80 = 100$
$30 + 70 = 100$
$40 + 60 = 100$
$50 + 50 = 100$
$60 + 40 = 100$
$70 + 30 = 100$
$80 + 20 = 100$
$90 + 10 = 100$
$100 + 0 = 100$

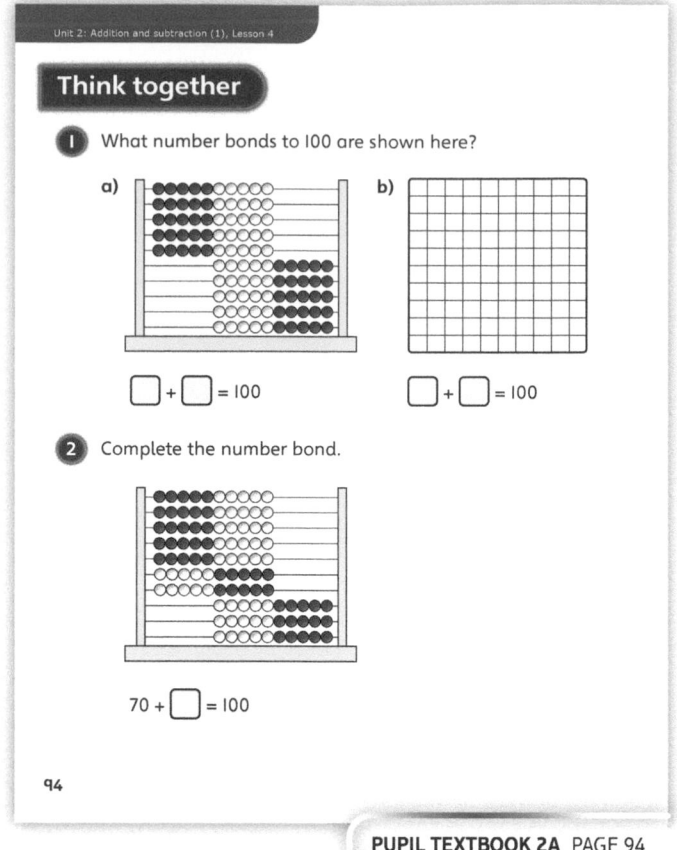

Practice

WAYS OF WORKING Independent thinking

IN FOCUS Question ❶ allows children to practise addition of multiples of 10 to 100. In question ❷ they explore the composition of 100 with part-whole models. Question ❹ requires children to match up complements to 100 and in question ❺ children use the number line to solve missing number calculations.

STRENGTHEN Use place value equipment to ensure all children have a good understanding of the fact that 100 is 10 tens.

DEEPEN Ask children to discuss and explain strategies for learning all multiples of 10 number bonds to 100 systematically.

THINK DIFFERENTLY Question ❸ introduces fact families. Children use the part-whole model to identify which number facts belong together.

ASSESSMENT CHECKPOINT Use question ❹ to assess whether children know which multiples of 10 are complements to 100.

ANSWERS Answers for the **Practice** part of the lesson can be found in the *Power Maths* online subscription.

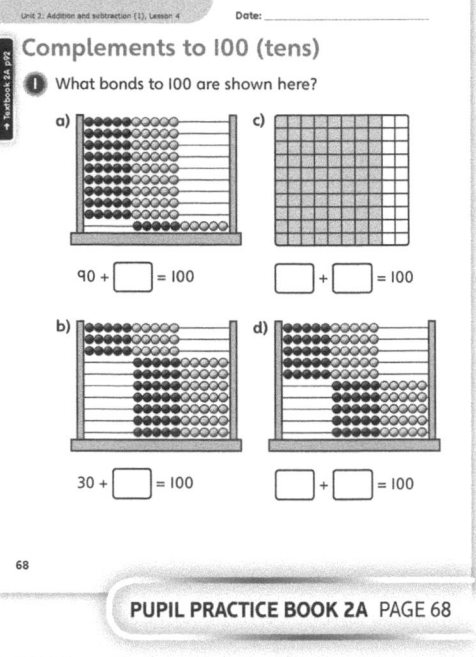

Unit 2: Addition and subtraction (1), Lesson 4   Date: _____

Complements to 100 (tens)

❶ What bonds to 100 are shown here?

a) 90 + ☐ = 100

c) ☐ + ☐ = 100

b) 30 + ☐ = 100

d) ☐ + ☐ = 100

68

PUPIL PRACTICE BOOK 2A PAGE 68

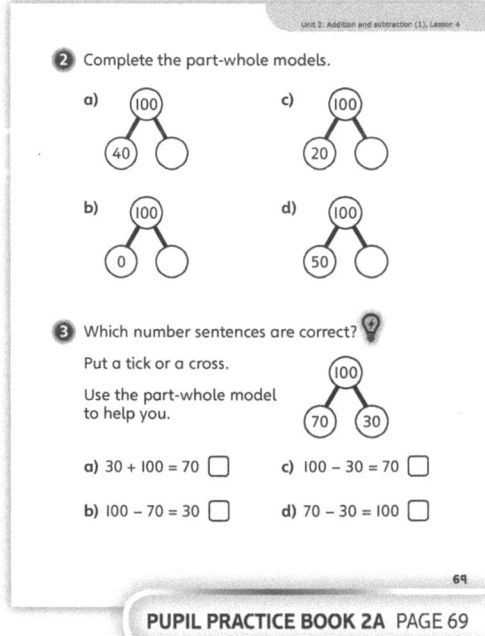

Unit 2: Addition and subtraction (1), Lesson 4

❷ Complete the part-whole models.

a) 100 / 40 / ☐

c) 100 / 20 / ☐

b) 100 / 0 / ☐

d) 100 / 50 / ☐

❸ Which number sentences are correct?

Put a tick or a cross.

Use the part-whole model to help you.

100 / 70 / 30

a) 30 + 100 = 70 ☐

c) 100 – 30 = 70 ☐

b) 100 – 70 = 30 ☐

d) 70 – 30 = 100 ☐

69

PUPIL PRACTICE BOOK 2A PAGE 69

Reflect

WAYS OF WORKING Independent thinking

IN FOCUS The **Reflect** part of the lesson requires children to explain the link between bonds to 10 and multiple of 10 complements to 100.

ASSESSMENT CHECKPOINT Can children explain how to add 10s to make 100 and describe how place value helps them to do this?

ANSWERS Answers for the **Reflect** part of the lesson can be found in the *Power Maths* online subscription.

Unit 2: Addition and subtraction (1), Lesson 4

❹ Draw lines to join the number bonds to 100.

| 90 | 50 | 100 | 20 | 60 | 70 |

| 50 | 40 | 80 | 0 | 30 | 10 |

❺ Complete the missing number.   CHALLENGE

– 40

0 ——— ☐ ——— 100

Reflect

Discuss with a partner how you can use your number bonds to 10 to work out your 10 bonds to 100.

70

PUPIL PRACTICE BOOK 2A PAGE 70

After the lesson

• If you write a number bond to 10 on the board, can children give the related number bond for 100?

Add and subtract Is

Learning focus

In this lesson, children will add and subtract 1s to or from a 2-digit number without exchanging, using number bonds to help them.

Before you teach

- Can children confidently recall number bonds?
- Which misconception do you think will be the most likely?

NATIONAL CURRICULUM LINKS

Number – addition and subtraction

Add and subtract numbers using concrete objects, pictorial representations, and mentally, including: a 2-digit number and 1s.

Solve problems with addition and subtraction: using concrete objects and pictorial representations, including those involving numbers, quantities and measures.

ASSESSING MASTERY

Children can identify the number of 10s and 1s in a number, can add and subtract an additional number of 1s without exchange and notice that only the digit in the 1s column changes. Children can verbalise the changes that occur and use known number bonds to calculate the answer, rather than counting on or back in 1s.

COMMON MISCONCEPTIONS

Children may not know their number bonds and may resort to counting on or back, perhaps incorrectly. Ask:
- *What number bond could you use for this calculation?*

Children may add the 1s to the 10s of the other number when calculating mentally. Ask:
- *Can you explain to me in words what you are doing?*

STRENGTHENING UNDERSTANDING

Give children a list of number bonds and encourage children to use and memorise them. Ask them to identify the fact they are using each time to help them see the link between the known fact and the calculation they are doing.

GOING DEEPER

Challenge children to record any calculation that they write in as many different ways as possible, practising skills that they have learned in previous lessons. For example, can they use the commutative law and change the location of the equals sign?

KEY LANGUAGE

In lesson: adding, subtracting, +, −, =, ones (1s), in total, number sentence, count on, tens (10s)

Other language to be used by the teacher: digit, altogether, take away

STRUCTURES AND REPRESENTATIONS

Number line, place value chart

RESOURCES

Mandatory: base 10 equipment

 In the eTextbook of this lesson, you will find interactive links to a selection of teaching tools.

Quick recap

Work together to partition these 2-digit numbers into 10s and 1s:

32, 43, 64, 77, 20

134

Discover

Add and subtract 1s

Discover

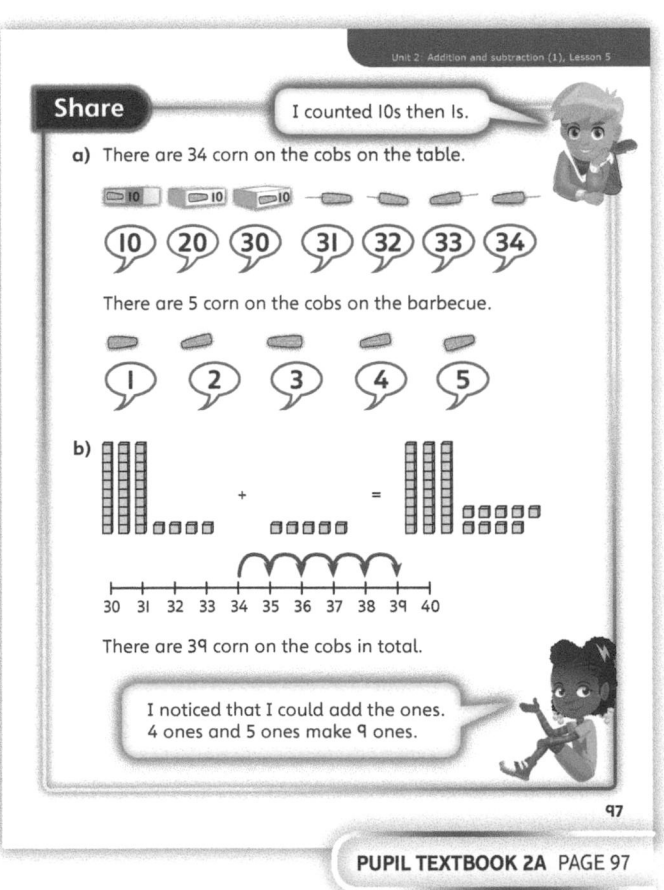

1 a) How many corn on the cobs are on the table?
How many corn on the cobs are on the barbecue?

b) How many corn on the cobs are there altogether?

96

PUPIL TEXTBOOK 2A PAGE 96

WAYS OF WORKING Pair work

ASK

• Question 1 a): *What does the box of corn on the cobs represent?*
• Question 1 b): *Is there more than one way to find how many altogether?*

IN FOCUS Question 1 requires children to differentiate between a box of corn on the cobs, representing 10, and an individual corn on the cob, representing 1, in order to successfully create number sentences.

PRACTICAL TIPS Provide base 10 equipment so that children can model the two numbers, paying particular attention to the 10s and the 1s.

ANSWERS

Question 1 a): There are 34 corn on the cobs on the table.
There are 5 corn on the cobs on the barbecue.

Question 1 b): There are 39 corn on the cobs altogether.
$34 + 5 = 39$

Share

WAYS OF WORKING Whole class teacher led

ASK

• Question 1 a): *What resources could you use to represent the corn cobs? What is the most efficient way to make the correct number?*
• Question 1 b): *Is there a quicker way to calculate the total than just counting on?*

IN FOCUS Ask children to share how they calculated the number of corn on the cobs in both parts of question 1. Discuss both methods in each case. *What is the same and what is different about the methods? Which method is more efficient?*

Share

I counted 10s then 1s.

a) There are 34 corn on the cobs on the table.

10   20   30   31   32   33   34

There are 5 corn on the cobs on the barbecue.

1   2   3   4   5

b)

+   =

30 31 32 33 34 35 36 37 38 39 40

There are 39 corn on the cobs in total.

I noticed that I could add the ones.
4 ones and 5 ones make 9 ones.

97

PUPIL TEXTBOOK 2A PAGE 97

Think together

Whole class teacher led (I do, We do, You do)

ASK

- Question **1**: *How many 10s are there in each number? How many 1s?*
- Question **2**: *What is the most efficient way to calculate the solution?*

IN FOCUS Question **3** exposes the common issue of choosing the wrong operation when trying to solve a word problem. Ask children to explain why Astrid is incorrect.

STRENGTHEN Encourage children to make each number of objects using concrete manipulatives. They should use the manipulatives to check their answers.

DEEPEN Ask children to describe what is the same and what is different about the numbers in each of the calculations in question **2**. Can they demonstrate how knowing the answer to one calculation can help them find the answer to the next?

ASSESSMENT CHECKPOINT Assess whether children can identify the number fact they can use to calculate each solution. Children should also be able to verbalise the steps they are completing mentally (for example, 2 ones plus 5 ones is equal to 7 ones).

ANSWERS

Question **1** a): $41 + 6 = $ **47**

Question **1** b): $52 + 4 = $ **56**

Question **2** a): $42 + 5 = $ **47** $45 + 2 = $ **47**

Question **2** b): $47 - 2 = $ **45** $47 - 5 = $ **42**

Question **3** a): 23 eggs are left

Question **3** b): $18 - 3 = 15$
$ 28 - 3 = 25$
$ 38 - 3 = 35$
$ 48 - 3 = 45$
Children should notice the 10s digit going up by 1 each time.

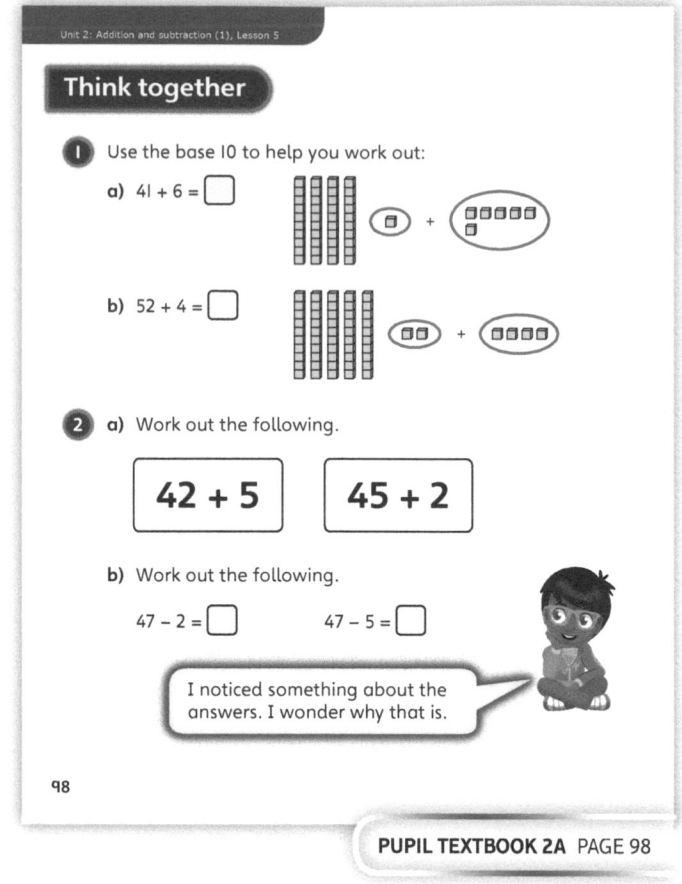

PUPIL TEXTBOOK 2A PAGE 98

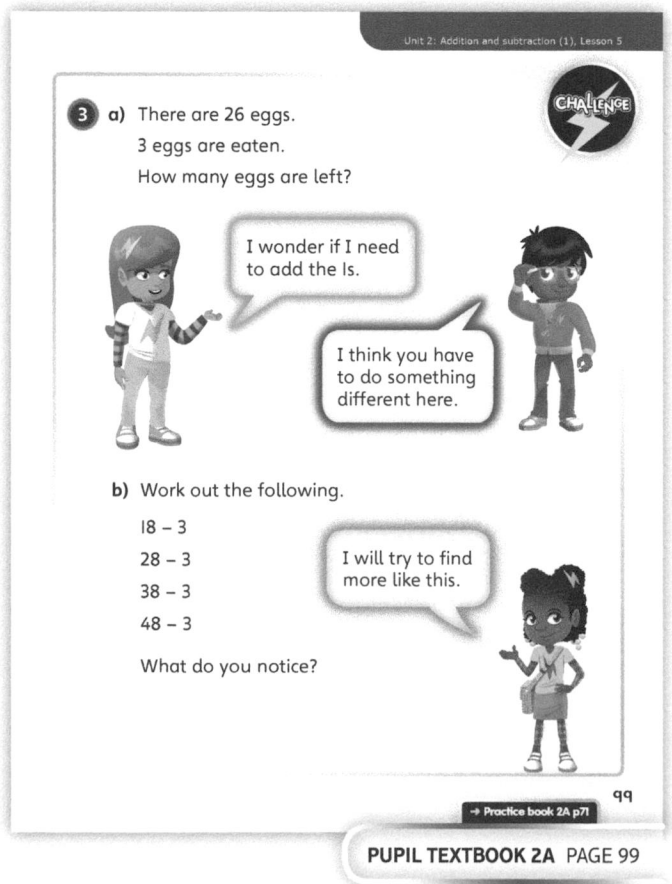

PUPIL TEXTBOOK 2A PAGE 99

Practice

WAYS OF WORKING Independent thinking

IN FOCUS Question ❶ shows numbers represented using base 10; in question ❷, children can draw their own base 10 representations to encourage understanding.

Question ❺ requires children to use the answer to an initial calculation with 1s to help them determine the answers for other related calculations that include 10s.

STRENGTHEN Encourage children to make concrete representations of the base 10 drawings and discuss what happens to the 10s and the 1s each time. They should notice that in these calculations, the 10s do not change, but the 1s do.

DEEPEN Provide children with a number sentence such as 2☐ + ☐ = 27 and ask them to explore all the different ways that this question could be answered. Encourage children to work systematically to find all possible ways.

ASSESSMENT CHECKPOINT Assess whether children are able to identify how many 10s and 1s there are in a number. Can children instantly recall number bonds within 10 and relate them to the starting number or are they relying on counting on or counting back in 1s?

ANSWERS Answers for the **Practice** part of the lesson can be found in the *Power Maths* online subscription.

Reflect

WAYS OF WORKING Pair work

IN FOCUS In the **Reflect** part of the lesson, children verbalise to their partner different ways to work out an addition and a subtraction. They could use resources to help them if required.

ASSESSMENT CHECKPOINT Assess whether children can confidently explain more than one way to answer each question. Do they use accurate mathematical language? Can they tell you which of the methods their partner explained was the more efficient, and why?

ANSWERS Answers for the **Reflect** part of the lesson can be found in the *Power Maths* online subscription.

After the lesson

- Did children link the use of number bonds to becoming more efficient?
- Can children confidently and accurately explain how they calculated their solutions?

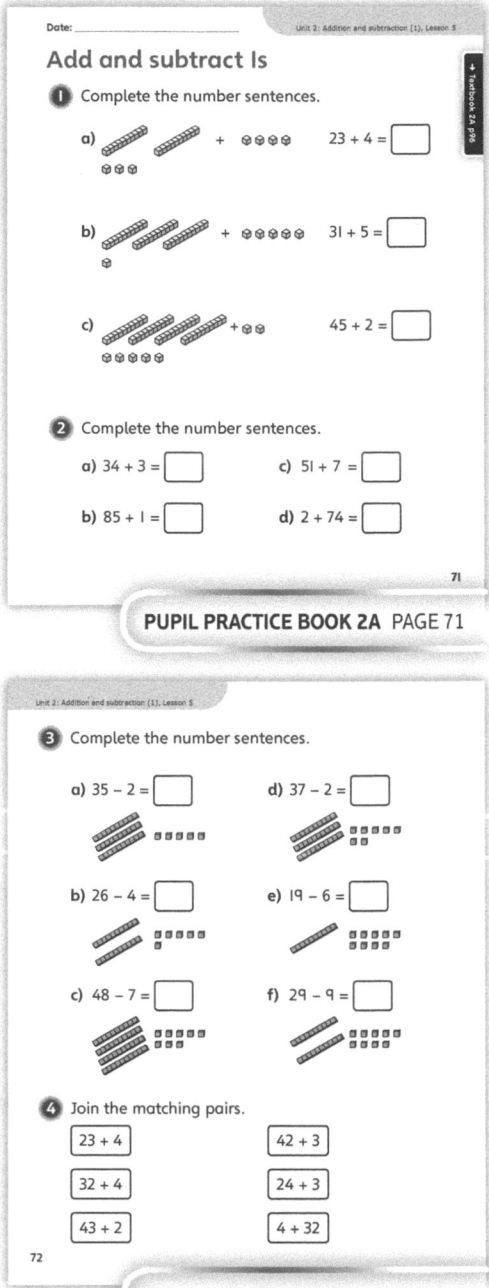

PUPIL PRACTICE BOOK 2A PAGE 71

PUPIL PRACTICE BOOK 2A PAGE 72

PUPIL PRACTICE BOOK 2A PAGE 73

Add by making 10

Learning focus

In this lesson, children will add two single-digit numbers that total more than 10, by breaking one number into two parts to bridge the 10.

Before you teach

- Are children confident partitioning a single-digit number?
- Can children add a single-digit number to 10 without counting on?

NATIONAL CURRICULUM LINKS

Number – addition and subtraction

Add and subtract numbers using concrete objects, pictorial representations, and mentally, including: two 2-digit numbers.

Solve problems with addition and subtraction: applying their increasing knowledge of mental and written methods.

ASSESSING MASTERY

Children can decide correctly how to partition one of the single-digit numbers to bridge 10 when adding. Children can add a single digit to 10 without counting, for example 10 + 4 = 14.

COMMON MISCONCEPTIONS

Children may partition the first number randomly rather than using the correct bond to 10. Ask:
- *What do you add to this number to make 10? How should we split this number up to help us use the bond to 10?*

STRENGTHENING UNDERSTANDING

Ten frames are an excellent visual support for bonds to 10. Practise by adding different single-digit numbers to 9, for example, showing that the partition you need will always be 1 and another number. A part-whole diagram can also be a useful tool to explore the partitions of a number, and to understand how choosing one part allows us to know what the other part must be.

GOING DEEPER

Encourage children to explain to each other how to decide which partitions to use, and to justify their reasoning using apparatus such as a ten frame and part-whole models. Challenge children to explain why the method works for additions such as 7 + 5, but not for 4 + 5.

KEY LANGUAGE

In lesson: make 10, add, altogether, break up

Other language to be used by the teacher: total, greatest, smallest, break apart, predict

STRUCTURES AND REPRESENTATIONS

Ten frame

RESOURCES

Mandatory: counters or cubes, ten frames

 In the eTextbook of this lesson, you will find interactive links to a selection of teaching tools.

Quick recap 🔄

Play 'Say the next number'. Say a number between 0 and 99 and ask children to call out the next number.

Play 'Say the previous number'. Say a number between 1 and 100 and ask children to call out the previous number.

Discover

WAYS OF WORKING Pair work

ASK

- Question ❶ b): *How could you write the addition?*
- Question ❶ b): *Before working out the answer, do you think it will total more than 10 or less than 10?*
- Question ❶ b): *How could you represent the two numbers to be added?*

IN FOCUS Children have already learnt methods to find the answer, including count all and count on. This lesson will teach a more efficient method: using known bonds to 10. To begin the discussion, ask children to predict whether the total will be greater or less than 10.

PRACTICAL TIPS Provide ten frames and counters in two colours so that children can make, compare and combine the two numbers.

ANSWERS

Question ❶ a): Sam has found 7 stars. Eva has found 5 stars.

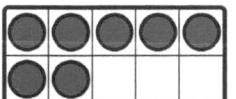 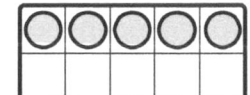

Question ❶ b): They have found 12 stars altogether.

Share

WAYS OF WORKING Whole class teacher led

ASK

- Question ❶ a): *How do we know the answer will be greater than 10?*
- Question ❶ b): *What do you add to 7 to make 10?*
- Question ❶ b): *What is left from 5 if you have used 3 to make 10?*

IN FOCUS This part of the lesson highlights splitting 5 into 3 add 2 to make 10 with the 7, with 2 left over to make 12 in total. Ensure children understand why 3 is moved from the 5 to the 10 (7 + 3 = 10), and why the 5 has not been lost, but rather just split into two parts. Emphasise that choosing 3 and 2 as the parts for 5 is not a random choice but is based on number bonds to 10.

PUPIL TEXTBOOK 2A PAGE 100

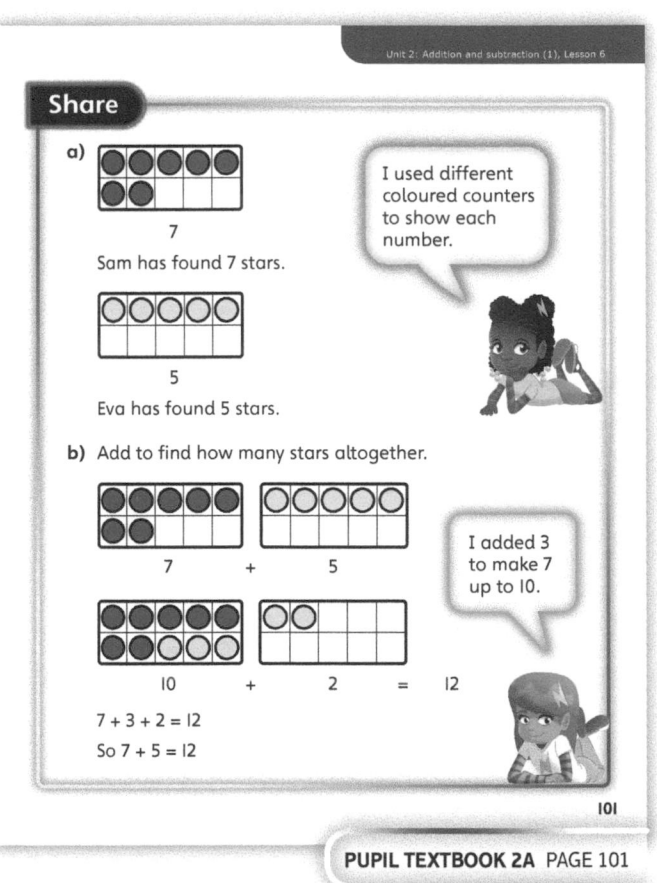

PUPIL TEXTBOOK 2A PAGE 101

Think together

WAYS OF WORKING Whole class teacher led (I do, We do, You do)

ASK

• Question ❶: *Why do you only need to split one of the numbers?*
• Question ❶: *How do you know which parts to use?*

IN FOCUS This part of the lesson builds an understanding of splitting one number to make 10, and choosing the parts with that in mind. The ten frame models it very clearly. Children may rely on counting, so they should be encouraged to use their understanding of number bonds. Can they predict the two parts of the number before checking using counters on a ten frame? Question ❸ uses variations on adding 5, to encourage children to build confidence in how the decisions are made.

STRENGTHEN To be efficient using this method, children will need to understand both bonds to and bonds within 10. They will need to work in stages. Support children to break the task into two stages: first, find the bond to bridge 10; second, find the remaining part. Children can use ten frames and part-whole models to support their reasoning.

DEEPEN Challenge children to demonstrate the concept using a range of manipulatives and to build confidence by predicting the part that will be left, before checking with counters or cubes on a ten frame. Can children explain why the answer to question ❷ is one more than the answer to question ❶? Can they also explain the pattern of answers in question ❸?

ASSESSMENT CHECKPOINT Question ❸ will demonstrate if children have understood how to use the different bonds for 5, and which to choose based on the calculations needed to bridge 10.

ANSWERS

Question ❶: 7 + 3 + 3 = 13

Question ❷: They have 14 apples altogether.

Question ❸ a): 6 + 5 = 11

Question ❸ b): 8 + 5 = 13

Question ❸ c): 9 + 5 = 14

Question ❸ d): 5 + 7 = 12

PUPIL TEXTBOOK 2A PAGE 102

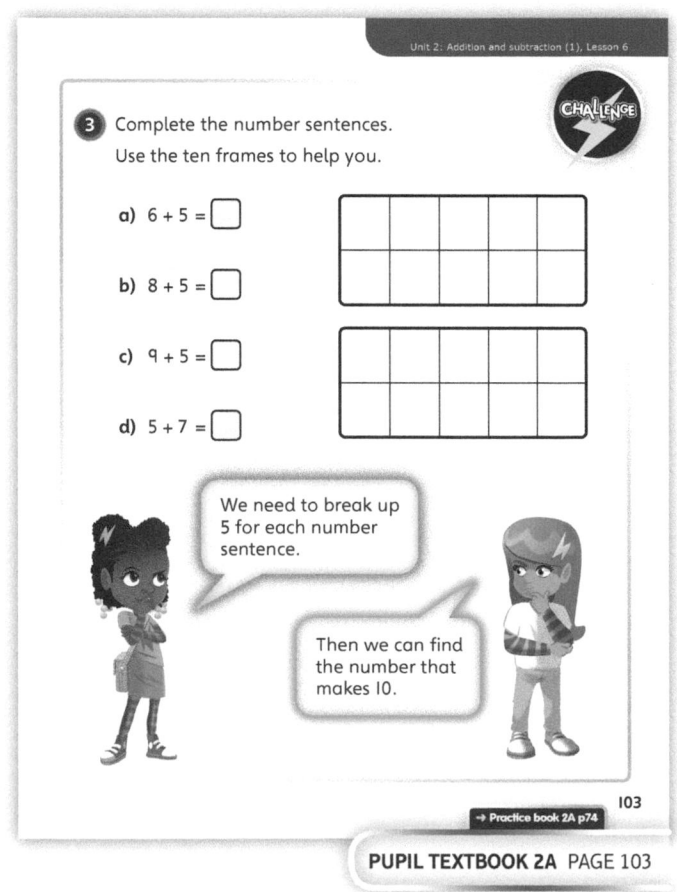

PUPIL TEXTBOOK 2A PAGE 103

Practice

WAYS OF WORKING Independent thinking

IN FOCUS Questions ❶ and ❷ structure the thought process for bridging 10 very clearly in stages. Children should complete each stage of the calculation to ensure understanding of how the bridging is made. Question ❻ links with understanding about comparing numbers. Engage children with making decisions about how to bridge the 10 at every stage, and provide plenty of practice so they become secure.

STRENGTHEN Use ten frames and counters to support reasoning. Encourage children to see the gaps left in one ten frame to support their decision about how many to move across to bridge the 10. At this stage, understanding the concept and the process is more important than quickly or mentally finding an answer. The next lesson will be developing the same skill, so in this lesson focus on building understanding of why the bridging 10 method works for these additions.

DEEPEN After completing question ❻, can children generalise about how to find the greatest and smallest totals from a set of numbers? How did they approach the problem?

ASSESSMENT CHECKPOINT Questions ❺ and ❻ will demonstrate if children are confident using the method when the parts are not provided and they have to make the decisions about partitioning for themselves.

ANSWERS Answers for the **Practice** part of the lesson can be found in the *Power Maths* online subscription.

Reflect

WAYS OF WORKING Pair work

IN FOCUS The important point of this activity is to be able to explain how making 10 involves breaking apart one of the numbers, rather than simply being able to find the answer. The ten frames allow children to either break the 5 or the 7, although the lesson has taught to break the second number.

ASSESSMENT CHECKPOINT Assess whether children can explain how to decide which parts to break 5 (or 7) into.

ANSWERS Answers for the **Reflect** part of the lesson can be found in the *Power Maths* online subscription.

After the lesson ⏸

- Are children able to decide how to break apart one of the numbers?
- Can children explain why they do not need to count all the counters to find the answer?
- Have they understood that this method can be more efficient than simply using a counting strategy?

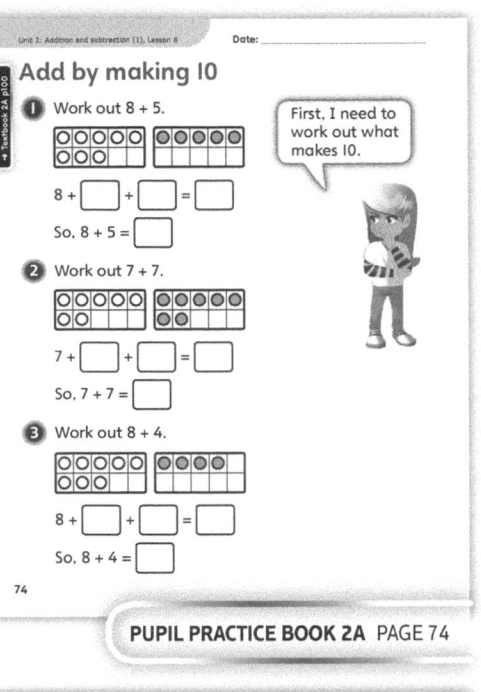

PUPIL PRACTICE BOOK 2A PAGE 74

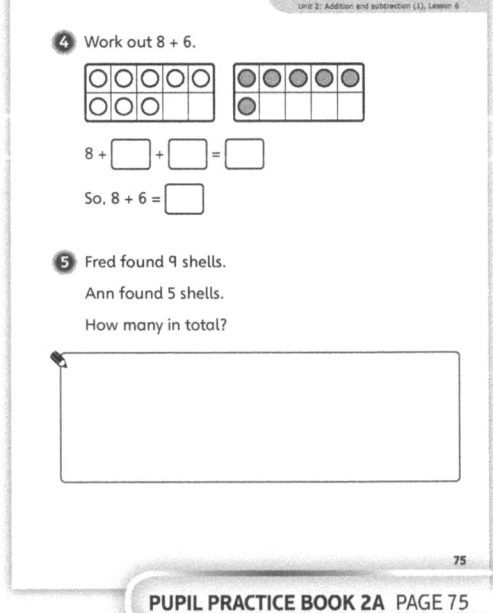

PUPIL PRACTICE BOOK 2A PAGE 75

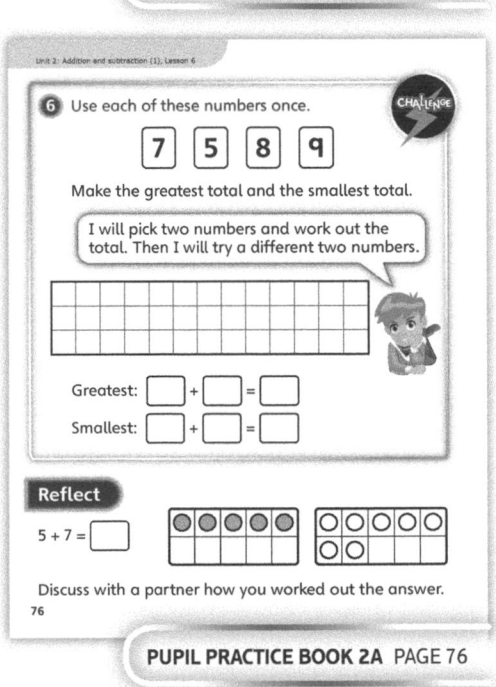

PUPIL PRACTICE BOOK 2A PAGE 76

Add using a number line

Learning focus

In this lesson, children will deepen their understanding and proficiency in adding two single-digit numbers by bridging 10.

Before you teach

- Can children show all the partitions of a given single-digit number?
- Can children explain what is the same and what is different about 2 + 4 and 4 + 2?
- Can children recognise which will give a total greater than 10: 2 + 7 or 7 + 5?

NATIONAL CURRICULUM LINKS

Number – addition and subtraction

Add and subtract numbers using concrete objects, pictorial representations, and mentally, including: adding 3 one-digit numbers.

Solve problems with addition and subtraction: applying their increasing knowledge of mental and written methods.

ASSESSING MASTERY

Children can decide which partitions to use to add by making 10. Children can represent the process using ten frames and number lines.

COMMON MISCONCEPTIONS

Some children may think this method can be used for adding any two numbers, even when they add up to less than 10. Ask:
- *How can you tell if the total will be greater than 10?*

STRENGTHENING UNDERSTANDING

Use a range of manipulatives, including bead strings and ten frames to support the concept of breaking apart a number to make 10 plus the extra 1s. The bead string can be particularly effective in making this concept clear, especially where children can compare it with using a ten frame. Alternatively, use two towers of interlocking cubes and break one apart to add to the other to make 10. Encourage children to use known facts wherever possible, rather than a counting method.

GOING DEEPER

Ask children to reason about the order of the calculation. If they were to work out 4 + 9, which number will they choose to break apart? Challenge children to see the links between calculations. For example, if they have solved 5 + 8, can they use that knowledge to solve 5 + 9 or 4 + 8 instead of actually working it out?

KEY LANGUAGE

In lesson: ten frames, number line, make 10, in total, altogether, answer

Other language to be used by the teacher: partition, calculation, addition

STRUCTURES AND REPRESENTATIONS

Ten frame, number line

RESOURCES

Mandatory: ten frames, counters

Optional: bead string, interlocking cubes

 In the eTextbook of this lesson, you will find interactive links to a selection of teaching tools.

Quick recap

Show children a ten frame with a certain number of counters in it.

Challenge them to say how many more they would need to add to make 10. Repeat with different starting numbers.

Discover

WAYS OF WORKING Pair work

ASK

- Question **1** b):*Will there be more than 10 jumpers in total?*
- Question **1** b):*Why can you not just count all the jumpers by pointing at them?*
- Question **1** b):*Can you think of two different ways to solve this addition?*

IN FOCUS The jumpers in the box cannot be counted, so this problem is not suited to a count all strategy. As 9 is one less than 10, children may see that if one of the remaining jumpers were to go in the box, it would make a complete 10. They could then work out the total from the remaining jumpers for the 1s digit. Some children may revert to a count on method, given the 4 jumpers to be counted. However, this lesson builds on the previous lesson for adding to make 10 plus any remaining, so encourage this method.

PRACTICAL TIPS Provide ten frames and counters in two colours so that children can make and compare the numbers, and can see how a counter is moved from one frame to the other to make a 10 and 3 more.

ANSWERS

Question **1** a): There are 9 jumpers in the box.
There are 4 jumpers on the ground.

Question **1** b): There are 13 jumpers altogether.

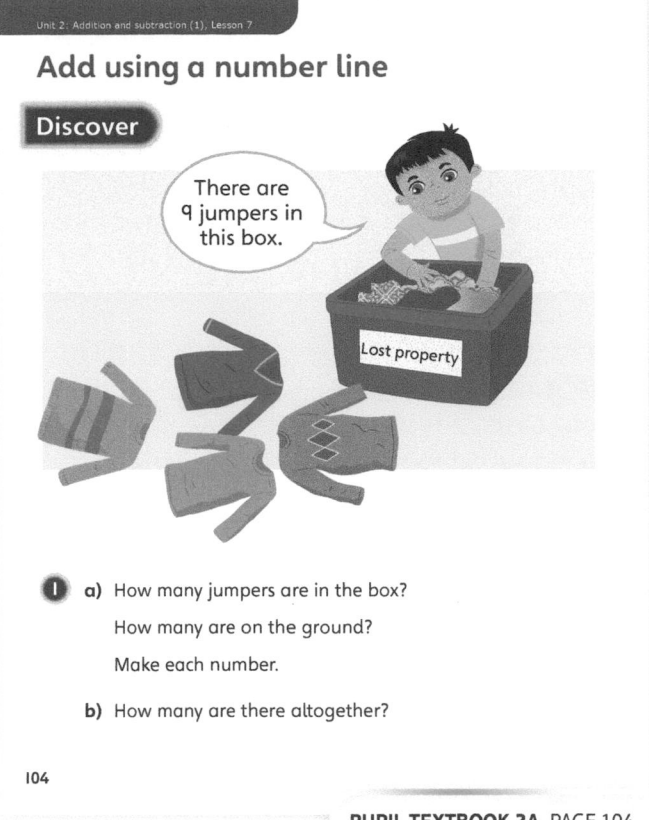

PUPIL TEXTBOOK 2A PAGE 104

Share

WAYS OF WORKING Whole class teacher led

ASK

- Question **1** b): *Why has 4 been split into 1 and 3?*
- Question **1** b): *Does the number line show the same method as the ten frame?*
- Question **1** b): *Why are the jumps different sizes on the number line?*

IN FOCUS This lesson builds deeper understanding of the concept of bridging to 10 to add, by continuing the same approach as the previous lesson. In this part of the lesson, a number line is used to support children's developing understanding of how the addition is broken into two parts.

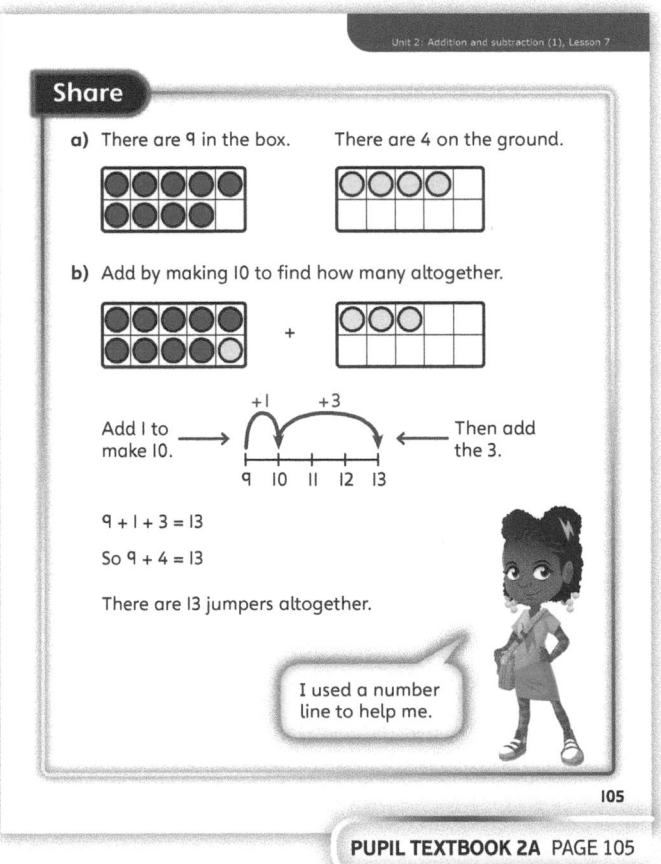

PUPIL TEXTBOOK 2A PAGE 105

Think together

WAYS OF WORKING Whole class teacher led (I do, We do, You do)

ASK

- Question ❶: *Why are the jumps on number lines sometimes different sizes?*
- Question ❸: *How do you know which numbers to choose for the parts?*
- Question ❸: *Can you check this answer using a ten frame?*

IN FOCUS Children are becoming more proficient at choosing the parts to make 10, by thinking about known bonds, and using the number line to support their reasoning. Question ❸ challenges children to make a decision about why 5 + 7 and 7 + 5 produce the same total.

STRENGTHEN Encourage children to use counters in a part-whole model to support their reasoning about which parts to use when making 10.

DEEPEN Challenge children to justify whether they think 5 + 7 or 7 + 5 is easier to solve by making 10. Can they explain and illustrate their reasoning using a range of manipulatives?

ASSESSMENT CHECKPOINT Questions ❶ and ❷ will show if children can find the parts of a number by using recall rather than counting. Question ❸ will demonstrate children's emerging understanding of commutativity.

ANSWERS

Question ❶: 8 + 5 = 13

Question ❷: 7 + 6 = 13
8 + 3 = 11
9 + 5 = 14
6 + 8 = 14

Question ❸: If children calculate 5 + 7 they will start at 5 and count on 7 to reach 12. If children calculate 7 + 5 they will start at 7 and count on 5 to also reach 12.

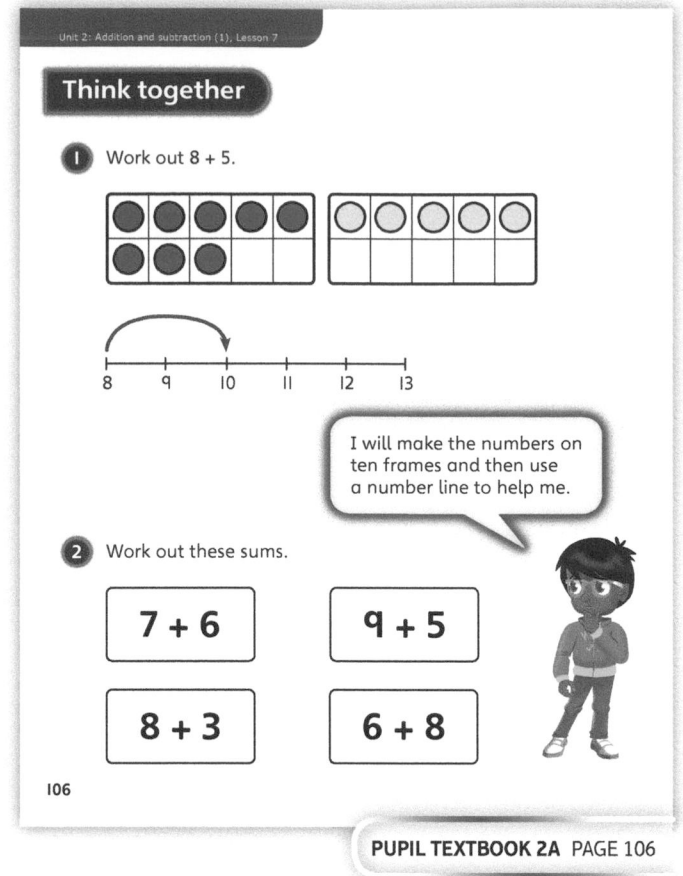

PUPIL TEXTBOOK 2A PAGE 106

PUPIL TEXTBOOK 2A PAGE 107

Practice

WAYS OF WORKING Independent thinking

IN FOCUS This part of the lesson builds confidence and fluency in adding by making 10 first. In questions ❶ and ❷ children use number lines to make the jump to 10 then on to the total. Questions ❸ and ❹ require children to work out the addition shown on the number line.

STRENGTHEN Children could use counters and ten frames to support their reasoning. For children who struggle making jumps greater than 1 on a number line, stress that they do not need to count the jumps one by one, as the method is about jumping to 10 (which they know from number bonds) or jumping from 10 (which they know because of place value of numbers between 10 and 20). Practise 10 + 7; 10 + 2 until children see the pattern.

DEEPEN Challenge children to explore whether or not it is always true that you can add two numbers in any order. Can they show on number lines that they always produce the same total? Can they explain to a partner why they think it works?

THINK DIFFERENTLY In question ❸, children are shown two jumps on a number line and use this to work out which calculation is being done. Encourage them to describe what each jump represents and to show how they used this to find the correct calculation.

ASSESSMENT CHECKPOINT Are children using the correct method of making 10 first or are they simply counting on? Are they able to use number bonds to 10 as the first jump and identify the second jump needed to find the total? In question ❸, are children able to identify the calculation easily or do they need to draw number lines to see which one matches?

ANSWERS Answers for the **Practice** part of the lesson can be found in the *Power Maths* online subscription.

Reflect

WAYS OF WORKING Independent thinking

IN FOCUS In this part of the lesson, children explain how they would find the answer to 5 + 9.

ASSESSMENT CHECKPOINT Assess whether children can explain clearly how to use making 10 to solve the addition. Can they explain how to choose the parts that will make 10? They may also discuss which representations they prefer (number line, ten frame) and in which order they would work out the addition.

ANSWERS Answers for the **Reflect** part of the lesson can be found in the *Power Maths* online subscription.

After the lesson ⏸

- Can children use number lines to show how to add by making 10 first?
- Can children make choices about which number to split apart?
- Can children use one answer to find another related answer?

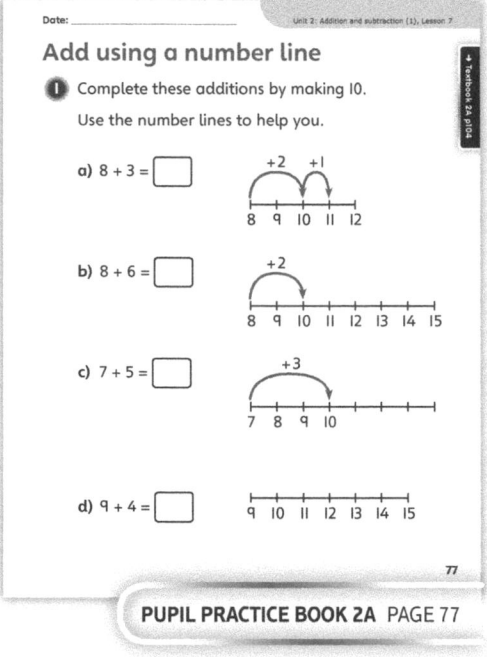

PUPIL PRACTICE BOOK 2A PAGE 77

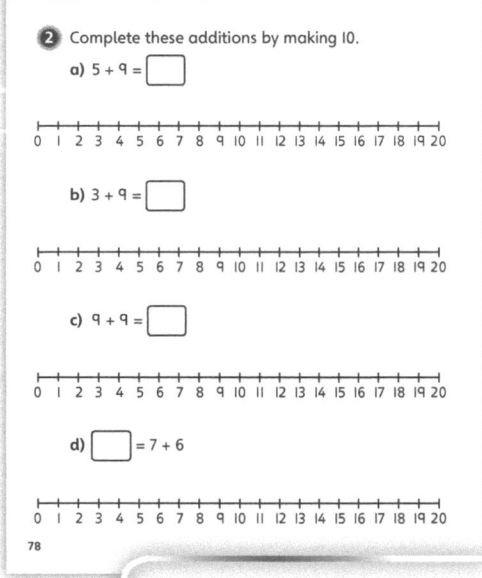

PUPIL PRACTICE BOOK 2A PAGE 78

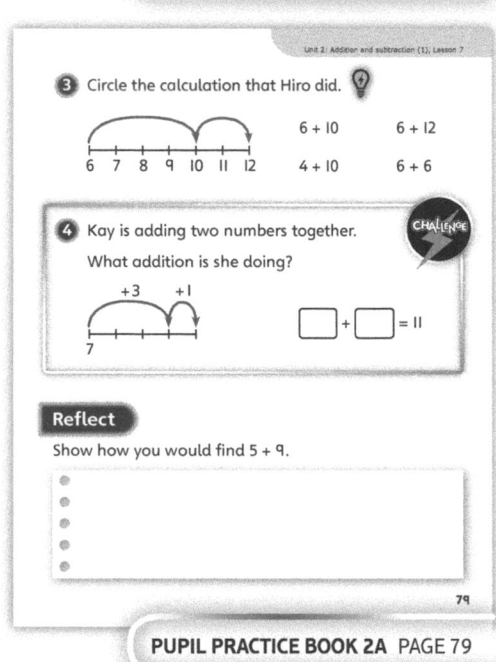

PUPIL PRACTICE BOOK 2A PAGE 79

Add three I-digit numbers

Learning focus

In this lesson, children will add three numbers presented in a variety of ways, including concrete and pictorial representations. They will select the most appropriate resource to help them and rearrange the numbers to add efficiently.

Before you teach

- How secure are children recalling known number facts?
- How can you enable children to critique approaches used by others?

NATIONAL CURRICULUM LINKS

Number – addition and subtraction

Add and subtract numbers using concrete objects, pictorial representations, and mentally, including: two 2 digit numbers.

Solve problems with addition and subtraction: applying their increasing knowledge of mental and written methods.

ASSESSING MASTERY

Children can use their knowledge of number bonds to make decisions regarding the order in which to complete mental addition. Children can understand that the order does not affect the final total.

COMMON MISCONCEPTIONS

Children may think the numbers have to be added in the order that they are given. Make each number using cubes and ask:
- *Does the order in which you add the groups of cubes change the total number of cubes?*

Children may confuse addition and subtraction when they are asked to work out a missing number, as in $19 = 8 + 5 + \square$. Ask:
- *Have you read through the completed calculation to check you have not made any mistakes? Do the three parts add up to equal the whole?*

STRENGTHENING UNDERSTANDING

Give children a completed set of number bonds within 20 to increase their confidence and emphasise recall, rather than allowing them to calculate the answer by counting in 1s using their fingers.

GOING DEEPER

Challenge children to justify why they have chosen to add numbers in a specific order.

KEY LANGUAGE

In lesson: add, +, =, altogether, missing number, number bonds, method

Other language to be used by the teacher: digits, partition, rearrange, total, compare

STRUCTURES AND REPRESENTATIONS

Ten frame, part-whole model

RESOURCES

Mandatory: ten frames, counters

Optional: number bonds within 20, counters in three different colours

 In the eTextbook of this lesson, you will find interactive links to a selection of teaching tools.

Quick recap 🔁

Show children a 0 to 10 number track and ask them to find pairs of numbers that total 10.

Discover

Unit 2: Addition and subtraction (1), Lesson 8

Add three 1-digit numbers

WAYS OF WORKING Pair work

ASK

- Question **1** b): *What number did you choose to start with?*
- Question **1** b): *What strategies could you use to effectively add numbers mentally?*

IN FOCUS Encourage children to discuss different ways to calculate the answer to question **1** b). Which way is easiest? Which way is most efficient?

PRACTICAL TIPS Provide ten frames and counters in three colours. Support children in making the three numbers and identifying the bond to 10.

ANSWERS

Question **1** a): Kendi is holding up 7, Malik is holding up 3 and Lily is holding up 5.

Question **1** b): 15 fingers and thumbs altogether.

Discover

1 a) How many fingers and thumbs is each child holding up?

   b) How many fingers and thumbs are being held up altogether?

108

PUPIL TEXTBOOK 2A PAGE 108

Share

WAYS OF WORKING Whole class teacher led

ASK

- Question **1** b): *Which two children make a number bond to 10 with the fingers they are holding up?*
- Question **1** b): *Is it easier to look for a number bond to 10 first, instead of just adding up the three numbers? Why?*

IN FOCUS Throughout question **1**, encourage children to use what they know, rather than using their fingers to count on in 1s. For example, they could start by working out 7 + 5 if they are comfortable doing so; if not, they may prefer to look for a number bond to 10 (7 + 3) as their initial strategy.

DEEPEN Challenge children to find three different numbers that make a total of 15. How many different ways can they find?

Share

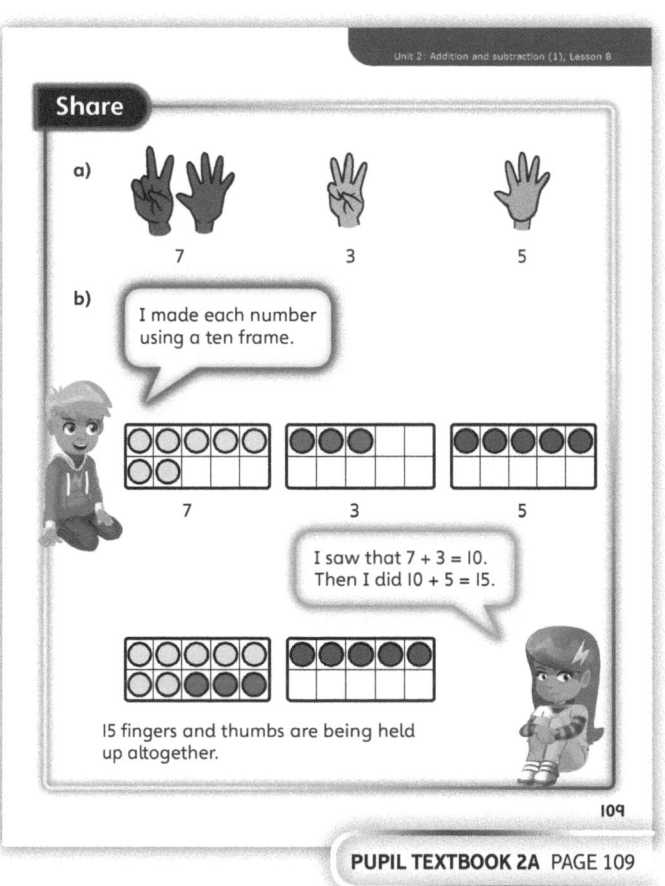

I made each number using a ten frame.

I saw that 7 + 3 = 10. Then I did 10 + 5 = 15.

15 fingers and thumbs are being held up altogether.

109

PUPIL TEXTBOOK 2A PAGE 109

Think together

WAYS OF WORKING Whole class teacher led (I do, We do, You do)

ASK

- Question **1**: *What strategies could you use to answer the question?*
- Question **2**: *How can you use resources to help you decide in what order to add the numbers mentally?*
- Question **3**: *Who used Astrid's method to answer the question? Who used Flo's method?*

IN FOCUS In question **2**, it is not possible to use number bonds to 10. Encourage children to visualise making 10 to help them partition and add larger numbers, such as 8 + 5.

STRENGTHEN Ensure children understand that the order in which the numbers to be added are given does not change the total. You can do this by giving them a part-whole model with three parts and asking them to complete it in as many different ways as possible for a given whole.

DEEPEN The answer to question **3** is 17. Ask children to find as many other possible ways of making 17 with three numbers as they can. Can they put the different ways into categories: those that use number bonds to 10, those that use doubling or near doubling facts, and those that require partitioning?

ASSESSMENT CHECKPOINT Check whether children are using their fingers to count on in 1s. If they are, they may need extra practice in using and recalling number bonds within 20.

ANSWERS

Question **1** a): 8 fingers

Question **1** b): 13 fingers

Question **2**: 8 + 5 + **6** = 19

Question **3**: 5 + 7 + 5 = 17.

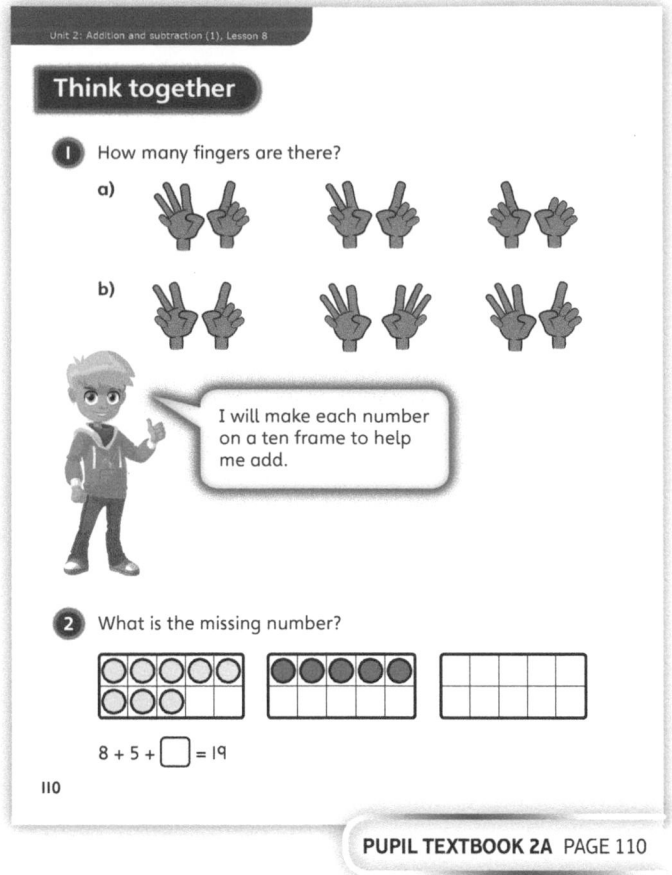

PUPIL TEXTBOOK 2A PAGE 110

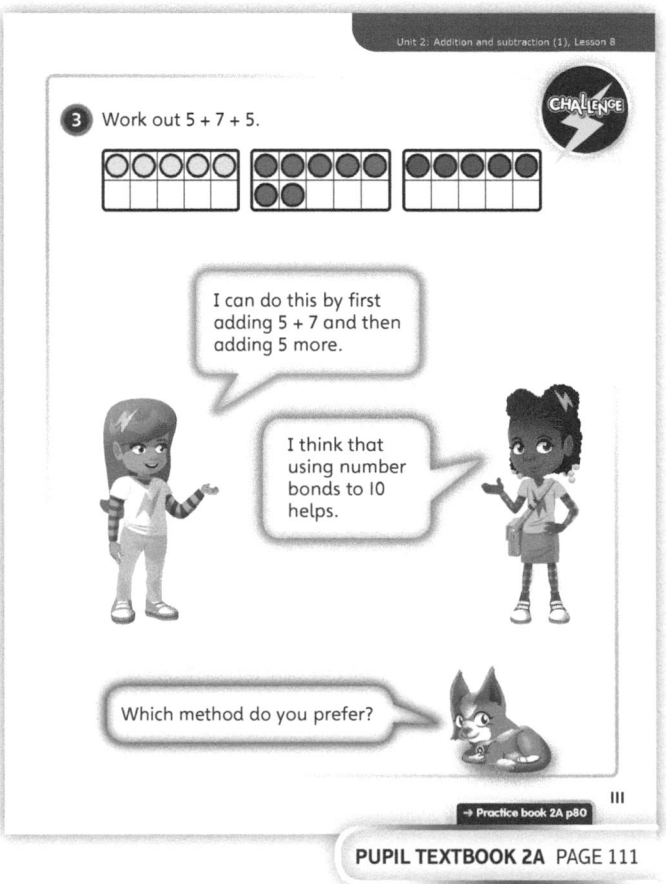

PUPIL TEXTBOOK 2A PAGE 111

Practice

WAYS OF WORKING Independent thinking

IN FOCUS Questions **1** and **2** include numbers that add to 10, or include doubles or near doubles. Children should make clear in what order they chose to add the numbers, for example using different colours in the ten frames, rather than simply drawing each number and then counting in 10s and 1s to find the total.

STRENGTHEN Using counters in three different colours will help children see which pairs of numbers work well together. Children can also use counters to see how a number can be partitioned to make 10, and to see what remains.

DEEPEN In question **4**, encourage children to find all the different ways of making 12 with three numbers. Can they prove, by the way that they have worked, that they have found all the possible ways?

ASSESSMENT CHECKPOINT Use question **5** to assess how well children can use known facts, including bonds to 10 and doubles, to find missing numbers. Can they explain which method they have used, and why?

ANSWERS Answers for the **Practice** part of the lesson can be found in the *Power Maths* online subscription.

Reflect

WAYS OF WORKING Independent thinking

IN FOCUS The **Reflect** part of the lesson requires children to use different strategies to add combinations of three numbers. Four different totals are possible. If pairs of children have both worked out all four totals correctly, they will find that they both made the biggest number possible. Encourage them to make the link between the size of the numbers they added and the size of the total: adding the three biggest numbers makes the biggest total; adding the three smallest numbers makes the smallest total.

ASSESSMENT CHECKPOINT Assess whether children can explain clearly how they found the different totals. Did they find some combinations easier than others?

ANSWERS Answers for the **Reflect** part of the lesson can be found in the *Power Maths* online subscription.

After the lesson ⏸

- Did children use known number facts or count on in 1s to calculate the answers?
- Did children understand the focus of the lesson, or did they just concentrate on finding the answer to each question?

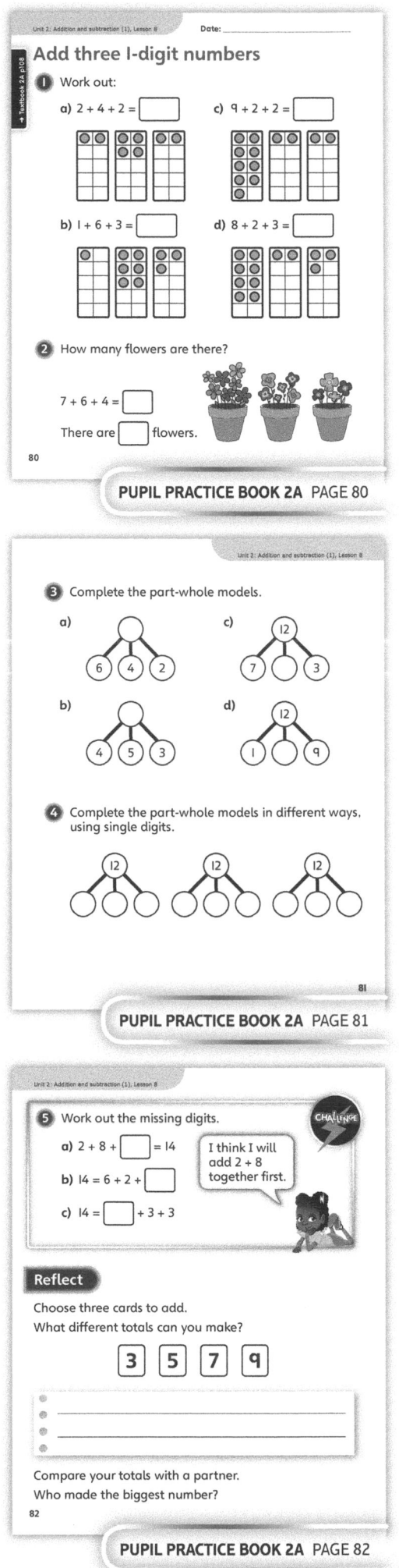

Add to the next 10

Learning focus

In this lesson, children learn to add from a 2-digit number to the next multiple of 10, in order to prepare for learning how to do additions that bridge 10s.

Before you teach

- Can children count on in 1s from a given 2-digit number?
- Are children able to identify the 10s on a 100 square?
- Can children identify 10s on a number line from 0 to 100?

NATIONAL CURRICULUM LINKS

Number – addition and subtraction

Add and subtract numbers using concrete objects, pictorial representations, and mentally, including: a 2-digit number and 1s.

ASSESSING MASTERY

Children can add from a 2-digit number to the next multiple of 10.

COMMON MISCONCEPTIONS

Children may rely on counting on in 1s rather than using jumps or recalling number facts. Ask:
- *How many to make 10? Does this remind you of a fact you have seen before?*

STRENGTHENING UNDERSTANDING

Work on securing children's knowledge of the number bonds to 10 in order for them to engage fully with this lesson. Some children may need to have a list of bonds to 10 to hand to support them in their working.

GOING DEEPER

Challenge children to explore and explain patterns such as 8 + 2 = 10, 18 + 2 = 20, 28 + 2 = 30, 38 + 2 = 40, and so on.

KEY LANGUAGE

In lesson: add, more, ten, next ten, number bond

Other language to be used by the teacher: altogether

STRUCTURES AND REPRESENTATIONS

Ten frames, number line, part-whole model

RESOURCES

Mandatory: ten frames

 In the eTextbook of this lesson, you will find interactive links to a selection of teaching tools.

Quick recap

Children take it in turns to roll three dice. Add the scores together to find the total. Repeat this a few times. Discuss the order that children choose to add the numbers in.

Discover

WAYS OF WORKING Pair work

ASK

• Question ➊: What do you notice about each of the stacks?
• Question ➊ a): How many cups are in each stack?

IN FOCUS In order to count the cups, children will count in 10s and then 1s. They then carry out an addition to the next multiple of 10.

PRACTICAL TIPS Make towers using safe stacking resources, for example paper cups or wooden blocks. Explore the 4-3-2-1 pyramid pattern that makes the number 10.

ANSWERS

Question ➊ a): There are 23 cups in total.

Question ➊ b): 7
$$23 + 7 = 30$$

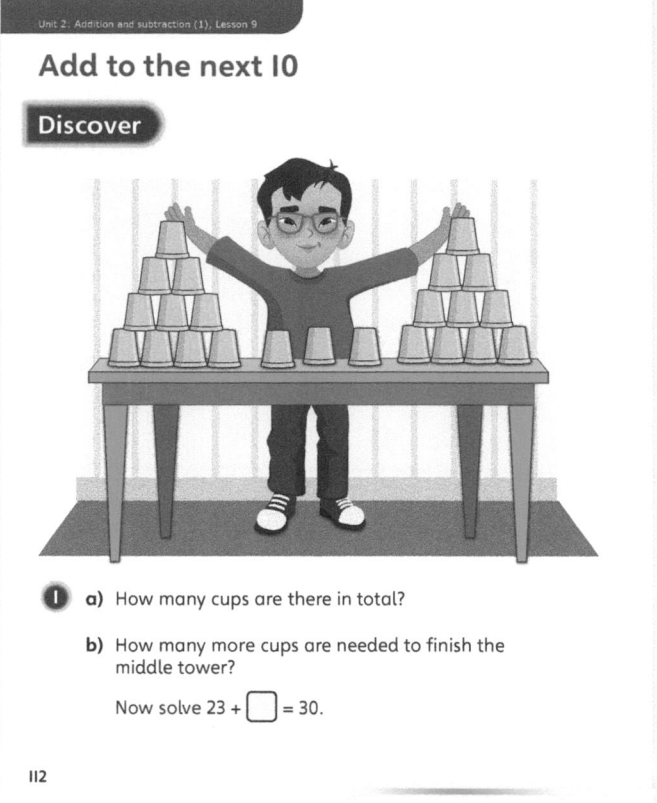

Add to the next 10

Discover

a) How many cups are there in total?

b) How many more cups are needed to finish the middle tower?

Now solve $23 + \boxed{} = 30$.

112

PUPIL TEXTBOOK 2A PAGE 112

Share

WAYS OF WORKING Whole class teacher led

ASK

• Question ➊ a): *Can you describe a good way to count the total number?*
• Question ➊ a): *Why do we often count in 10s then 1s?*
• Question ➊ b): *Can you predict how many it will take to finish the final tower?*
• Question ➊ b): *How does this help solve the calculation?*

IN FOCUS Children first explore the pattern of 10s and 1s in a 2-digit number. They then identify which number to add on to reach the next multiple of 10. Look out for children using their knowledge of number bonds to do this.

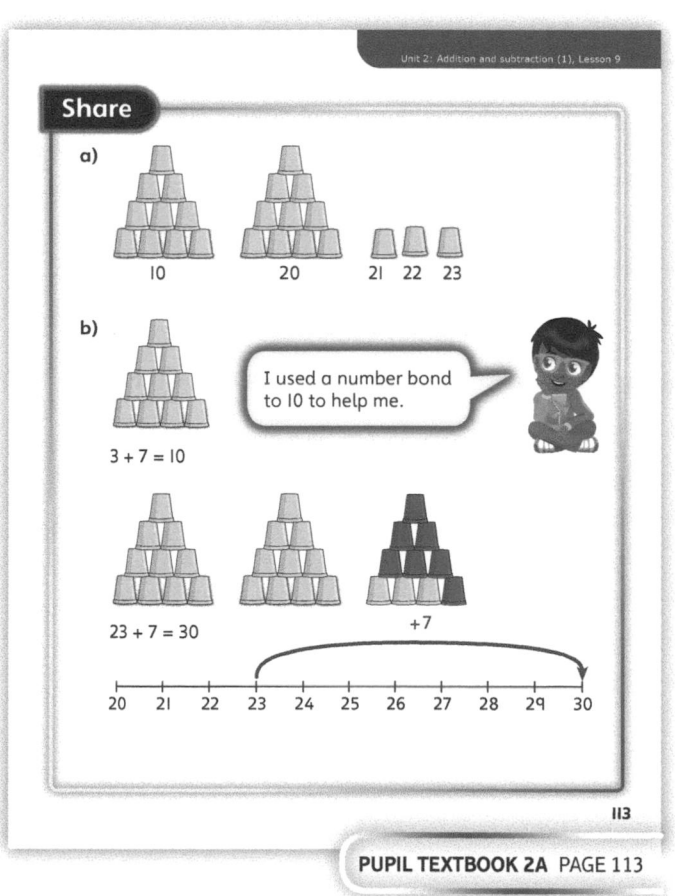

Share

a)

10   20   21 22 23

b)

I used a number bond to 10 to help me.

3 + 7 = 10

23 + 7 = 30   +7

20 21 22 23 24 25 26 27 28 29 30

113

PUPIL TEXTBOOK 2A PAGE 113

Think together

WAYS OF WORKING Whole class teacher led (I do, We do, You do)

ASK

- Question **1**: *What is the same and what is different about the calculations in this question?*
- Question **2**: *If you know 3 + 7 = 10, what else do you know?*
- Question **3**: *Do you recognise any of these models? How would you explain them to someone who had never seen them before*

IN FOCUS In questions **1** and **2**, children are using ten frames to support them with adding to the next multiple of 10. Question **3** requires them to interpret several different representations of adding to the next 10.

STRENGTHEN Provide ten frames for children to use to support their working throughout.

DEEPEN Challenge children to investigate patterns in calculations such as 2 + 8, 12 + 8, 22 + 8, and so on. What do they notice?

ASSESSMENT CHECKPOINT Use question **2** to assess whether children can correctly identify which number to add to reach the next multiple of 10. Do they notice the pattern in their answers and make the link to number bonds?

ANSWERS

Question **1**: 8 + **2** = 10
28 + **2** = 30

Question **2**: 3 + **7** = 10
33 + **7** = 30
43 + **7** = 50
73 + **7** = 80

Question **3** a): 14 + **6** = 20

Question **3** b): 55 + **5** = 60

Question **3** c): 86 + **4** = 90

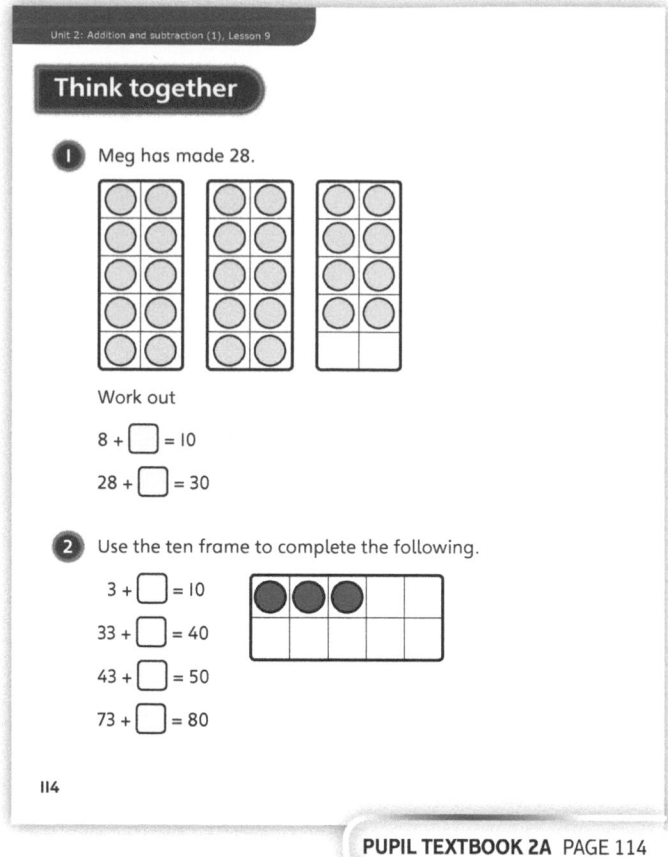

PUPIL TEXTBOOK 2A PAGE 114

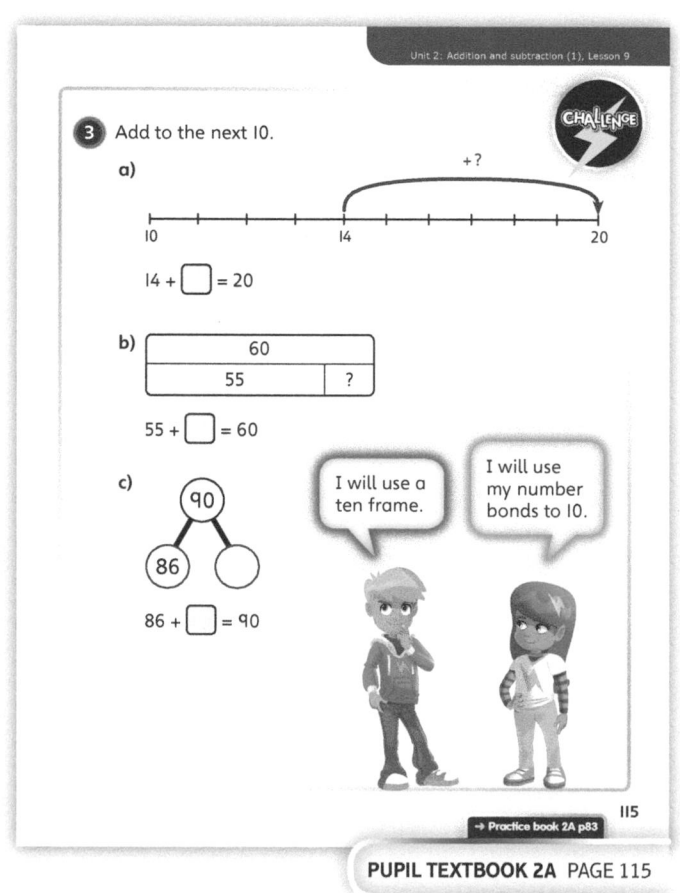

PUPIL TEXTBOOK 2A PAGE 115

Practice

WAYS OF WORKING Independent thinking

IN FOCUS In questions ❶ and ❷, children are using ten frames to support them in adding to the next 10. In question ❸ they use the number line representation. Question ❹ has a variety of more abstract questions about adding to the next ten. In question ❺, children explore variations in calculations by finding the missing 2-digit numbers. In question ❻, children add three parts and begin bridging a ten.

STRENGTHEN Children can use ten frames to support them in visualising the 10s and 1s that make up a number, and how many 1s it will take to make the next 10.

DEEPEN Use question ❻ as a prompt to explore bridging a 10. Can children draw other part-whole models like the one in the question? Can they explain how they know their answer will bridge a 10?

ASSESSMENT CHECKPOINT Use question ❹ to assess whether children can use number bonds to find the number that bridges to the next multiple of 10.

ANSWERS Answers for the **Practice** part of the lesson can be found in the *Power Maths* online subscription.

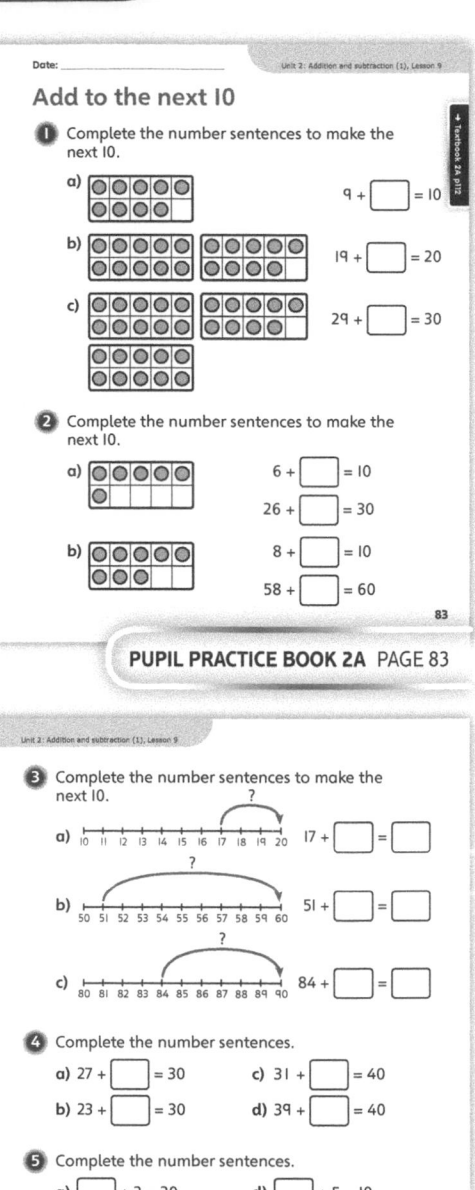

PUPIL PRACTICE BOOK 2A PAGE 83

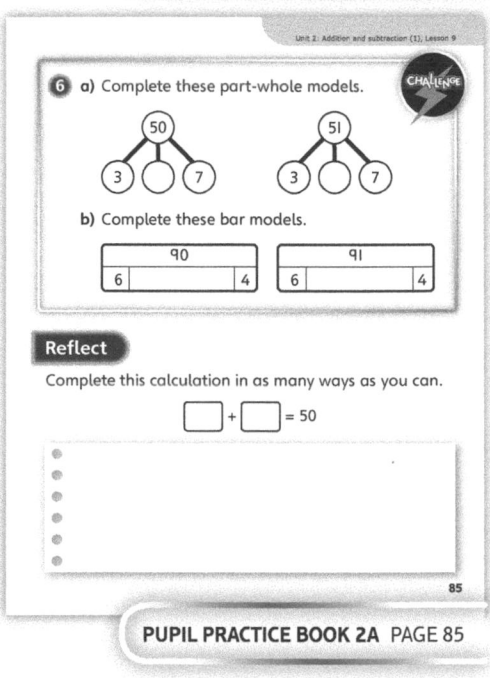

PUPIL PRACTICE BOOK 2A PAGE 84

Reflect

WAYS OF WORKING Independent thinking

IN FOCUS The **Reflect** part of the lesson requires children to explore questions that have many possible answers.

ASSESSMENT CHECKPOINT Can children justify how to add a number to reach the next multiple of 10?

ANSWERS Answers for the **Reflect** part of the lesson can be found in the *Power Maths* online subscription.

After the lesson

- If you choose a 2-digit number at random, can children say what to add to it to make the next multiple of 10?

PUPIL PRACTICE BOOK 2A PAGE 85

Add across a 10

Learning focus

In this lesson, children will add 2-digit and 1-digit numbers together, with the focus on bridging 10. They will represent this using ten frames and jumps on a number line.

Before you teach

- Are children confident about adding and subtracting 10s?
- Are children secure using the 'make 10' strategy?

NATIONAL CURRICULUM LINKS

Number – addition and subtraction

Add and subtract numbers using concrete objects, pictorial representations, and mentally, including: a 2-digit number and 1s.

Solve problems with addition and subtraction: using concrete objects and pictorial representations, including those involving numbers, quantities and measures.

ASSESSING MASTERY

Children can make links to previous lessons and concepts and recognise that 10 ones is the same as 1 ten. Children understand how this is represented when writing numbers in digits and when using resources to make different numbers.

COMMON MISCONCEPTIONS

Children may count on in 1s from a starting number, rather than using their understanding of number. Ask:
- *Is there a more efficient way to calculate the answer? Can you use what you know about numbers to calculate the answer?*

Children may try to partition the 1-digit number in a way that is not helpful to aid their mental calculation. Ask:
- *How does the way that you have partitioned the 1-digit number help with your mental calculation?*

STRENGTHENING UNDERSTANDING

Children should be encouraged to use the 'make 10' strategy, as this will help them understand how to use number bonds to calculate the answers; using ten frames throughout the lesson will highlight this strategy. Using counters of one colour to make the starting number, and counters of another colour to make 10 and then count on to the final answer, will clearly highlight the partitioning of the 1-digit number.

GOING DEEPER

Encourage children to record calculations in as many different ways as possible, to show the different methods that they could use to find the answers, and to consolidate learning from previous lessons in this unit.

KEY LANGUAGE

In lesson: adding, +, =, digit, total, whole, parts, number sentence

Other language to be used by the teacher: ones (1s), tens (10s), partition

STRUCTURES AND REPRESENTATIONS

Ten frames, number line

RESOURCES

Mandatory: counters (double-sided or two colours, if possible), ten frames

Optional: 1–100 number line, pictures of completed ten frames

 In the eTextbook of this lesson, you will find interactive links to a selection of teaching tools.

Quick recap 🔁

Count on together in 1s from a given 2-digit number. Say any 2-digit number at random and ask children to count on in 1s in time while you clap the rhythm.

Discover

Unit 2: Addition and subtraction (1), Lesson 10

WAYS OF WORKING Pair work

ASK

- Question ❶ a): *What number does four complete ten frames represent?*
- Question ❶ b): *How many more do you need to fill the ten frame?*
- Question ❶ b): *How could you represent the different parts on a part-whole model?*

IN FOCUS Question ❶ a) requires children to make a 2-digit number (45) using counters and ten frames. This will help them see how many complete 10s and how many 1s make 45. They then build on this in question ❶ b), which is designed to encourage them to use the 'make 10' strategy to add on a further 7. It is important that children see that five of the additional chairs can be used to fill the fifth ten frame from question ❶ a).

PRACTICAL TIPS Provide base 10 equipment for children to model the groups of 10s and some 1s. Encourage them to describe what happens when they add enough 1s to create a new group of 10.

ANSWERS

Question ❶ a): There are 45 stacked chairs.

Question ❶ b): 45 + 7 = 45 + 5 + 2 = 52
There are 52 chairs in total.

Add across a 10

Discover

❶ a) There are 10 chairs in each full stack.
How many chairs are stacked?
Show this number on ten frames.

b) How many chairs are there in total?

116

PUPIL TEXTBOOK 2A PAGE 116

Share

WAYS OF WORKING Whole class teacher led

ASK

- Question ❶ b): *Would it be useful to partition the 1-digit number?*
- Question ❶ b): *Why is it better not to count on in 1s seven times?*

IN FOCUS Make sure children understand what the different representations used in question ❶ b) each show, and how they are linked. In particular, children need to understand how the ten frames can help them see how to partition the 1-digit number.

Share

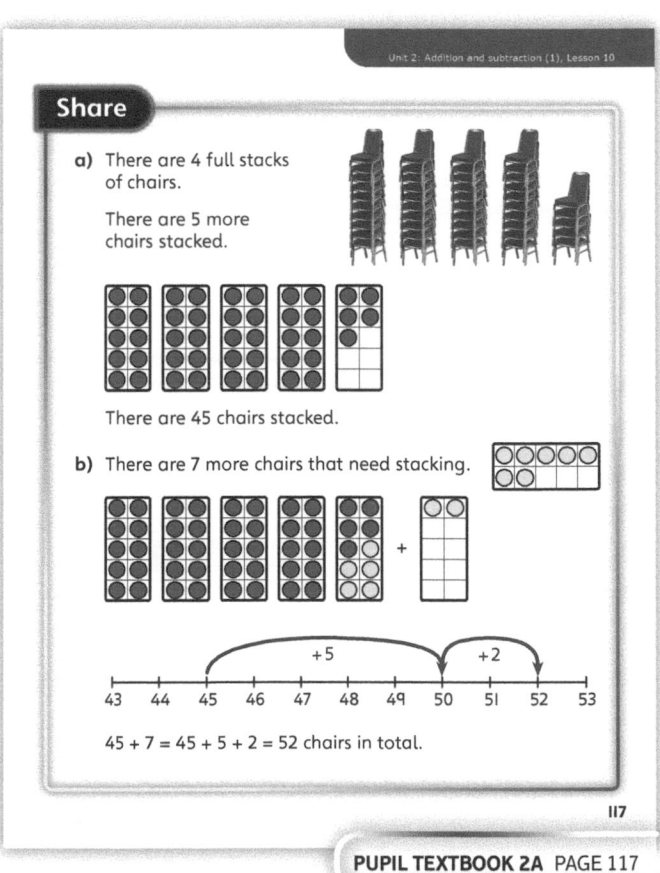

a) There are 4 full stacks of chairs.

There are 5 more chairs stacked.

There are 45 chairs stacked.

b) There are 7 more chairs that need stacking.

45 + 7 = 45 + 5 + 2 = 52 chairs in total.

117

PUPIL TEXTBOOK 2A PAGE 117

155

Think together

Think together

Unit 2: Addition and subtraction (1), Lesson 10

WAYS OF WORKING Whole class teacher led (I do, We do, You do)

ASK

- Question **1**: *How can you decide what to partition the 1-digit number into?*

IN FOCUS Make sure children understand why the 1s in questions **1** and **2** are being partitioned. Encourage them to explore this using different representations, such as ten frames and number lines.

Question **3** tests children's recall of bonds to 10 and number facts. Discuss with children what working out they needed to do and why, or if there were some answers they knew just by looking.

STRENGTHEN Children can continue to use ten frames to see how many more they need to add on to the starting number to make 10. To save time, you could give them pictures of completed ten frames to represent the 10s in the starting number, but children should make the 1s themselves before adding on the 1-digit number.

DEEPEN Challenge children to write several number sentences to represent a calculation, partitioning the 1-digit number in as many different ways as possible (for example, 42 + 9 can be written as 42 + 8 + 1 = 51, but also 42 + 7 + 2 = 51, 42 + 6 + 3 = 51, and so on). Doing this will both highlight and increase children's flexibility of number.

ASSESSMENT CHECKPOINT Ask children if they can explain why they have chosen to partition the 1-digit number as they have. This will allow you to assess their recall of number bonds and their ability to apply them to different concepts.

ANSWERS

Question **1**: 27 + 8 = 27 + 3 + 5 = 35
There are 35 stars in total.

Question **2**: 34 + 8 = 34 + 6 + 2 = 42

Question **3**: 42 + 6 = 48 (not greater that 50)
43 + 6 = 49 (not greater than 50)
47 + 6 = 47 + 3 + 3 = 53 (greater than 50)
49 + 6 = 49 + 1 + 5 = 55 (greater than 50)

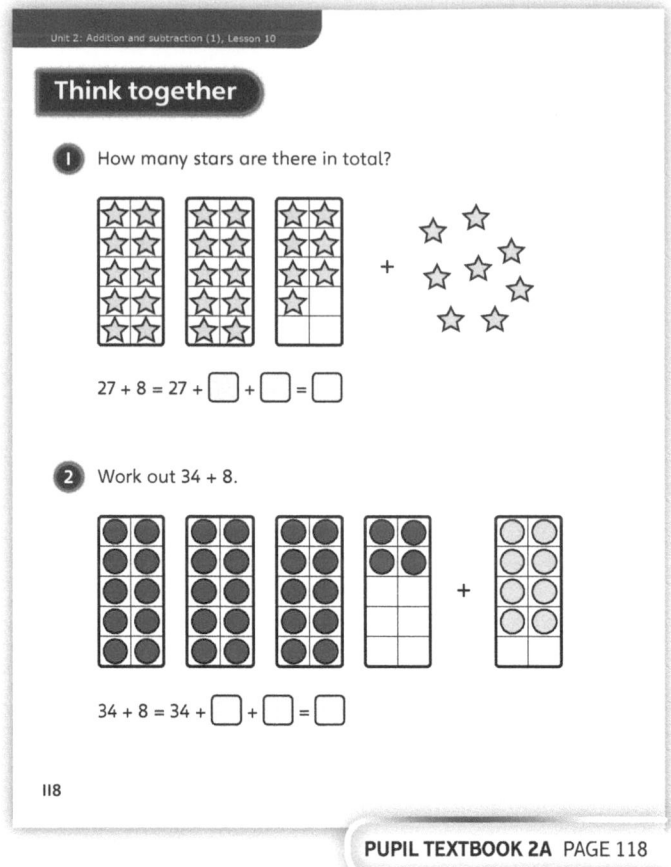

1 How many stars are there in total?

$$27 + 8 = 27 + \square + \square = \square$$

2 Work out 34 + 8.

$$34 + 8 = 34 + \square + \square = \square$$

118

PUPIL TEXTBOOK 2A PAGE 118

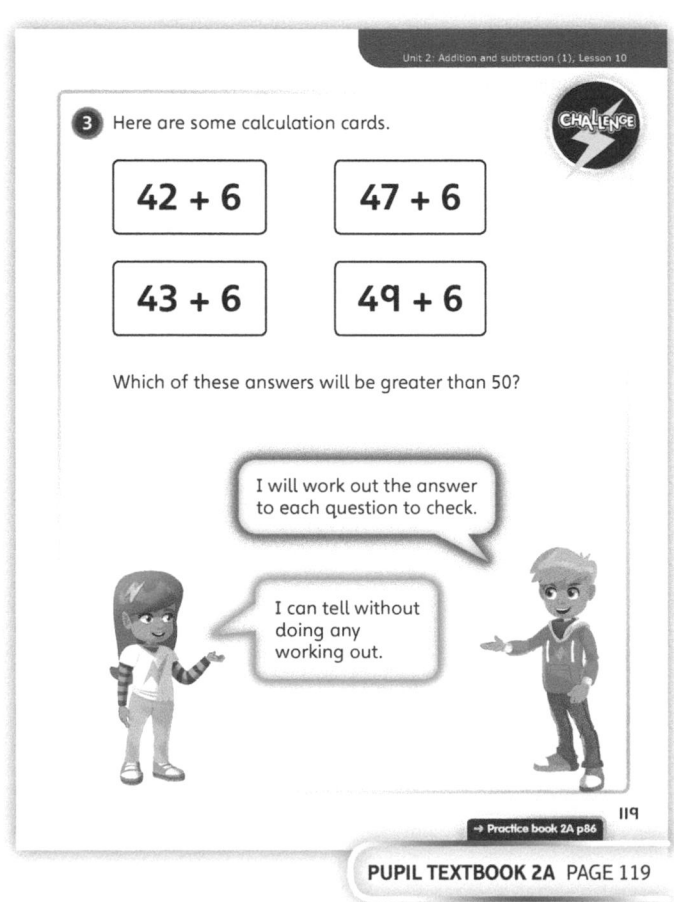

3 Here are some calculation cards.

CHALLENGE

| 42 + 6 | 47 + 6 |
| 43 + 6 | 49 + 6 |

Which of these answers will be greater than 50?

I will work out the answer to each question to check.

I can tell without doing any working out.

119

→ Practice book 2A p86

PUPIL TEXTBOOK 2A PAGE 119

Practice

WAYS OF WORKING Independent thinking

IN FOCUS The **Practice** questions provide progressively less scaffolding for children. Question ❸ models the calculation on ten frames and on a number line. Question ❹ requires children to decide which number to partition and then to draw this as jumps on a number line.

THINK DIFFERENTLY Question ❺ requires children to identify the best way to partition the 1-digit number in a calculation and to write this as a number sentence before finding the answer.

STRENGTHEN Encourage children to continue to use ten frames to help them see what to partition the 1-digit number into. They might be able to draw a pictorial representation of the ten frames as a step on from using concrete resources to make the number.

DEEPEN Present children with a number sentence with one unknown, such as $67 + \boxed{} = 75$. Can children apply what they have learnt in this lesson to calculate the unknown? Can they work backwards to see that $75 - 5 = 70$ and $70 - 3 = 67$, and therefore that the missing number is 8? Help children make the link by encouraging them to show their working on a number line.

ASSESSMENT CHECKPOINT Assess whether children can explain why they have chosen to partition the 1-digit number in the way that they have and how this helps their mental calculation. Ensure children are not simply counting on in 1s to find the solution.

ANSWERS Answers for the **Practice** part of the lesson can be found in the *Power Maths* online subscription.

Reflect

WAYS OF WORKING Independent thinking

IN FOCUS By the **Reflect** part of the lesson, children should be able to visualise the ten frames and identify that 6 needs to be added to 54 to 'make 10'; the remaining 2 can then be added to 60. If any children struggle with this, ask: *If you visualise the number 54, what can you see? How can this help you with your mental calculation?*

ASSESSMENT CHECKPOINT Check to see what explanations children give and how comfortable they are at describing them.

ANSWERS Answers for the **Reflect** part of the lesson can be found in the *Power Maths* online subscription.

After the lesson

- Were children able to identify the correct way to partition the 1-digit number?
- Did children continue to count in 1s on their fingers to calculate the answer?
- Do they understand the limitations this method will have when they start working with larger numbers?

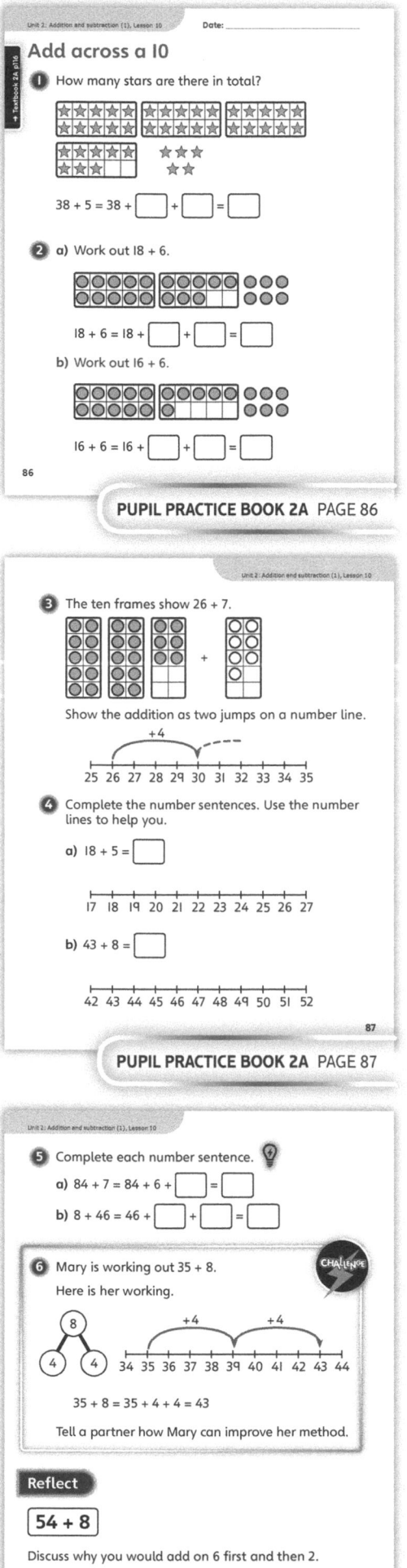

PUPIL PRACTICE BOOK 2A PAGE 86

PUPIL PRACTICE BOOK 2A PAGE 87

PUPIL PRACTICE BOOK 2A PAGE 88

Subtract across a 10

Learning focus

In this lesson, children learn how to subtract by crossing 10. The range stays within 20, so the subtractions cross the 10 barrier. This is an important part of the progression of this sequence of lessons.

Before you teach

- Do children know how to make the numbers 11 to 20 on ten frames?
- Can children find the numbers 11 to 20 on a number line?
- Can they partition the numbers 11 to 20 into 10s and 1s?

NATIONAL CURRICULUM LINKS

Number –addition and subtraction

Add and subtract numbers using concrete objects, pictorial representations, and mentally, including: two 2-digit numbers.

Solve problems with addition and subtraction: applying their increasing knowledge of mental and written methods.

ASSESSING MASTERY

Children can solve subtractions that involve crossing 10, for example 13 – 4.

COMMON MISCONCEPTIONS

Some children may still rely on counting back in 1s, rather than using number bonds and finding complements to 10. Ask:
- *Could we use fewer steps to find the answer? Are there any facts that you already know?*

STRENGTHENING UNDERSTANDING

Ask children throughout to model each question using ten frames to support their understanding of the calculations and the steps that are taken to solve them.

GOING DEEPER

Look for children who are able to link this learning to number facts within 20, for example doubles or near doubles. Can they explain how they will use this to find answers?

KEY LANGUAGE

In lesson: subtract, ten, tens (10s), ones (1s)

STRUCTURES AND REPRESENTATIONS

Ten frames

RESOURCES

Mandatory: ten frames

Optional: items that can be counted individually or grouped in 10s, such as pencils

 In the eTextbook of this lesson, you will find interactive links to a selection of teaching tools.

Quick recap

Show children a number line from 40 to 50 and ask children to count on from 40 to 50. Then ask children to count back from 50 to 40.

Repeat for different intervals between two multiples of 10.

Discover

Unit 2: Addition and subtraction (1), Lesson 11

Subtract across a 10

Discover

WAYS OF WORKING Pair work

ASK

• Question ① a): *How many pencils does the teacher have in total?*
• Question ① b): *How many pencils will the teacher take out of the box?*

IN FOCUS Children will need to demonstrate how they can count in 10s and 1s to make a teen number and can then subtract a given 1-digit number from this, crossing 10 to find the answer.

PRACTICAL TIPS Provide real boxes of pencils or similar that children can practise counting. Reinforce for children the idea that they can find the total without needing to count every individual item in a group of 10.

ANSWERS

Question ① a): There are 13 pencils.

Question ① b): There are 8 pencils left.

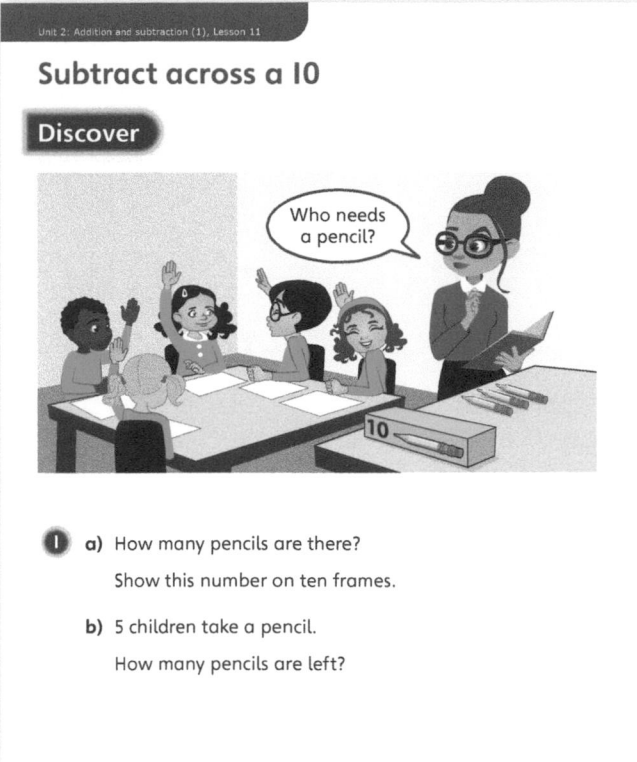

① a) How many pencils are there?
Show this number on ten frames.

b) 5 children take a pencil.
How many pencils are left?

120

PUPIL TEXTBOOK 2A PAGE 120

Share

WAYS OF WORKING Whole class teacher led

ASK

• Question ① a): *Explain how you can tell quickly that this shows 13.*
• Question ① a): *Why is this split up into – 3 and then – 2. Is that the same as – 5?*

IN FOCUS Children first explore the composition of numbers between 11 and 20 and understand that they are made of a 10 and some 1s. In question ① b), they carry out a subtraction of a 1-digit number from a teen number, that involves crossing 10. This is modelled by removing counters from the ten frame in two steps. The first step gets us to 10 and the second step takes us beyond 10 to reach the answer.

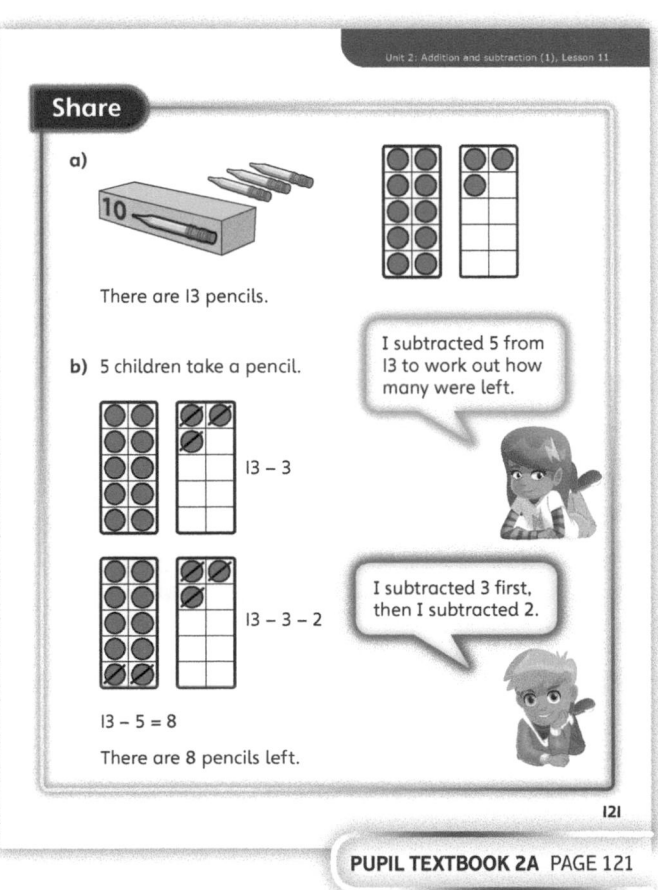

Unit 2: Addition and subtraction (1), Lesson 11

Share

a)

There are 13 pencils.

b) 5 children take a pencil.

13 – 3

13 – 3 – 2

13 – 5 = 8
There are 8 pencils left.

I subtracted 5 from 13 to work out how many were left.

I subtracted 3 first, then I subtracted 2.

121

PUPIL TEXTBOOK 2A PAGE 121

Think together

WAYS OF WORKING Whole class teacher led (I do, We do, You do)

ASK

- Questions **1** and **2**: *How can you decide what to subtract first?*
- Questions **1** and **2**: *Can you work out the next part to take away?*
- Question **3**: *How could the ten frame help you?*

IN FOCUS Questions **1** and **2** both involve children doing two steps to solve subtractions that involve crossing 10. Make sure they understand that the role of the first step is to get to 10. Question **3** uses different representations to demonstrate subtracting through 10. Can children explain how the calculation is shown on the ten frame and on the number line?

STRENGTHEN Ask children to subtract from a given teens number to get to 10. Then ask them to use this to subtract from the same teens number to get to 9. For example:

13 – 3 = 10

So, 13 – 3 – 1 = 9

So, 13 – 4 = 9

Discuss the link between these two calculations.

DEEPEN Challenge children to investigate how they can solve 14 – 7 either as 14 – 4 – 3 or by knowing double 7. Which do they prefer? Why?

ASSESSMENT CHECKPOINT Use question **3** to assess whether children can understand and use a variety of representations of subtractions crossing 10 and can explain each step that is shown.

ANSWERS

Question **1**: 12 – 2 – 3 = 7; 12 – 5 = 7

Question **2**: 14 – 8 = 6; 14 – 4 – 4 = 6

Question **3**: 13 – 7 = 6

Dan needs to make another jump of 4.

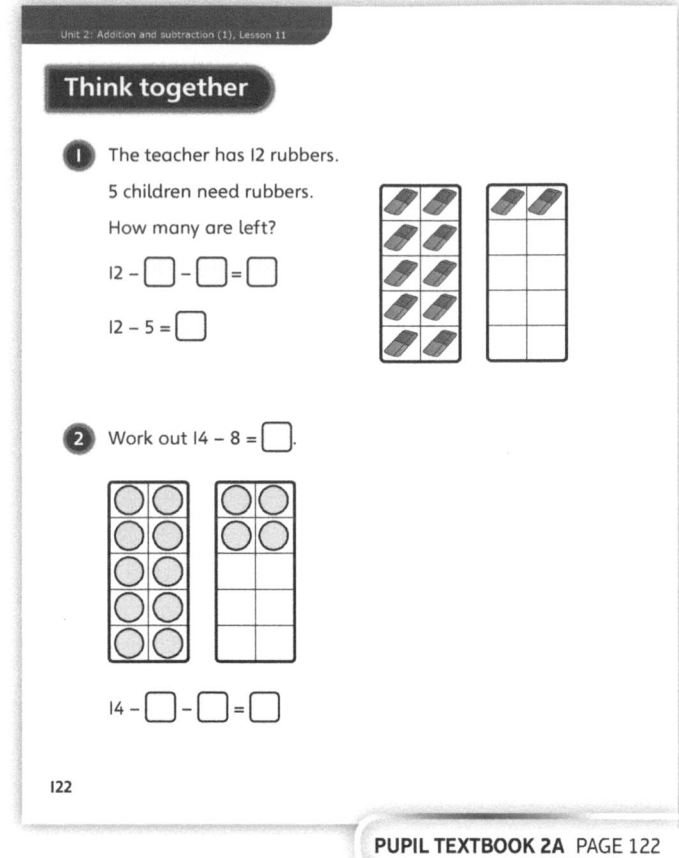

Think together

1 The teacher has 12 rubbers.

5 children need rubbers.

How many are left?

12 – ☐ – ☐ = ☐

12 – 5 = ☐

2 Work out 14 – 8 = ☐.

14 – ☐ – ☐ = ☐

122

PUPIL TEXTBOOK 2A PAGE 122

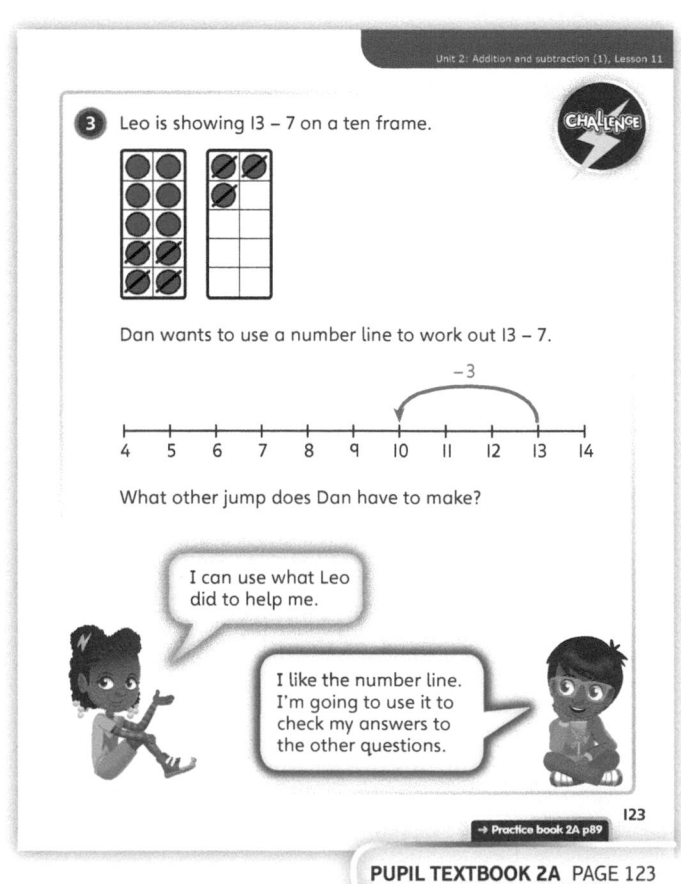

3 Leo is showing 13 – 7 on a ten frame.

CHALLENGE

Dan wants to use a number line to work out 13 – 7.

– 3

4 5 6 7 8 9 10 11 12 13 14

What other jump does Dan have to make?

I can use what Leo did to help me.

I like the number line. I'm going to use it to check my answers to the other questions.

123

→ Practice book 2A p89

PUPIL TEXTBOOK 2A PAGE 123

Practice

WAYS OF WORKING Independent thinking

IN FOCUS In questions ❶, ❷ and ❸, children are carrying out subtractions that cross 10 with scaffolding from ten frames. In questions ❹ and ❺ scaffolding comes from the number line. Question ❻ requires children to solve abstract calculations, where no representation or scaffolding are given.

STRENGTHEN Provide ten frames for children to use to support their working throughout.

DEEPEN Challenge children to investigate the following sequence: 11 – 2, 12 – 3, 13 – 4, 14 – 5. Can they continue and explain the pattern?

ASSESSMENT CHECKPOINT Question ❸ will allow you to assess whether children can identify the steps needed to subtract a 1-digit number, crossing 10.

ANSWERS Answers for the **Practice** part of the lesson can be found in the *Power Maths* online subscription.

Reflect

WAYS OF WORKING Independent thinking

IN FOCUS The **Reflect** part of the lesson gives children the chance to explain the possible methods they could use to solve a subtraction which crosses 10.

ASSESSMENT CHECKPOINT Assess whether children can justify their decisions about the methods they have used.

ANSWERS Answers for the **Reflect** part of the lesson can be found in the *Power Maths* online subscription.

After the lesson

- Are children able to solve missing number problems such as 14 – ☐ = 8?
- Are there any underlying misconceptions that should be addressed before the next lesson?

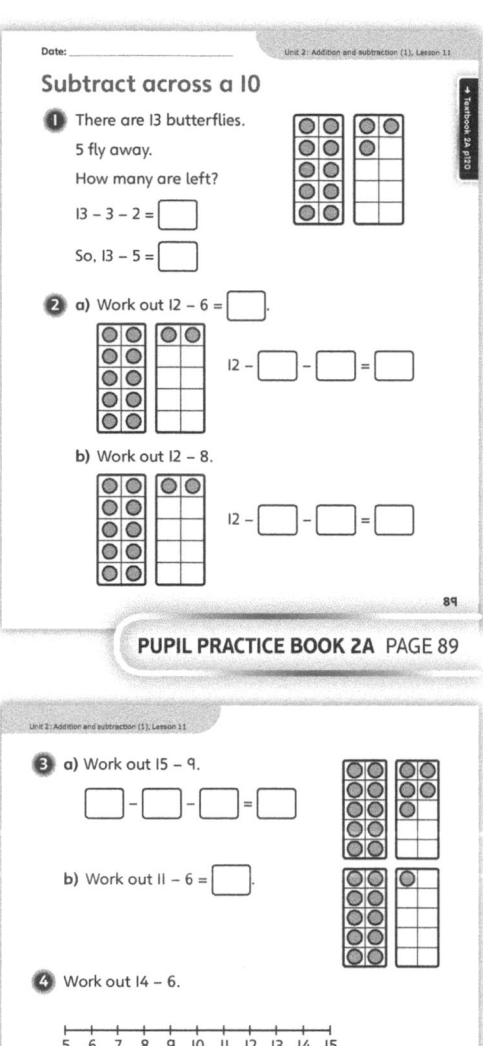

PUPIL PRACTICE BOOK 2A PAGE 89

PUPIL PRACTICE BOOK 2A PAGE 90

PUPIL PRACTICE BOOK 2A PAGE 91

161

Subtract from a 10

Learning focus

In this lesson, children begin to build up their understanding of how to subtract from 2-digit numbers more flexibly, by subtracting from a given multiple of 10.

Before you teach

- Can children count in 10s to 100?
- Have children had the opportunity to rehearse the language of 10s? For example, 3 tens is 30, 30 is 3 tens.
- Can children count back from a given multiple of 10 to the previous multiple of 10?

NATIONAL CURRICULUM LINKS

Number – addition and subtraction

Add and subtract numbers using concrete objects, pictorial representations, and mentally, including: a 2-digit number and 1s.

Solve problems with addition and subtraction: using concrete objects and pictorial representations, including those involving numbers, quantities and measures.

ASSESSING MASTERY

Children can subtract a 1-digit number from any given multiple of 10 up to 100.

COMMON MISCONCEPTIONS

Some children may still rely on counting back in 1s to subtract, rather than using their knowledge of number bonds. Ask:
- *Is there a quicker way? Does this remind you of any facts that you have seen before?*

STRENGTHENING UNDERSTANDING

Provide ten frames for children to use throughout, to help them visualise each calculation and build understanding.

GOING DEEPER

Challenge children to explore, predict, continue and explain the patterns that they find in questions that show variation, for example 10 – 9, 20 – 9, 30 – 9, 40 – 9, and so on.

KEY LANGUAGE

In lesson: tens (10s), ones (1s), subtract

STRUCTURES AND REPRESENTATIONS

Ten frames, number line

RESOURCES

Mandatory: ten frames, counters

 In the eTextbook of this lesson, you will find interactive links to a selection of teaching tools.

Quick recap

Count down together from 20 to 0.

Discover

WAYS OF WORKING Pair work

ASK

- Question **1** a): *How many pencils are in each pack?*
- Question **1** a): *How could you use number bonds to help you answer the question?*
- Question **1** b): *Could you represent the number using any other equipment or drawings?*

IN FOCUS This activity introduces subtraction related to number bonds within 10. Children are then asked to subtract a 1-digt number from three different multiples of 10.

PRACTICAL TIPS Ensure that children start by representing each of the multiples of 10 on ten frames.

ANSWERS

Question **1** a): There are 7 pencils left in the pack.

Question **1** b): There will be 27 pens left. There will be 47 rubbers left.

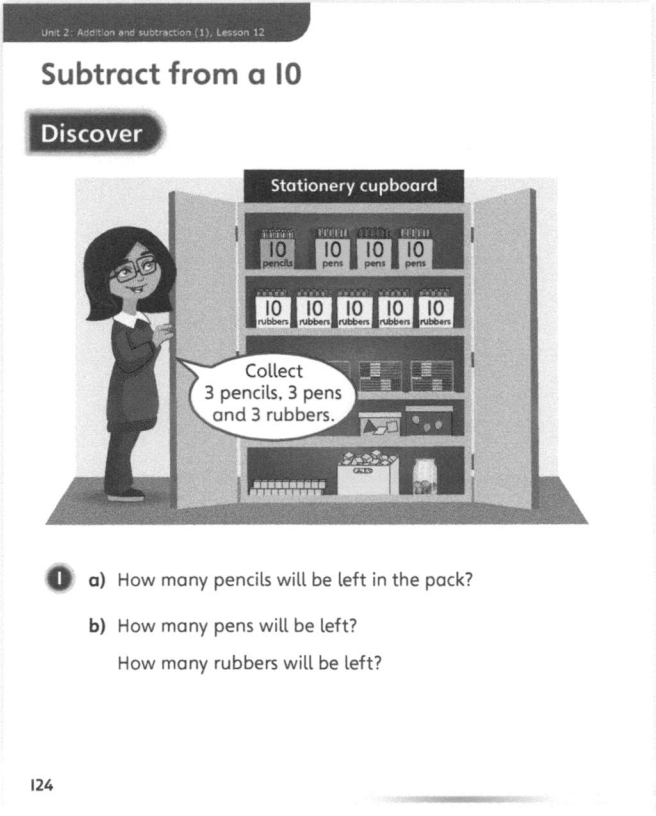

Subtract from a 10

Discover

Collect 3 pencils, 3 pens and 3 rubbers.

1 a) How many pencils will be left in the pack?

b) How many pens will be left?
How many rubbers will be left?

124

PUPIL TEXTBOOK 2A PAGE 124

Share

WAYS OF WORKING Whole class teacher led

ASK

- Question **1** a): *How can you use number bonds to work out 10 − 3?*
- Question **1** b): *What is the same and what is different about each calculation?*

IN FOCUS In question **1** a), children are encouraged to use number bonds to subtract a 1-digit number from 10. In question **1** b), they subtract the same 1-digit number from other multiples of 10, using their knowledge of number bonds to calculate efficiently.

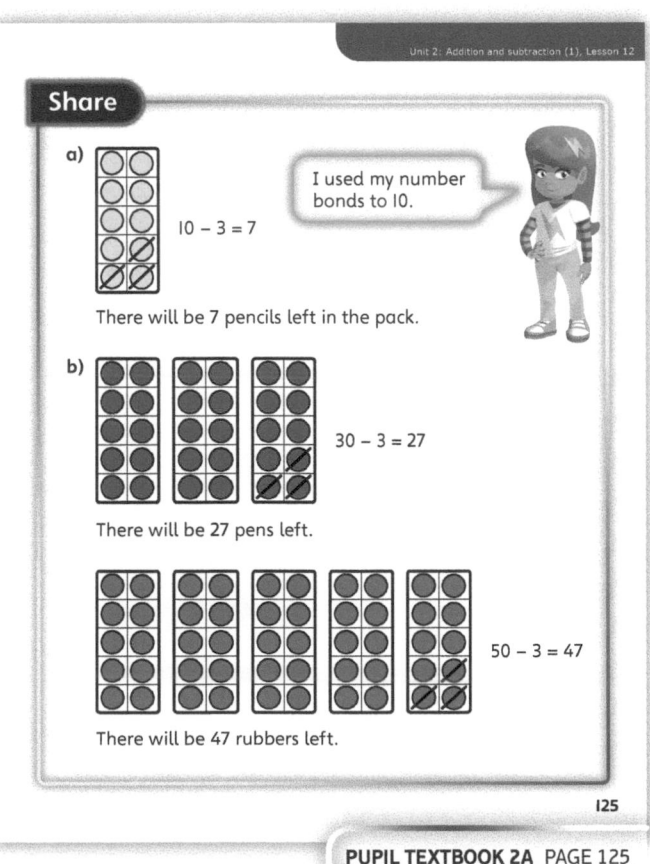

Share

a)

$10 - 3 = 7$

I used my number bonds to 10.

There will be 7 pencils left in the pack.

b)

$30 - 3 = 27$

There will be 27 pens left.

$50 - 3 = 47$

There will be 47 rubbers left.

125

PUPIL TEXTBOOK 2A PAGE 125

Think together

WAYS OF WORKING Whole class teacher led (I do, We do, You do)

ASK

- Question **1**: *How many counters are there in total in the ten frames? How many counters do you need to cross out?*
- Question **2**: *What numbers are in the word problem? Where do those numbers go in the number sentence?*
- Question **3**: *How can number bonds help you to find the missing numbers?*

IN FOCUS In question **1** children are using number bonds to subtract a given 1-digit number from a multiple of 10. Question **2** presents subtractions as word problems in a real-life context. The subtractions in question **3** are represented on a number line.

STRENGTHEN Check regularly that children are focusing on the use of number bonds to 10 in order to calculate efficiently.

DEEPEN Challenge children to explore patterns in a series of related subtractions such as 10 – 5, 20 – 5, 30 – 5, 40 – 5, and so on. What do they notice?

ASSESSMENT CHECKPOINT Use question **2** to assess whether children can interpret subtractions in the context of a word problem and can use number bonds to find answers efficiently.

ANSWERS

Question **1**: 40 – 8 = 32

Question **2** a): 30 – 5 = 25
There are 25 apples left on the tree.

Question **2** b): 30 – 7 = 23
7 apples have fallen off in total.

Question **3**: 10 – 2 = **8**
50 – 2 = **48**
80 – 2 = 78

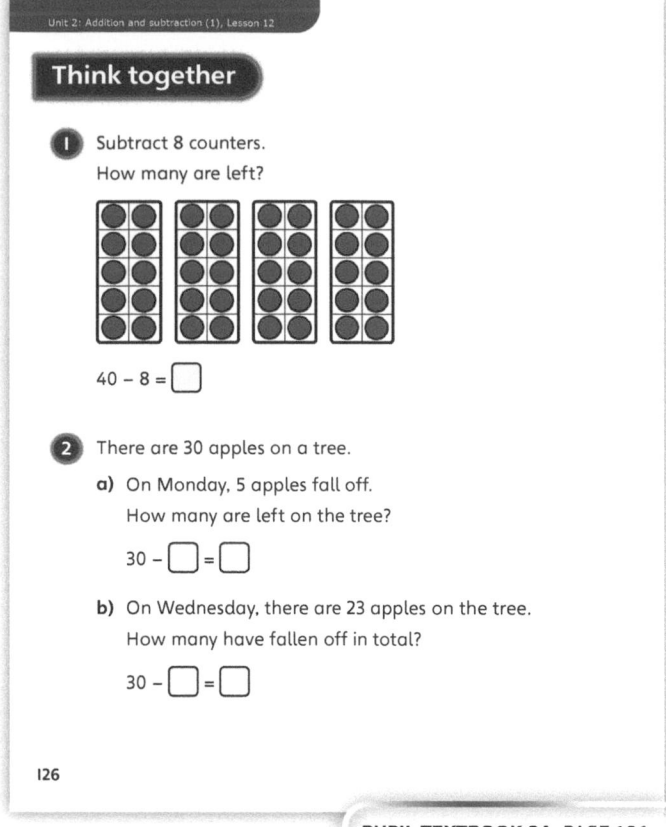

PUPIL TEXTBOOK 2A PAGE 126

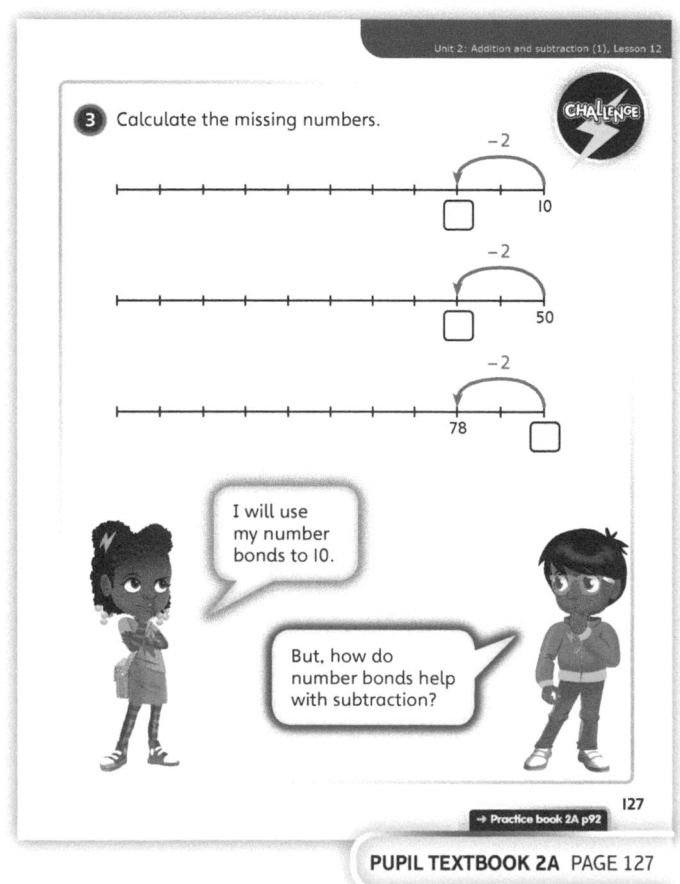

PUPIL TEXTBOOK 2A PAGE 127

Practice

WAYS OF WORKING Independent thinking

IN FOCUS In question ①, the subtractions are represented using ten frames. Questions ② and ③ require children to use their knowledge of number bonds to subtract. In questions ④, ⑤ and ⑥ there are patterns in the variation in the exercises. These patterns can prompt further insight into the number system and offer opportunities to revisit as a whole class discussion.

STRENGTHEN Ensure that children use ten frames or other equipment to secure their knowledge of number bonds to 10.

DEEPEN Ask children to explore and explain the patterns that they see in the answers to the missing number problems in question ③.

ASSESSMENT CHECKPOINT Use question ② to assess whether children can make efficient use of number bonds to subtract a series of 1-digit numbers from given multiples of 10.

ANSWERS Answers for the **Practice** part of the lesson can be found in the *Power Maths* online subscription.

Unit 2: Addition and subtraction (1), Lesson 12   Date: _____

Subtract from a 10

① Complete the number sentences.

a) $10 - 5 = \boxed{}$

b) $20 - 5 = \boxed{}$

c) $30 - 5 = \boxed{}$

② Complete the number sentences.

a) $10 - 1 = \boxed{}$

b) $10 - 2 = \boxed{}$

c) $10 - 3 = \boxed{}$

d) $10 - 4 = \boxed{}$

e) $50 - 1 = \boxed{}$

f) $30 - 3 = \boxed{}$

g) $90 - 4 = \boxed{}$

h) $70 - 1 = \boxed{}$

i) $40 - 2 = \boxed{}$

j) $80 - 3 = \boxed{}$

92

PUPIL PRACTICE BOOK 2A PAGE 92

---

③ Find the missing numbers.

a) $\boxed{}$ —6→ 10 $10 - 6 = \boxed{}$

b) $\boxed{}$ —6→ 60 $60 - 6 = \boxed{}$

④ Complete the number sentences.

a) $40 - \boxed{} = 32$

b) $40 - \boxed{} = 38$

c) $100 - \boxed{} = 96$

d) $100 - \boxed{} = 91$

⑤ Complete the number sentences.

a) $\boxed{} - 7 = 23$

b) $\boxed{} - 7 = 33$

c) $\boxed{} - 7 = 53$

d) $\boxed{} - 7 = 83$

e) $\boxed{} - 9 = 21$

f) $\boxed{} - 9 = 41$

g) $\boxed{} - 9 = 61$

h) $\boxed{} - 9 = 81$

93

PUPIL PRACTICE BOOK 2A PAGE 93

Reflect

WAYS OF WORKING Pair work

IN FOCUS The **Reflect** part of the lesson requires children to use number bonds to subtract efficiently in order to complete the grid. Look for children who are working systematically in order to identify and explain patterns.

ASSESSMENT CHECKPOINT Can children draw on their learning to explain the patterns that they encounter in the columns of the grid?

ANSWERS Answers for the **Reflect** part of the lesson can be found in the *Power Maths* online subscription.

Unit 2: Addition and subtraction (1), Lesson 12

⑥ Draw a line to join each matching pair.   CHALLENGE

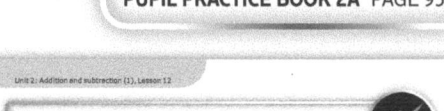

| 50 + 7 | 50 + 4 | 50 + 3 | 60 + 4 | 60 + 6 | 60 + 2 |

| 70 – 6 | 70 – 8 | 70 – 4 | 60 – 3 | 60 – 7 | 60 – 6 |

Reflect

Work out the subtractions to complete the grid.
Talk about any patterns you notice with a partner.

10 – 0	10 – 2	10 – 4	10 – 6	10 – 8	10 – 10
20 – 0	20 – 2	20 – 4	20 – 6	20 – 8	20 – 10
40 – 0	40 – 2	40 – 4	40 – 6	40 – 8	40 – 10
60 – 0	60 – 2	60 – 4	60 – 6	60 – 8	60 – 10
80 – 0	80 – 2	80 – 4	80 – 6	80 – 8	80 – 10
100 – 0	100 – 2	100 – 4	100 – 6	100 – 8	100 – 10

94

PUPIL PRACTICE BOOK 2A PAGE 94

After the lesson ⏸

- If you choose a multiple of 10 at random, can children subtract 5 from that number?

Subtract a 1-digit number from a 2-digit number – across 10

Learning focus

In this lesson, children build on their subtraction skills to subtract across a multiple of 10.

Before you teach

- Do children know how to partition 2-digit numbers into 10s and 1s?
- Can children locate 2-digit numbers on 0 to 100 number lines?
- Are children able to subtract from a 2-digit number to the previous multiple of 10?

NATIONAL CURRICULUM LINKS

Number – addition and subtraction

Add and subtract numbers using concrete objects, pictorial representations, and mentally, including: a 2-digit number and 1s.

Solve problems with addition and subtraction: using concrete objects and pictorial representations, including those involving numbers, quantities and measures.

ASSESSING MASTERY

Children can subtract across a multiple of 10 by making two jumps. For example, representing 32 – 5 as 32 – 2 – 3.

COMMON MISCONCEPTIONS

Some children may still be relying on counting back in 1s, rather than using known number facts. Ask:
- *How many in this jump? Do you know a quicker way to find out?*

STRENGTHENING UNDERSTANDING

Provide ten frames for children to use to support their understanding throughout.

GOING DEEPER

Link this learning to number bonds within 20. For example, if we know 12 – 3 = 9, we also know 22 – 3, 32 – 3, 62 – 3, and so on. Can children give any other examples?

KEY LANGUAGE

In lesson: subtract, ten, tens (10s), ones (1s)

STRUCTURES AND REPRESENTATIONS

Ten frames

RESOURCES

Mandatory: ten frames

Optional: place value equipment

 In the eTextbook of this lesson, you will find interactive links to a selection of teaching tools.

Quick recap

Ask children to count back in 1s from a given 2-digit number. Say a 2-digit number at random and ask children to count back in 1s in time to you clapping the rhythm, until they reach 0.

Discover

Unit 2: Addition and subtraction (1), Lesson 13

Subtract a 1-digit number from a 2-digit number – across 10

WAYS OF WORKING Pair work

ASK

- Question **1**: *What information do you think is written on the board?*
- Question **1**: *What other questions could we ask about this information?*

IN FOCUS Children explore the composition of 2-digit numbers as 10s and 1s. They then consider a question that involves subtracting a 1-digit number from a 2-digit number, crossing the multiple of 10.

PRACTICAL TIPS Ask children to make the number using ten frames or other place value equipment.

ANSWERS

Question **1** a):

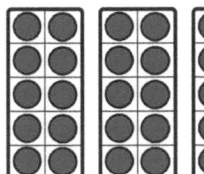 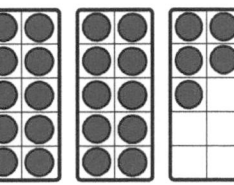

Question **1** b): There are 29 children in Class B today.

Discover

1 a) Show 35 on ten frames.

   b) Today 6 children are away.
     How many children are in Class B today?

128

PUPIL TEXTBOOK 2A PAGE 128

Share

WAYS OF WORKING Whole class teacher led

ASK

- Question **1** a): *How does this show the 10s and the 1s?*
- Question **1** b): *What do you notice about the steps in this calculation?*
- Question **1** b): *Does this remind you of subtractions that cross 10?*

IN FOCUS The ten frame is used to show the composition of a 2-digit number that is made up of 10s and 1s. Counters are then removed from the ten frame in two steps to show a subtraction that crosses the previous multiple of 10. There is a jump back to the multiple of 10 and then another jump to the answer.

Share

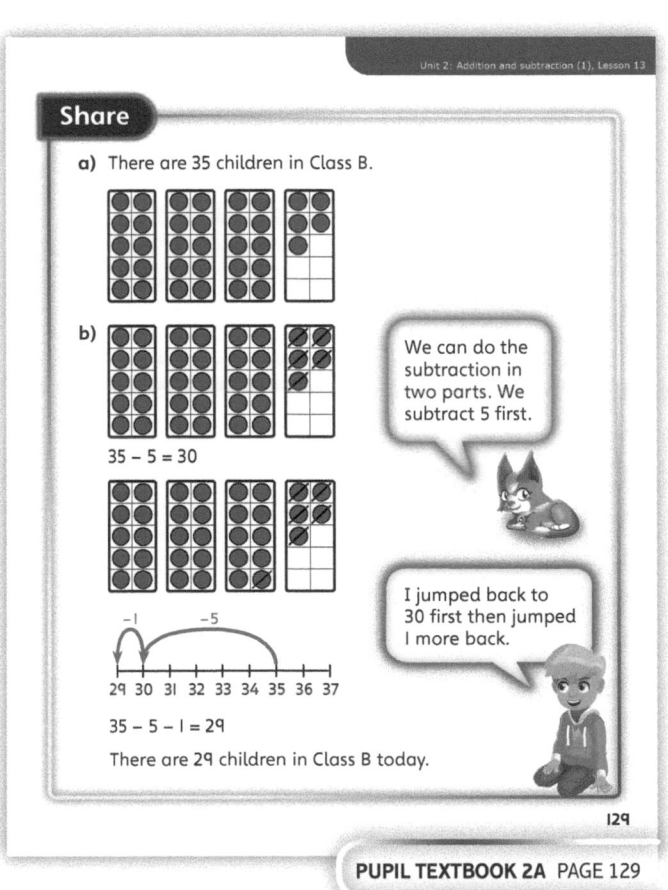

a) There are 35 children in Class B.

b) 35 – 5 = 30

We can do the subtraction in two parts. We subtract 5 first.

35 – 5 – 1 = 29

I jumped back to 30 first then jumped 1 more back.

There are 29 children in Class B today.

129

PUPIL TEXTBOOK 2A PAGE 129

Think together

Whole class teacher led (I do, We do, You do)

ASK

- Question ❶: *What decisions are you making when deciding how to partition the 1-digit number?*
- Question ❷: *Are you ensuring that you are being efficient with your calculations?*

IN FOCUS Question ❷ requires children to subtract 7 from three different starting numbers. This means they need to partition 7 in three different ways, subtracting a different number each time to get back to a multiple of 10. Discuss what is the same and what is different about the three calculations, and help children see the link between the starting number and how they partition the 7.

STRENGTHEN Children who find the concept difficult should continue to work using ten frames as well.

DEEPEN Ask children to explain in writing what is the same and what is different about the two calculations in question ❸. This will allow them to demonstrate their deepening understanding using correct mathematical vocabulary.

ASSESSMENT CHECKPOINT Assess whether children can explain the strategy that they are using, how this uses their knowledge of number bonds and why their method is efficient.

ANSWERS

Question ❶: $24 - 4 - 2 = 18$
$24 - 6 = 18$

Question ❷ a): $34 - 7 = 34 - 4 - 3 = 27$

Question ❷ b): $46 - 7 = 46 - 6 - 1 = 39$

Question ❷ c): $55 - 7 = 55 - 5 - 2 = 48$

Question ❸: Both calculations use the digits 3, 7 and 4 and both involve a subtraction.
In the first, you start at 37 and jump back 4 to get 33. This does not cross a 10.
In the second, you start at 34 and jump back 7 to give 27. This calculation crosses a 10.

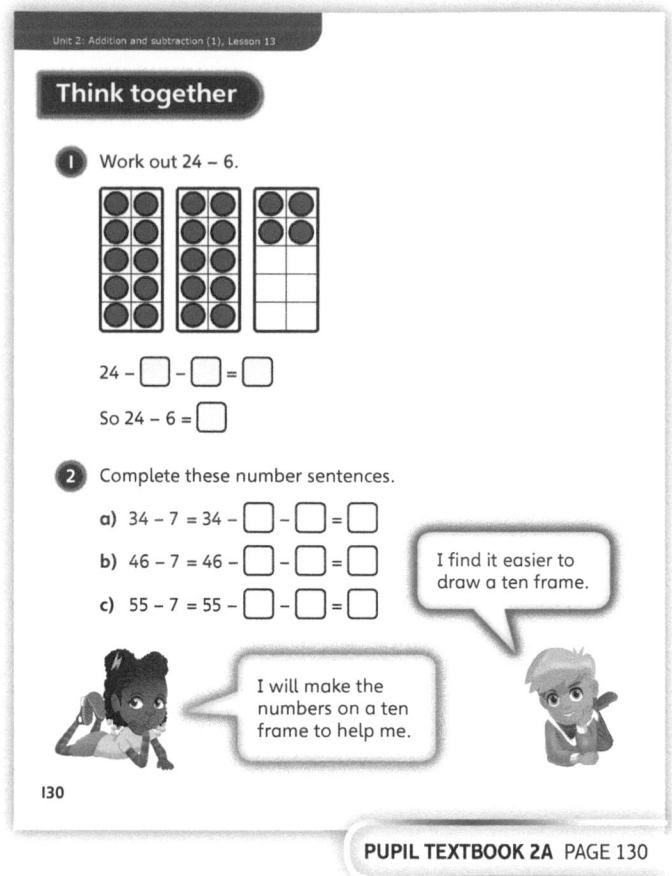

PUPIL TEXTBOOK 2A PAGE 130

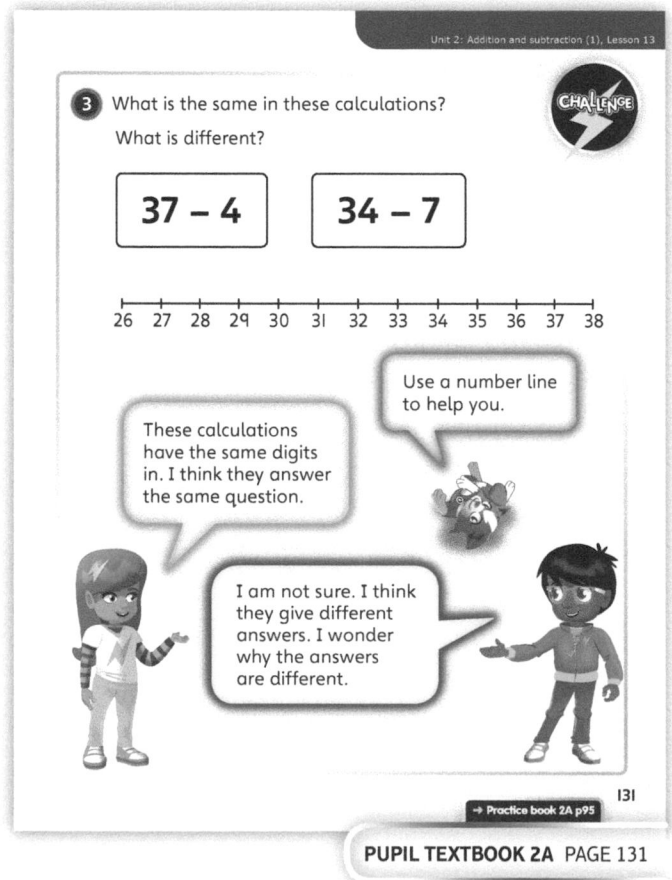

PUPIL TEXTBOOK 2A PAGE 131

Practice

WAYS OF WORKING Independent thinking

IN FOCUS In questions **1** and **2**, children are subtracting using ten frames to support their reasoning. In question **4** the scaffolding comes from the number line. Question **5** presents subtractions that are not scaffolded with a visual representation. In question **6**, children explore the patterns that they find in a series of subtractions that have one variation each time.

STRENGTHEN Children who find the concept difficult should continue to work with concrete resources. If they are struggling to identify how many 1s they need to subtract to get to the nearest 10, give them practice by showing them number cards and asking them to quickly recall how many more 1s the number has than the nearest 10.

DEEPEN Challenge children to explore a method for subtracting 9. For example, *first I subtract 10, then I add 1 back on*. Can they explain why this works and why it can be an efficient method to use?

ASSESSMENT CHECKPOINT Use question **5** to assess whether children can apply the skills they have learnt to subtract across a multiple of 10. What representation do they choose to use? Are they using it accurately?

ANSWERS Answers for the **Practice** part of the lesson can be found in the *Power Maths* online subscription.

Reflect

WAYS OF WORKING Independent thinking

IN FOCUS In the **Reflect** part of the lesson, discuss what Hanna has partitioned 9 into and how effective this is. Hanna gets the correct answer, but children should see that it is not an effective strategy. Hanna's method is likely to result in mistakes as the nearest 10 has not been made and therefore the mental calculation is more difficult.

ASSESSMENT CHECKPOINT Assess children's ability to reflect on the method used, including its efficiency and effectiveness. Can they suggest a more suitable method?

ANSWERS Answers for the **Practice** part of the lesson can be found in the *Power Maths* online subscription.

After the lesson

- Can children describe what is the same and what is different about each of these calculations; 33 – 3, 33 – 1, 31 – 3?
- Do children need more practise of recalling number bonds and facts?

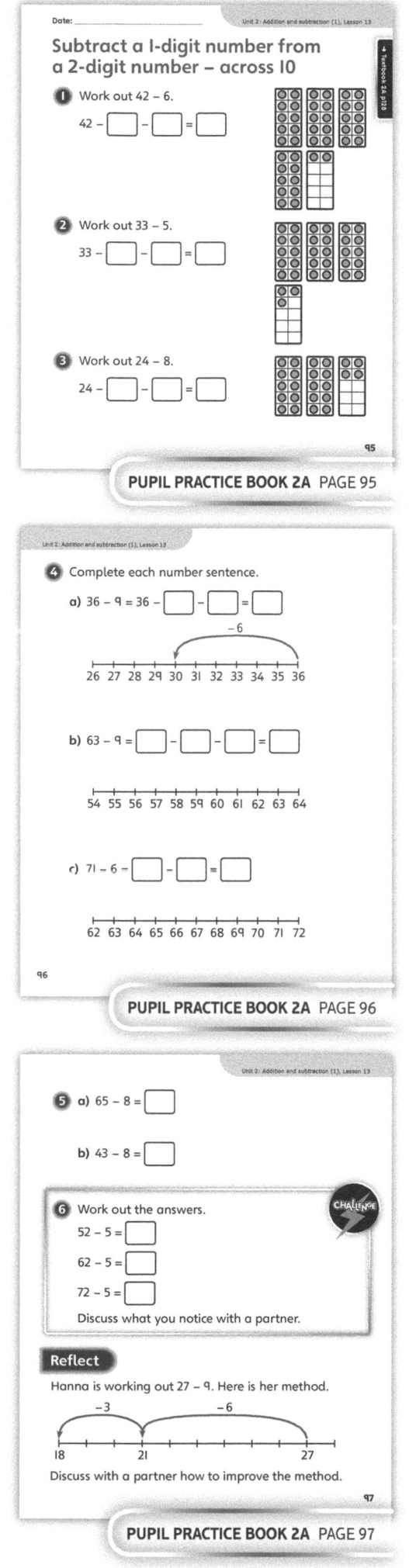

PUPIL PRACTICE BOOK 2A PAGE 95

PUPIL PRACTICE BOOK 2A PAGE 96

PUPIL PRACTICE BOOK 2A PAGE 97

End of unit check

Don't forget the unit assessment grid in your *Power Maths* online subscription.

WAYS OF WORKING Group work teacher led

IN FOCUS These questions assess the links that children can make between numbers and how they can use known facts to calculate unknowns. Children must understand the links between addition and subtraction and when each operation is needed as a result of the context that they are provided with.

Think!

WAYS OF WORKING Pairs or small groups

IN FOCUS This question allows children to think about the methods they would use for different addition and subtraction calculations.

Children should be able to use mathematical language to describe the method they would use, rather than just stating what the correct answer is.

ANSWERS AND COMMENTARY Children who have mastered this unit will be able to relate each number in a calculation to what it represents within a given context. Children will be able to use a variety of manipulatives to represent addition and subtraction. Children will also be fluent at recalling and applying their number bonds within 20 to addition and subtraction calculations.

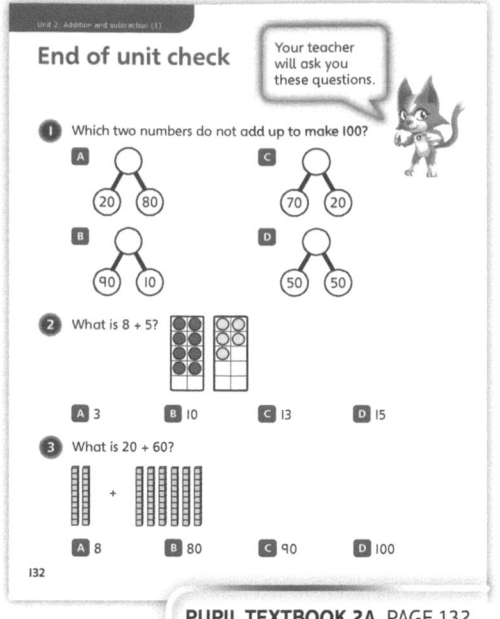

PUPIL TEXTBOOK 2A PAGE 132

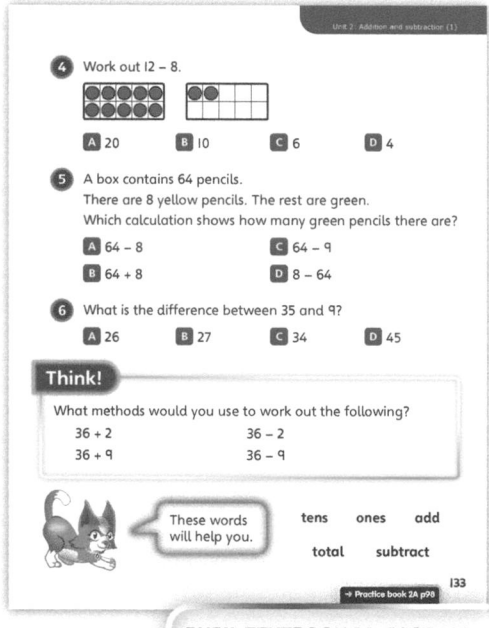

PUPIL TEXTBOOK 2A PAGE 133

Q	A	WRONG ANSWERS AND MISCONCEPTIONS	STRENGTHENING UNDERSTANDING
1	C	Wrong answers suggest the child should practise number bonds to 10.	Children who are still finding it difficult to identify which operation links with the context that they are provided with should be given the opportunity to create their own maths stories for the different operations, or sort problems into either subtraction or addition categories.

Children should also practise writing all possible calculations from a complete part-whole diagram (fact family), understanding that the equals sign can occur in different places and means 'is the same as'. |
2	C	A suggests they have subtracted rather than added.	
3	B	A suggests they have not understood that they should add 10s rather than 1s. C and D show basic number bond errors.	
4	D	A suggests they have added rather than subtracted. B or C suggests they need to work on using ten frames to subtract.	
5	A	C and D both suggest that the child does not understand the concept of difference and its link to subtraction.	
6	A	B shows subtraction error. C and D show lack of understanding of relationship between subtraction and find the difference.	

My journal

WAYS OF WORKING Independent thinking

ANSWERS AND COMMENTARY Children will probably choose to count on and back to add and subtract 2. To add and subtract 9, they may choose to add and subtract 10 and then adjust or partition 9 in a way that they see fit. Encourage children to use the mathematical vocabulary they have been provided with to talk about the methods they used for each calculation and the reasoning behind their decisions.

Power check

WAYS OF WORKING Independent thinking

ASK

- *How well do you know your number bonds within 20?*
- *How confidently do you use these within additions and subtractions?*

Power puzzle

WAYS OF WORKING Pair work or small group

IN FOCUS It is possible to complete the **Power puzzle** by working sequentially through the problems. Children should be encouraged to record what they have calculated as they work, rather than trying to store this mentally. Children should also be encouraged not to complete the final stage until they have calculated all shapes.

ANSWERS AND COMMENTARY Children should count on from 63 to 68 to calculate 63 + **5** = 68. This could be shown on a number line.

Children then must apply what they have just calculated to form the calculation 5 + ☐ = 100. They should then count back from 100 to calculate 5 + **95** = 100.

22 + 50 = ☐ Children could start from 22 and count on 5 jumps of 10 (22, 32, 42, 52, 62, **72**), or start at 50, count on 2 and then count on 2 jumps of 10 (this will be explored as a strategy in the next unit, but some children may already see how this is possible using their understanding of commutativity, place value and the 100 square).

72 – 5 = ☐ Children should partition 5 into 2 and 3 and count back from 72 to 70 and then 70 to 67. This could be displayed on a number line.

Shapes from smallest to largest: star (5), triangle (67), square (72), diamond (95).

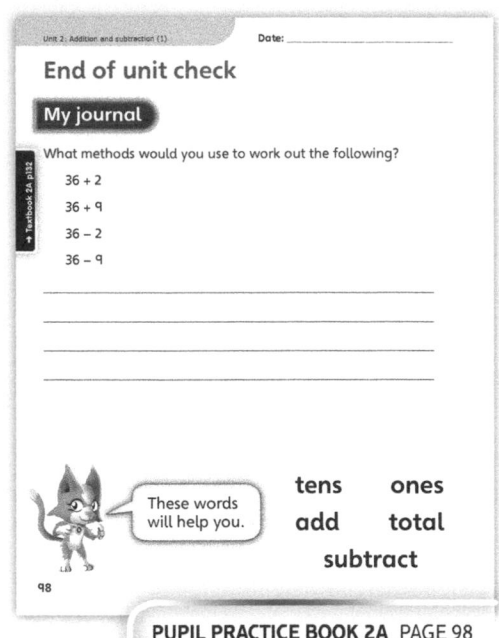

PUPIL PRACTICE BOOK 2A PAGE 98

PUPIL PRACTICE BOOK 2A PAGE 99

After the unit ⏸

- Could children spot the deliberate mistakes? Were they able to identify if they make similar mistakes in their own work?
- What are the most common misconceptions that still remain in the class that must be addressed before the content becomes more difficult in Unit 3?

Strengthen and **Deepen** activities for this unit can be found in the *Power Maths* online subscription.

Unit 3
Addition and subtraction ②

Mastery Expert tip! 'It is important for children to make links between different areas of maths and understand the progression of what they are learning. Children should view this unit as an opportunity for them to build on what they have previously learnt and extend their thinking to new content.'

Don't forget to watch the Unit 3 video!

WHY THIS UNIT IS IMPORTANT

This unit directly builds upon what children have learnt in Unit 2 and provides an opportunity for what they have understood to be applied to larger numbers. Children must understand this progression and see the importance of applying this learning, rather than seeing larger numbers as a different area of maths and therefore using inefficient methods as a result.

Within this unit children will progress to addition and subtraction involving two 2-digit numbers, again representing the steps within these calculations visually with different resources.

Children continue to use known number facts within mental calculations and use their understanding of the inverse as a way to check their calculations. The final stage of children's learning allows the bar model to be used to represent a word problem, to allow children to self-identify the operation needed to complete the calculation.

WHERE THIS UNIT FITS

→ Unit 2: Addition and subtraction (1)
→ **Unit 3: Addition and subtraction (2)**
→ Unit 4: Properties of shapes

This unit directly builds upon the content of the previous unit and exposes children to addition and subtraction involving two 2-digit numbers, where the 10s boundary is crossed and regrouping and exchange is required.

Before they start this unit, it is expected that children:

- know how to partition 2-digit numbers into 10s and 1s and place these onto a place value table
- understand the value of each digit within a 2-digit number and how these will change as a result of addition and subtraction
- know number bonds within 10 and 20 and how to apply these to mental addition and subtraction calculations.

ASSESSING MASTERY

Children who have mastered this unit will be able to differentiate between addition and subtraction problems, understanding how to represent the numbers provided within each context in different ways, using different resources. Children should be flexible with the methods they can use to calculate different problems depending on the problem's complexity in order to work efficiently.

COMMON MISCONCEPTIONS	STRENGTHENING UNDERSTANDING	GOING DEEPER
Children may confuse addition and subtraction and find it difficult to change between the two operations.	Children should have an increased opportunity to make numbers using different resources, and then be required to add and take away different quantities explaining the changes that occur to the number.	Use subtraction to check addition calculations and vice versa.
Children may interpret problems correctly but then make calculation errors resulting in a wrong answer.	Providing children with number facts that are useful to complete equations will increase the likelihood of children understanding the importance of memorisation.	Challenge children to calculate unknown quantities in the most efficient way possible and justify how they have been efficient and accurate.

UNIT STARTER PAGES

Use these pages to introduce the unit focus to children as a whole class. You can use the characters to explore different ways of working.

STRUCTURES AND REPRESENTATIONS

Part-whole model: This model helps children understand that a number can be partitioned in different ways and how changing the ways that it is partitioned suits different mental calculations.

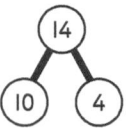

Number line: This model helps children to count on and back from a number. It is used to show jumps of different amounts to help children understand the 'make 10' strategy and the steps completed in the column method.

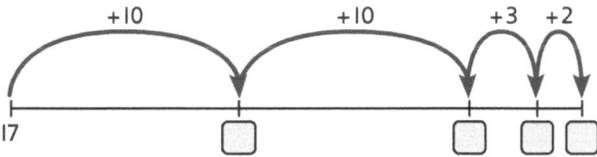

Bar model: This model is used to show parts and wholes. It can also be used to show comparisons between two numbers.

84	
50	?

KEY LANGUAGE

There is some key language that children will need to know as a part of the learning in this unit.

→ part, whole and part-whole, partition

→ add, added, plus, total, altogether, sum, calculation, +

→ count, count on, count back, left, difference

→ subtract, take away, minus, −

→ exchange, compare, greater than, less than, more, less, >, <, regroup, represent, difference

→ ones (1s), tens (10s), 10 more, 10 less, place value, column, 1-digit number, 2-digit number, bar model

→ number sentence, number bonds, known fact, fact family

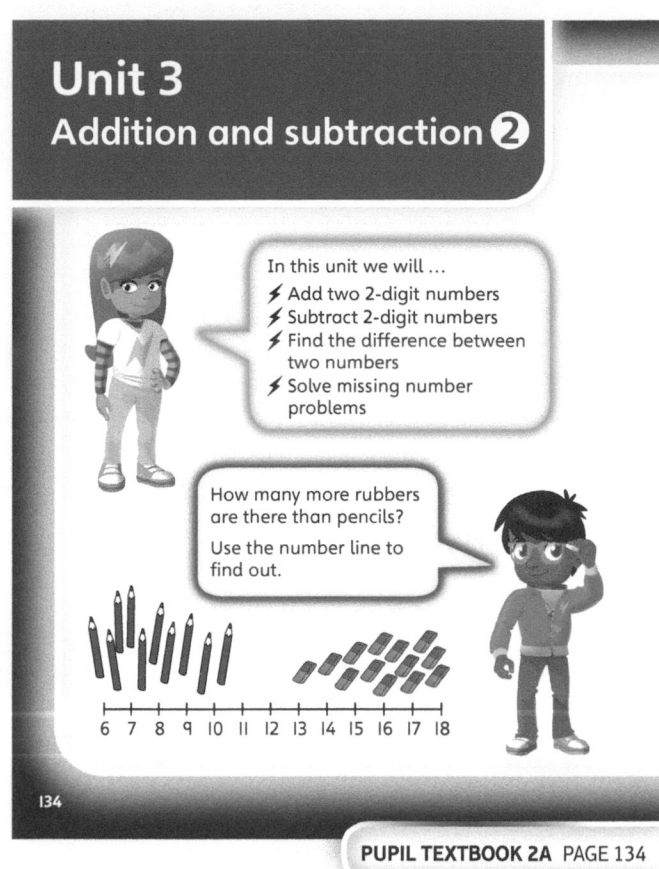

PUPIL TEXTBOOK 2A PAGE 134

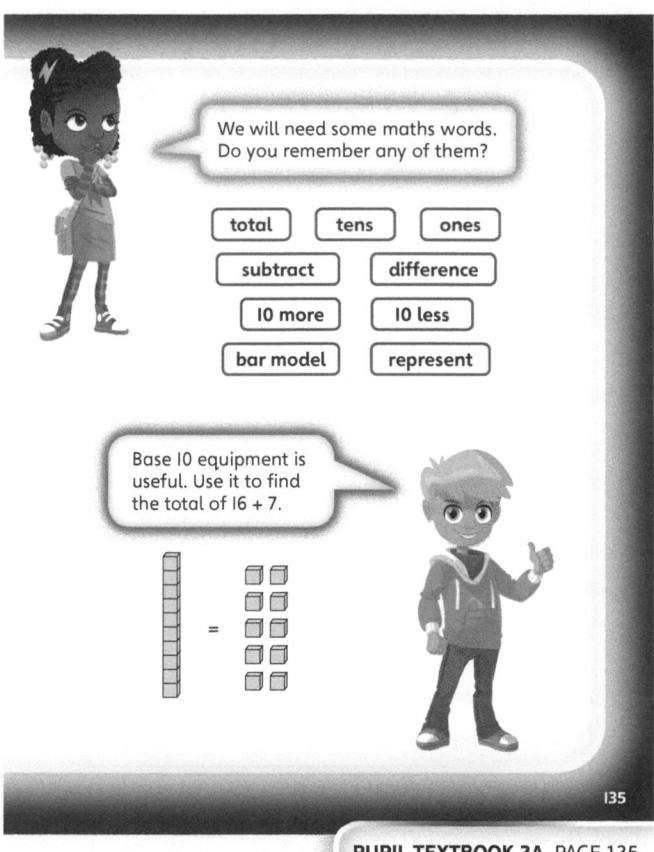

PUPIL TEXTBOOK 2A PAGE 135

10 more, 10 less

Learning focus

In this lesson, children will find 10 more and 10 less than a number and notice which digit changes during this process.

Before you teach

- Which representation do you think children will find hardest to use or understand?
- How will you support children in explaining the use of this representation?

NATIONAL CURRICULUM LINKS

Year 2 Number – addition and subtraction

Count in steps of 2, 3, and 5 from 0, and in 10s from any number, forward and backward.

Solve problems with addition and subtraction: using concrete objects and pictorial representations, including those involving numbers, quantities and measures.

ASSESSING MASTERY

Children can mentally add or subtract 10 to or from a 2-digit number (staying within 100) and can identify that only the digit in the tens column changes during this process. Children can show this visually in different representations and explain what each shows.

COMMON MISCONCEPTIONS

Children may add or subtract to or from the 1s digit rather than the 10s digit of a number. Ask:
- *What is the total number of 10s? Have you added/subtracted a 10 to/from the tens column?*

Use the sentence scaffold: ☐ *tens + 1 ten =* ☐ *tens or* ☐ *tens – 1 ten =* ☐ *tens.*

STRENGTHENING UNDERSTANDING

Encourage children to use concrete resources. Ask them to work in pairs, commenting on what they are doing during the calculation to their partner. Using the sentence scaffold above will help in this process.

GOING DEEPER

Children should notice that only the 10s digit changes when you add or subtract 10. Ask: *Is this always true, or only sometimes true?* Children might be able to see that if you add 10 to a number with 9 tens the digit in the hundreds column will also change (or will become 1).

KEY LANGUAGE

In lesson: 10 more, 10 less

Other language to be used by the teacher: digit, tens, ones, pattern, above, below, count on, count back

STRUCTURES AND REPRESENTATIONS

100 square, number track

RESOURCES

Mandatory: 100 square, base 10 equipment

Optional: completed number track (increasing in 10s)

 In the eTextbook of this lesson, you will find interactive links to a selection of teaching tools.

Quick recap

Chant together as a class: Count up in 10s from 0 to 100, and then count down in 10s from 100 to 0.

Then play 'Can you hear my mistake?'. Count up again, but this time make a deliberate error (for example, say '45' instead of '40', or say '70, 80, 90, 20'). Can children identify your mistake and explain what you should have said?

Discover

10 more, 10 less

Discover

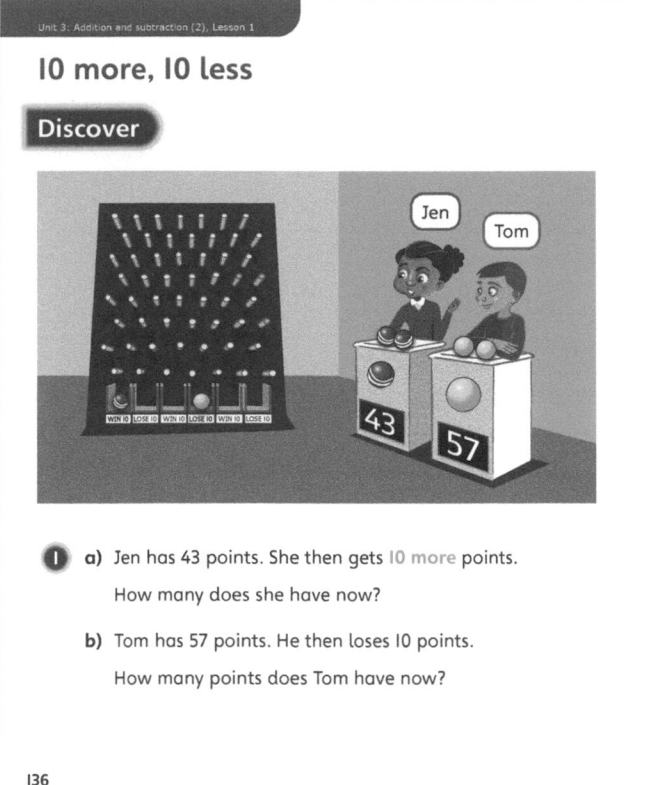

WAYS OF WORKING Pair work

ASK

- Question ① : *What is the same as yesterday's lesson? What is different?*
- Question ① : *Who can explain what is happening in this game?*

IN FOCUS Question ① introduces children to the concept of finding 10 more or 10 less than a number, using the context of a game. Encourage children to see the similarities between this concept and the previous lesson's work.

PRACTICAL TIPS Children can use base ten equipment to represent the numbers in the question and explore what happens when points are won or lost.

ANSWERS

Question ① a): Jen has 53 points now.

Question ① b): Tom has 47 points now.

① a) Jen has 43 points. She then gets 10 more points.
How many does she have now?

b) Tom has 57 points. He then loses 10 points.
How many points does Tom have now?

136

PUPIL TEXTBOOK 2A PAGE 136

Share

WAYS OF WORKING Whole class teacher led

ASK

- Question ① a): *What digit changes when you find 10 more or 10 less than a number? Is this always the case?*
- Question ① b): *What is 10 more than …? What is 10 less than …?*

IN FOCUS Encourage children to explain how they know how many points each player now has. Ask them to use resources and representations to illustrate their reasoning.

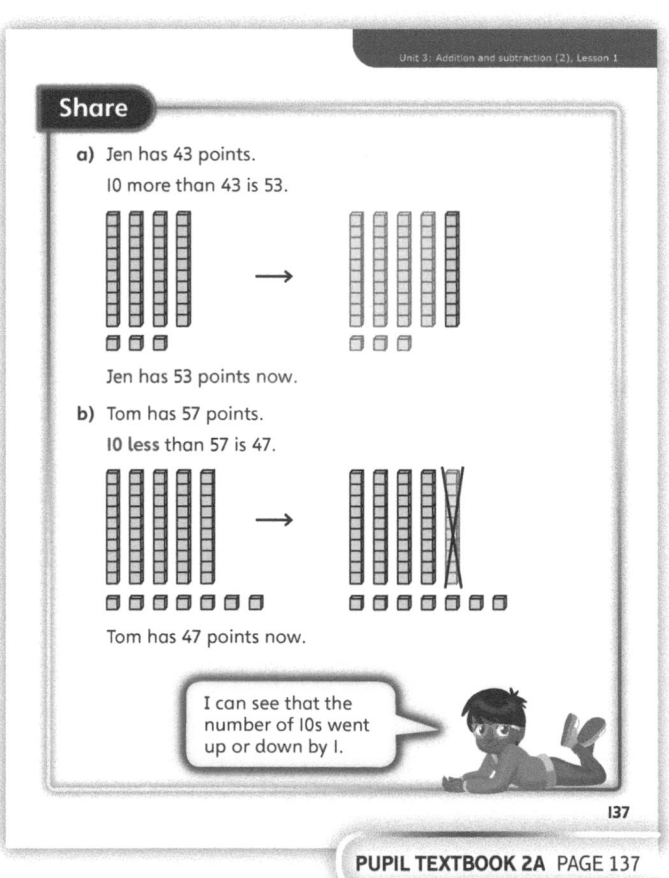

Share

a) Jen has 43 points.
10 more than 43 is 53.

Jen has 53 points now.

b) Tom has 57 points.
10 less than 57 is 47.

Tom has 47 points now.

I can see that the number of 10s went up or down by 1.

137

PUPIL TEXTBOOK 2A PAGE 137

Think together

WAYS OF WORKING Whole class teacher led (I do, We do, You do)

ASK

- Question ❶: *What digit changes when you add or subtract 10 to or from a number?*
- Question ❸: *What patterns can you spot in the arrangement of the numbers in the 100 square?*

IN FOCUS Question ❶ revisits the game context used in **Discover** and **Share**, while questions ❷ and ❸ use different ways of asking children to find 10 more or 10 less than given numbers. It is important that children continue to use a variety of different resources, rather than just the one that is clearest to them, in order to increase their understanding of this key concept.

STRENGTHEN If children make the starting number incorrectly, using resources such as base 10 equipment, this will need to be addressed at another time. In order to keep the focus of this lesson on finding 10 more or 10 less than a number, give children a pictorial representation of the starting number and ask them to draw another 10 items, or cross 10 out, as the question requires.

DEEPEN If children can calculate 10 more than a number, can they then also say the inverse? For example, '10 more than 33 is 43, so I know 10 less than 43 is 33'. Encourage children to verbalise or write down the inverse for each of the calculations they have completed.

ASSESSMENT CHECKPOINT Assess whether children can predict what 10 more or 10 less than a number will be before using resources. Can they explain how they know they are correct? Encourage them to prove it using a variety of different resources.

ANSWERS

Question ❶: 10 more than 76 is 86.

Question ❷ a):

23	33	43	53	63	73	83	93	103

Question ❷ b):

84	74	64	54	44	34	24	14	4

Question ❸ a): 10 more than 73 is 83.

On a 100 square, 10 more than a number is on the line below.

Question ❸ b): 10 less than 64 is 54.

On a 100 square, 10 less than a number is on the line above.

Question ❸ c): 10 more than 58 is **68**.

35 is 10 more than **25**.

10 less than 99 is **89**.

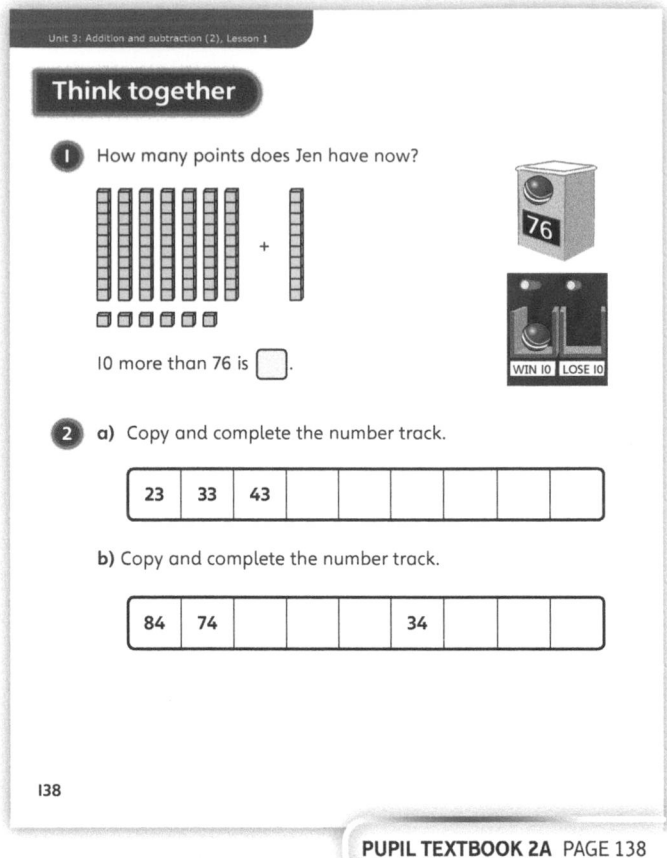

PUPIL TEXTBOOK 2A PAGE 138

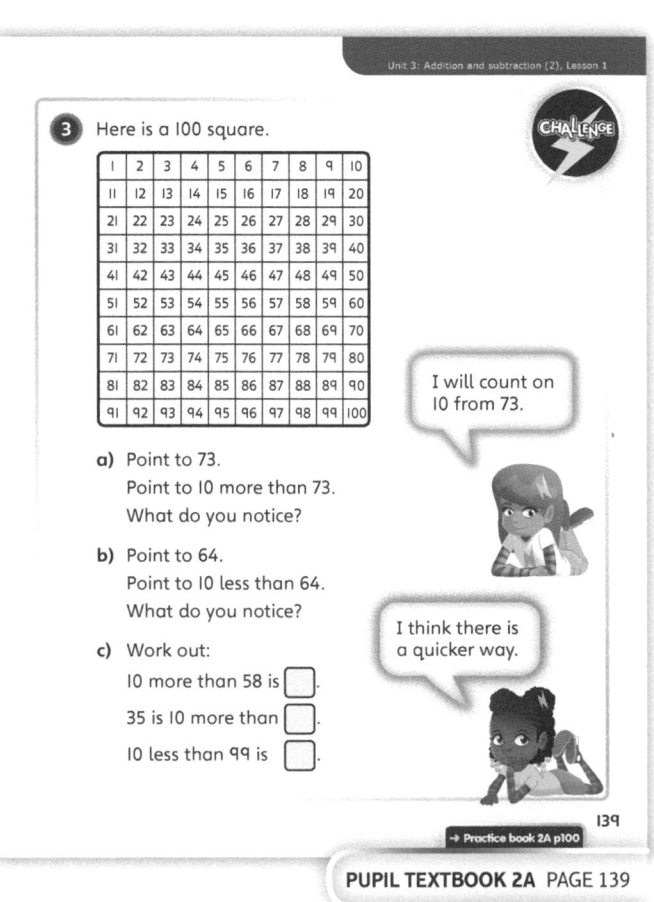

PUPIL TEXTBOOK 2A PAGE 139

Practice

WAYS OF WORKING Independent thinking

IN FOCUS In questions ❶ and ❷, children are required to interpret numbers represented in different ways, using different resources and representations. This variation will help strengthen their understanding of the topic.

STRENGTHEN Encourage children to continue working with concrete resources. Ask them to explain what they are doing. This will strengthen their understanding.

DEEPEN Question ❻ is a multi-step problem. Work with children to explore and plan the steps that will be needed, and decide how they can best make notes as they work through each step.

ASSESSMENT CHECKPOINT Check whether children can explain how they know their answer is correct. They should be able to explain why the digit in the tens column changes and why the digit in the ones column does not change.

Use questions ❹ to ❻, which have no pictorial support, to determine whether children understand the concept of 10 more and 10 less, or if they still need to make the numbers in order to see the answer.

ANSWERS Answers for the **Practice** part of the lesson can be found in the *Power Maths* online subscription.

Reflect

WAYS OF WORKING Pair work

IN FOCUS In the **Reflect** part of the lesson, children create their own questions for their partner to complete. Each child should use what they have learnt to find 10 more and 10 less, and to check each other's answers.

ASSESSMENT CHECKPOINT Assess what children say they would do to check their partner's answer. Can they accurately describe the digit in the tens column and how and why it changes as a result of the calculation?

ANSWERS Answers for the **Reflect** part of the lesson can be found in the *Power Maths* online subscription.

After the lesson

- Are children confident finding 10 more and 10 less?
- What resource did children find most or least easy to use?

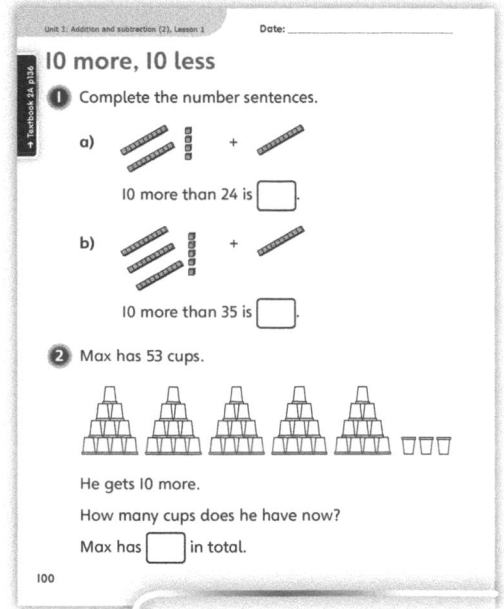

PUPIL PRACTICE BOOK 2A PAGE 100

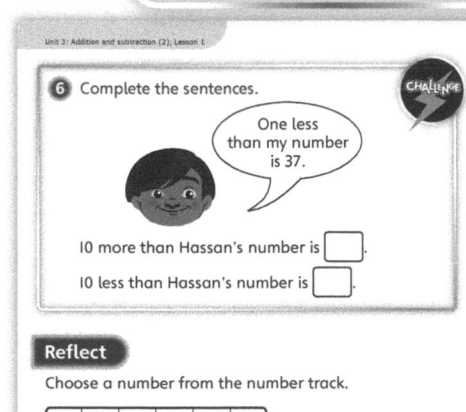

PUPIL PRACTICE BOOK 2A PAGE 101

Unit 3: Addition and subtraction (2), Lesson 1

❻ Complete the sentences. CHALLENGE

One less than my number is 37.

10 more than Hassan's number is ☐.

10 less than Hassan's number is ☐.

Reflect

Choose a number from the number track.

| 48 | 49 | 50 | 51 | 52 | 53 |

Ask a partner to say 10 more than the number. Are they correct?

Now ask your partner to choose a number.

Say 10 less than this number. Are you correct?

102

PUPIL PRACTICE BOOK 2A PAGE 102

Add and subtract 10s

Learning focus

In this lesson, children will build on what they learnt in the previous lesson, but will now focus on addition and subtraction of more than 1 ten to and from a 2-digit number.

Before you teach

- How will you promote the use of number bonds within this lesson?
- How can you help children critique each other's methods to see which is most efficient?

NATIONAL CURRICULUM LINKS

Year 2 Number – addition and subtraction

Add and subtract numbers using concrete objects, pictorial representations, and mentally, including a 2-digit number and 10s.

Solve problems with addition and subtraction, using concrete objects and pictorial representations, including those involving numbers, quantities and measures.

ASSESSING MASTERY

Children can make links to previous learning and identify that only the digit in the tens column changes when they add or subtract a multiple of 10. Children can recognise that the method they use (and visualise) is similar to the method for adding 1s to a 2-digit number.

COMMON MISCONCEPTIONS

Children may add tens to the ones column. Ask:
- *Have you added or subtracted 10 to or from the tens column?*

Children may also forget what operation they are using and may add when they should be subtracting and vice versa. Ask:
- *What operation does this question require? How do you know?*

STRENGTHENING UNDERSTANDING

Encourage children to work in pairs, with one child using the manipulatives and the other recording how many 10s and 1s there are at different points. Working in this way will help children conceptualise the numbers and strengthen their understanding of what the recorded numbers represent.

GOING DEEPER

Ask children to apply their understanding of adding and subtracting 10s to and from a number by representing or writing a calculation in as many different ways as possible. You could also ask them to make a whole in as many different ways as possible, with the restriction that one of the parts is always a multiple of 10.

KEY LANGUAGE

In lesson: adding, subtraction, +, –, =, total, ones, tens, columns, calculation

Other language to be used by the teacher: digit, multiple of 10, method, altogether, take away, minus

STRUCTURES AND REPRESENTATIONS

Place value chart, bar model, number wall

RESOURCES

Mandatory: base 10 equipment, place value chart

 In the eTextbook of this lesson, you will find interactive links to a selection of teaching tools.

Quick recap

Say a number and ask children to write or call out the number that is 10 more. Repeat, but this time ask children to say the number that is 10 less.

Discover

WAYS OF WORKING Pair work

ASK

- Question ❶: *What is the same as yesterday's lesson? What is different?*
- Question ❶: *Can you use number bonds to work in an efficient way?*

IN FOCUS To answer question ❶, children need to identify the number of 10s and 1s in the given number and add more than 1 ten to this number.

PRACTICAL TIPS Children can use base 10 equipment to represent the apples in the question. Support them in adding 10s, rather than counting every individual apple.

ANSWERS

Question ❶ a): There are 16 toffee apples on the table and 30 toffee apples on the ground.

Question ❶ b): There are 46 toffee apples in total.

Add and subtract 10s

Discover

❶ a) How many toffee apples are on the table?
How many toffee apples are on the ground?

b) How many toffee apples are there in total?

140

PUPIL TEXTBOOK 2A PAGE 140

Share

WAYS OF WORKING Whole class teacher led

ASK

- Question ❶ b): *How can you organise your work and the resources you are using? Have you organised your work in this way before?*
- Question ❶ b): *How many boxes of 10 are there in total? Can you think of a number bond to 10 which will help you find the total number of boxes? How can you use what you know about adding 1s to a 2-digit number to help you find the total number of toffee apples?*

IN FOCUS Encourage children to make links between what they are doing in this lesson and what they did when they added or subtracted 1s to or from a 2-digit number. Lead them towards arranging the resources on a place value chart in a way that can be related to the column method in future.

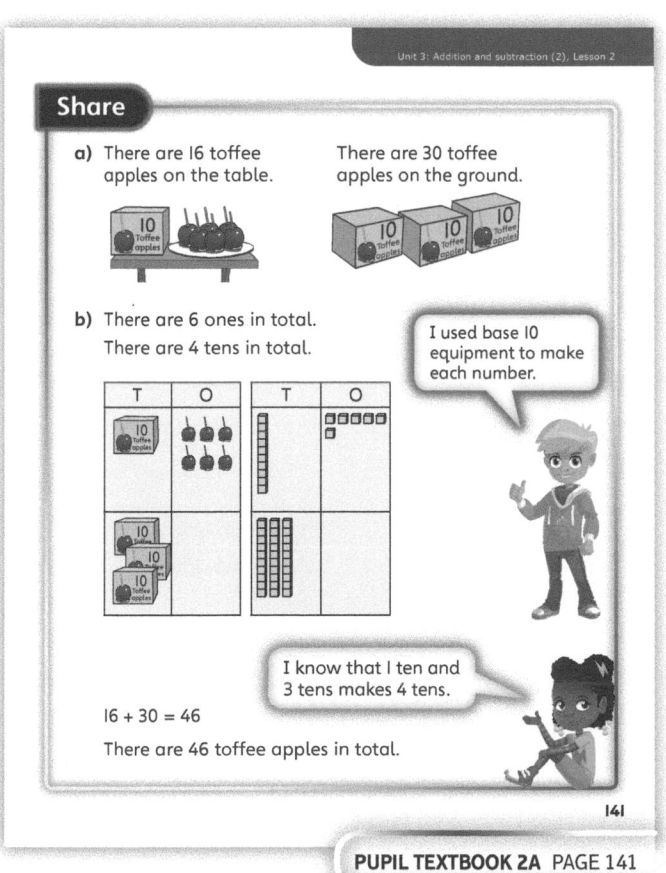

Share

a) There are 16 toffee apples on the table.

There are 30 toffee apples on the ground.

b) There are 6 ones in total.
There are 4 tens in total.

I used base 10 equipment to make each number.

I know that 1 ten and 3 tens makes 4 tens.

16 + 30 = 46
There are 46 toffee apples in total.

141

PUPIL TEXTBOOK 2A PAGE 141

Think together

WAYS OF WORKING Whole class teacher led (I do, We do, You do)

ASK

• Question **1**: *How do you know which operation to use in each question?*
• Question **3**: *Does the digit in the ones column change in any of these questions?*

IN FOCUS Encourage children to think about each question in different ways. How could it be represented using different resources? For example, in question **1**, prompt children to consider how each calculation could be represented on a number line.

STRENGTHEN Allow children to continue to use resources to calculate the answers. This will help them understand the concept of adding and subtracting a number of 10s from a given number.

DEEPEN Present children with some incorrect calculations and challenge them to explain how they know the given answers are not possible. For example, ask: *How do you know that 38 – 20 = 36 is not correct?*

ASSESSMENT CHECKPOINT Assess whether children can verbalise what they are doing with the resources and what they are doing mentally using the appropriate vocabulary. They should also be able to explain why the digit in the ones column does not change.

ANSWERS

Question **1** a): 25 + 30 = **55**
Question **1** b): 36 + 40 = **76**
Question **2** a): 51 – 20 = **31**
Question **2** b): 76 – 50 = **26**
Question **3** a): 36 + 20 = 56
Question **3** b): 35 + 20 = 55
    35 + 30 = 65
    35 + 40 = 75
    17 + 60 = 77
    24 + 60 = 84
    31 + 60 = 91

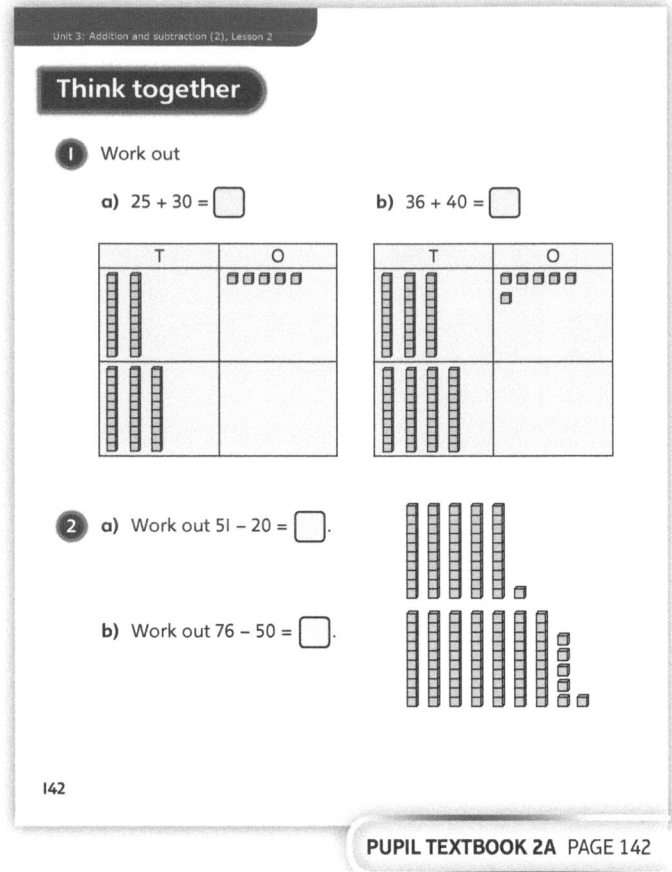

PUPIL TEXTBOOK 2A PAGE 142

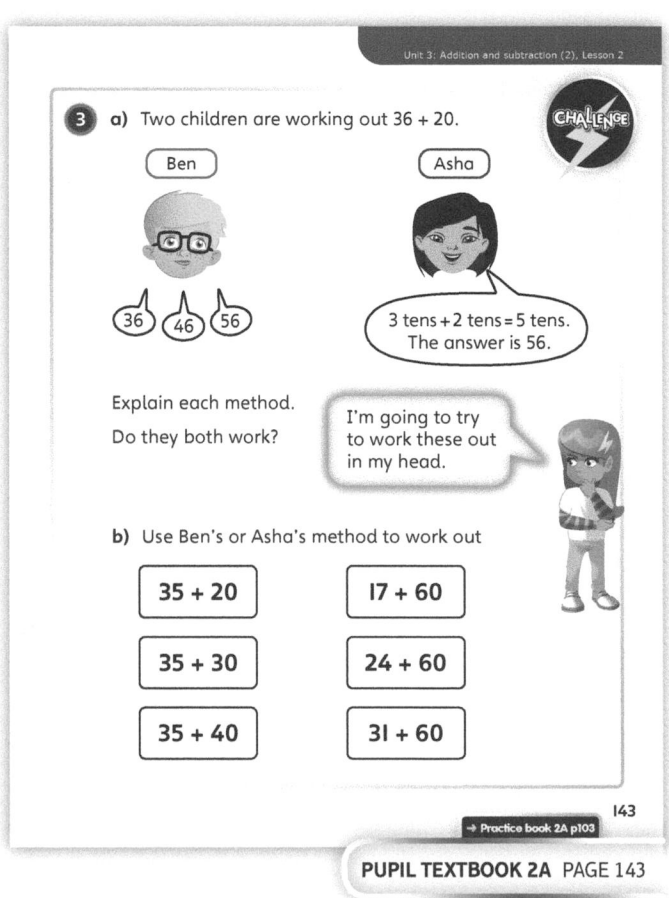

PUPIL TEXTBOOK 2A PAGE 143

Practice

Independent thinking

IN FOCUS Questions **1** and **3** include representations of some of the calculations using base 10 equipment. Question **4** uses the bar model to represent addition and subtraction of multiples of 10. Children may use base 10 equipment to represent the numbers in the bar model. This should help them to identify which operation is needed to find the missing number.

STRENGTHEN Encourage children to record 'skip counting' on a number track. For example, for 35 + 40, children would write 35 first, then 45, 55, 65, 75 in the next boxes. Doing this at the same time as using a resource such as base 10 equipment will help them understand how each 10 affects the total number.

DEEPEN Challenge children to choose a number and represent it on a number wall, as in question **6**, finding as many different ways to make it as possible. Note that one of the parts should always be a multiple of 10. For example, if children chose 72 as the whole, their number wall would show 10 + 62, 20 + 52, 30 + 42, and so on.

THINK DIFFERENTLY Question **5** requires children to fill in missing digits in calculations where an unknown multiple of 10 is being added to, or subtracted from, a given 2-digit number. They should draw on their knowledge of number bonds and place value to achieve this.

ASSESSMENT CHECKPOINT Children should be able to verbalise the steps that they are completing using the correct mathematical vocabulary. They should also be able to explain what each number they record represents.

ANSWERS Answers for the **Practice** part of the lesson can be found in the *Power Maths* online subscription.

Reflect

WAYS OF WORKING Pair work

IN FOCUS The **Reflect** part of the lesson requires children to find 2 tens digits that complete the number sentences. Children could be challenged to find more than one solution.

ASSESSMENT CHECKPOINT Assess whether children can explain how they found suitable numbers to complete the calculations. Can they tell you if more than one answer is possible, and why or why not?

ANSWERS Answers for the **Reflect** part of the lesson can be found in the *Power Maths* online subscription.

After the lesson ⏸

- Did all children understand the importance of place value in solving calculations like this?
- Do children need further practice with mental addition and subtraction of 10s to and from a given 2-digit number? For example, 55 + 10, 55 + 30, 55 + 40, 55 − 10, 55 − 30, 55 − 50.

181

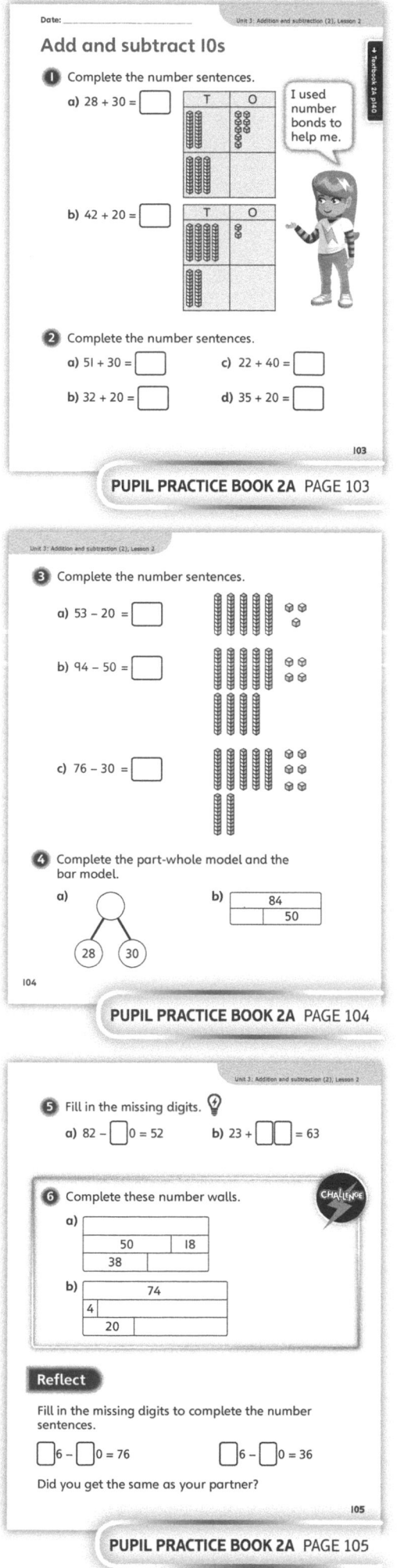

PUPIL PRACTICE BOOK 2A PAGE 103

PUPIL PRACTICE BOOK 2A PAGE 104

PUPIL PRACTICE BOOK 2A PAGE 105

Add two 2-digit numbers – add 10s and add 1s

Learning focus

In this lesson, children learn to add two 2-digit numbers by adding the 10s and the 1s separately, and then recombining.

Before you teach

- Can children add two multiples of 10 within 100?
- Can children make a 2-digit number from 10s and 1s?
- Can children partition 2-digit numbers into 10s and 1s?

NATIONAL CURRICULUM LINKS

Year 2 Number – addition and subtraction

Solve problems with addition and subtraction: applying their increasing knowledge of mental and written methods.

Add and subtract numbers using concrete objects, pictorial representations, and mentally, including two 2-digit numbers.

ASSESSING MASTERY

Children can use their knowledge of place value to add together two 2-digit numbers by adding the 10s, then adding the 1s and then recombining.

COMMON MISCONCEPTIONS

Children may revert to a count-all method, rather than using their number bond knowledge to help them find answers. Ask:
- *What number bond could help you add the 10s?*

STRENGTHENING UNDERSTANDING

Provide place value equipment for children to use throughout, to support and check their working and answers.

GOING DEEPER

Ask children to investigate pairs of calculations that will have the same total, for example 32 + 41, and 42 + 31. Can they explain how they chose the calculations and why the answers are the same?

KEY LANGUAGE

In lesson: tens, ones, total

STRUCTURES AND REPRESENTATIONS

Place value equipment

RESOURCES

Mandatory: place value equipment

 In the eTextbook of this lesson, you will find interactive links to a selection of teaching tools.

Quick recap

Work together as a class to collect and rehearse all the number bonds for 7. Repeat to find all the number bonds for 8 and for 9.

Discover

ASK

- Question **1**: *How would you describe what you can see in Milo's equipment?*
- Question **1**: *How would you describe what you can see in Seth's equipment?*
- Question **1** b): *What number bond could help with adding the 10s?*

IN FOCUS Children are shown a visual representation of base 10 equipment to support an addition of two 2-digit numbers where the 10s and the 1s will be added separately, and then recombined.

PRACTICAL TIPS Ask children to use or draw place value equipment to represent the numbers in the question.

ANSWERS

Question **1** a): Milo and Seth can find the total of their base 10 materials.

Question **1** b): 43 + 14 = 57

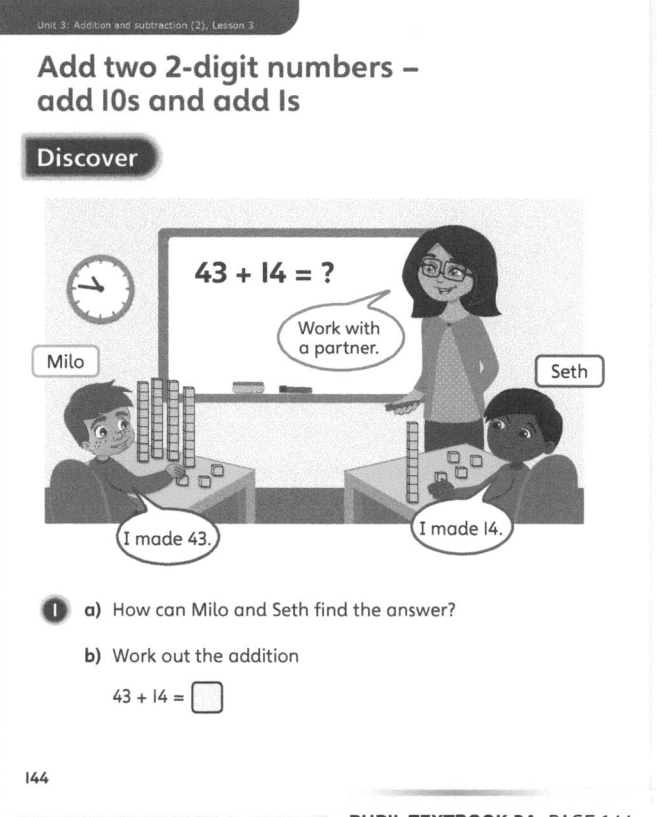

Share

WAYS OF WORKING Whole class teacher led

ASK

- Question **1** a): *Can you point to the group of 4 tens. Can you point to Seth's 1 ten?*
- Question **1** a): *What number bond can help you add the 10s? What number bond can help you add the 1s?*
- Question **1** b): *How is this related to your work in question **1** a)?*

IN FOCUS In question **1** a), children should add the 10s and 1s separately. In question **1** b), they recombine the 10s and 1s to complete the addition.

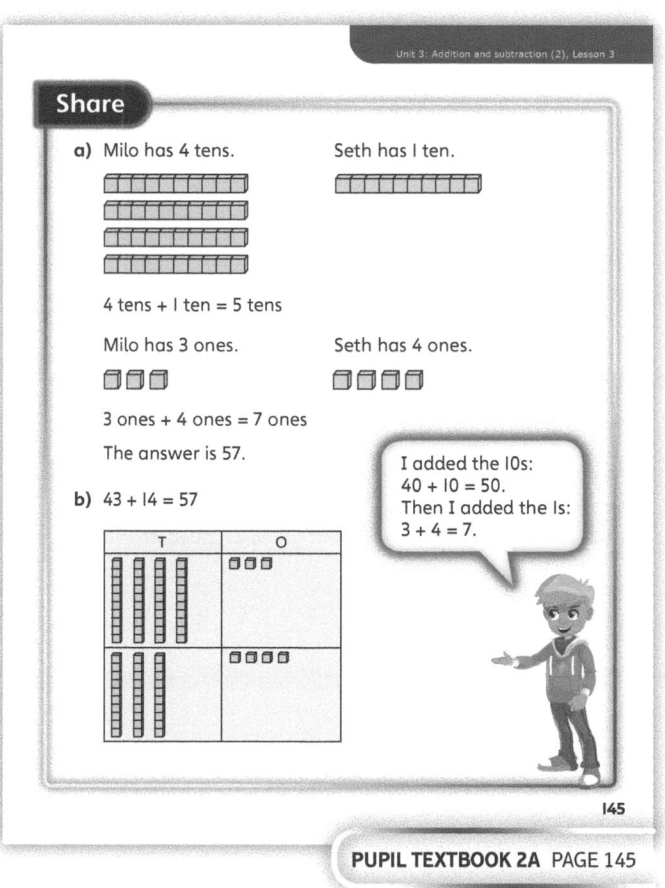

Think together

WAYS OF WORKING Whole class teacher led (I do, We do, You do)

ASK

- Question ❶: *What number bond could help you add the 10s? What number bond could help you add the 1s?*
- Question ❷: *Could you partition into 10s and 1s?*
- Question ❸: *What do you notice about the total of the 1s?*

IN FOCUS In question ❶, children practice solving a calculation where place value equipment representations are also shown. Question ❷ requires children to add two 2-digit numbers, this time without a visual representation. Encourage them to look for opportunities to use known number bonds to help them. In question ❸, children are solving additions where the 1s digits will add to 10 or more.

STRENGTHEN Encourage children to use or draw place value equipment to support them as they check their working and answers.

DEEPEN Ask children to compare these calculations: 45 + 54, 27 + 72 and 36 + 63.

Challenge them to find more calculations like these. What do they notice?

ASSESSMENT CHECKPOINT Question ❷ will allow you to assess whether children are able to partition and recombine 2-digit numbers in order to add, and whether they are making use of known number bonds to help them calculate.

ANSWERS

Question ❶ a): Children make or draw the numbers.

Question ❶ b): 42 + 25 = 67

Question ❷: 31 + 26 = 56
18 + 61 = 79
44 + 55 = 99

Question ❸ a): 35 + 27 = 62

Question ❸ b): 24 + 18 = 42
51 + 29 = 80
47 + 46 = 93

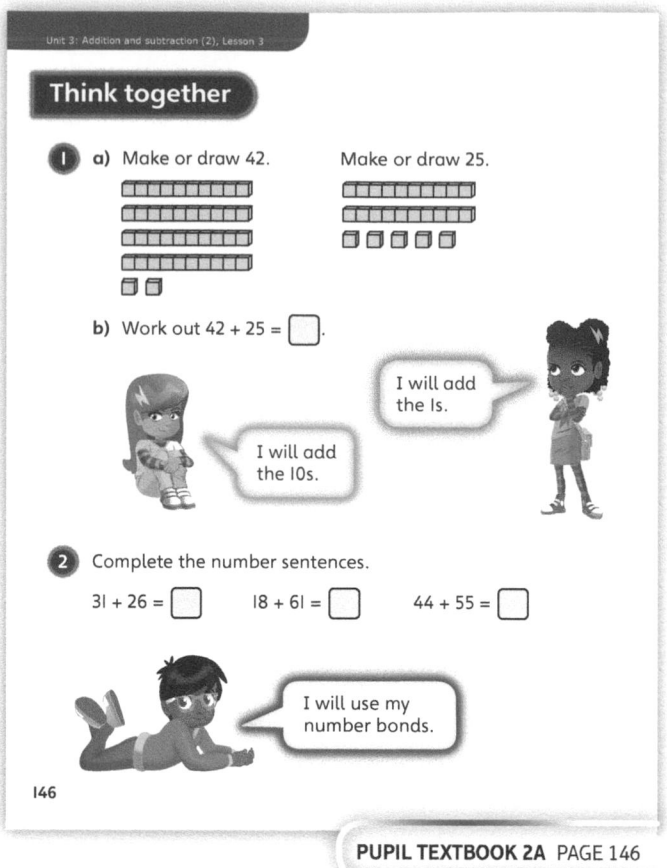

PUPIL TEXTBOOK 2A PAGE 146

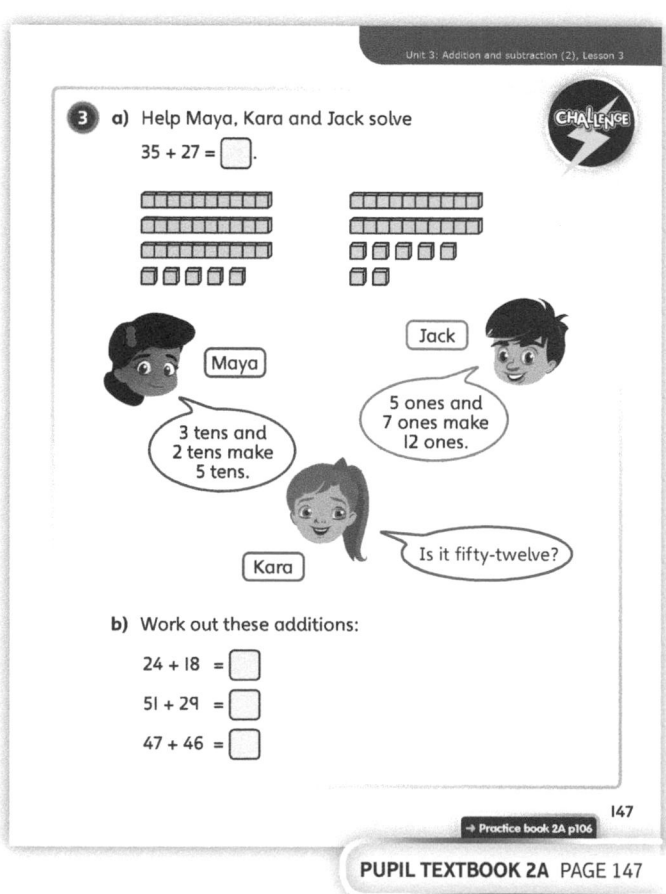

PUPIL TEXTBOOK 2A PAGE 147

Practice

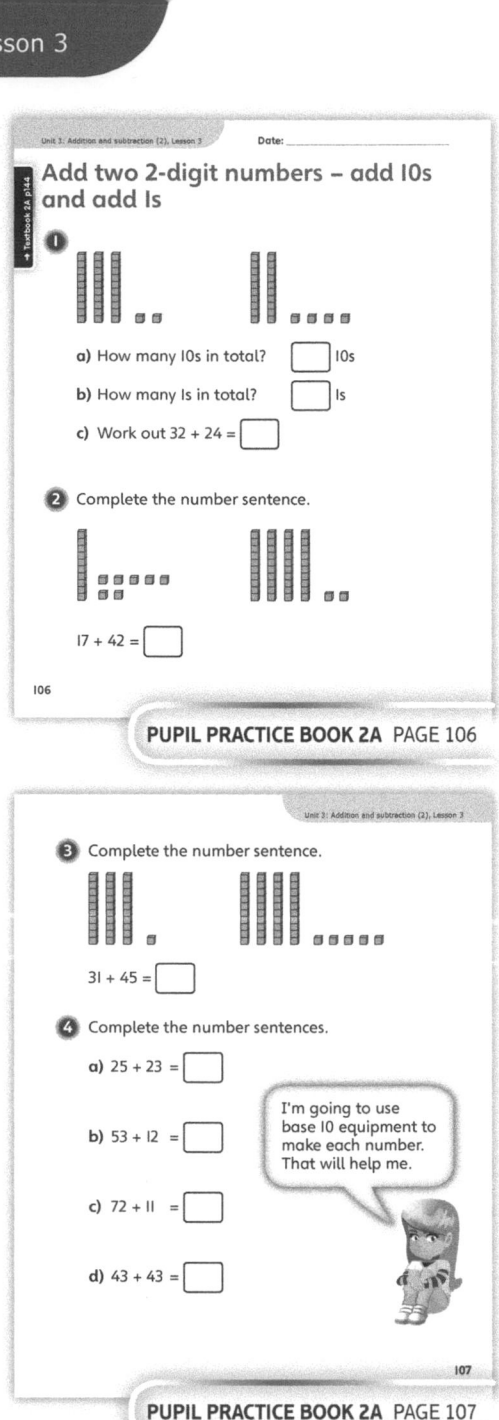

WAYS OF WORKING Independent thinking

IN FOCUS In questions ❶, ❷ and ❸, children are adding two 2-digit numbers and using place value equipment as a representation. The visual representation is not present in question ❹, but children can be prompted to make the numbers themselves if they find this helpful. The calculations in questions ❺ and ❻ will require children to make exchanges.

STRENGTHEN Children should be encouraged to draw place value equipment to represent the calculations throughout.

DEEPEN Challenge children to find as many calculations as they can with the same total as 44 + 45. Can they explain how they went about this?

ASSESSMENT CHECKPOINT Use question ❹ to assess whether children are able to partition and recombine to add in situations where a visual representation of the numbers is not provided.

ANSWERS Answers for the **Practice** part of the lesson can be found in the *Power Maths* online subscription.

> **PUPIL PRACTICE BOOK 2A** PAGE 106

> **PUPIL PRACTICE BOOK 2A** PAGE 107

Reflect

WAYS OF WORKING Pair work

IN FOCUS The **Reflect** part of the lesson prompts children to discuss the method they have used to add two 2-digit numbers. They should draw attention to the 10s and the 1s in each number and how these can be used to find the total.

ASSESSMENT CHECKPOINT Assess whether children can explain the process that they are using with reasoning about how it works and why it is a good method to use.

ANSWERS Answers for the **Reflect** part of the lesson can be found in the *Power Maths* online subscription.

> **PUPIL PRACTICE BOOK 2A** PAGE 108

After the lesson

- Can children add any pairs of 2-digit numbers, up to a total of 99?
- Did all children know how to use place value and number bonds to solve calculations like these?

Add two 2-digit numbers – add more 10s then more 1s

Learning focus

In this lesson, children add a 2-digit number to another 2-digit number by first adding on more 10s and then adding on more 1s.

Before you teach

- Can children add 1 to any given 2-digit number?
- Can children add 10 to any given 2-digit number?

NATIONAL CURRICULUM LINKS

Year 2 Number – addition and subtraction

Solve problems with addition and subtraction: applying their increasing knowledge of mental and written methods.

Add and subtract numbers using concrete objects, pictorial representations, and mentally, including two 2-digit numbers.

ASSESSING MASTERY

Children can partition in order to add two 2-digit numbers by adding on the 10s and then adding on the 1s.

COMMON MISCONCEPTIONS

Children may confuse this method with the one from the previous lesson, where they were adding the 10s and the 1s separately, and then recombining. Make sure you have made a clear decision about how to teach one or both methods. Ask:
- *How is this the same and how is it different to the previous lesson?*

STRENGTHENING UNDERSTANDING

The importance of this method is that it links to the subtraction method used in subsequent lessons. Make sure that children understand how to jump on 10 from a given 2-digit number, then how to count on in 10s from any given 2-digit number.

GOING DEEPER

Challenge children to compare this method with the partitioning and recombining method from the previous lesson and to explain the similarities and differences between them in detail.

KEY LANGUAGE

In lesson: add more, tens, ones

STRUCTURES AND REPRESENTATIONS

Place value equipment, open number lines

RESOURCES

Mandatory: place value equipment

 In the eTextbook of this lesson, you will find interactive links to a selection of teaching tools.

Quick recap 🔁

Ask children to count on together in 10s from any given 2-digit number. Repeat with different 2-digit numbers..

186

Discover

Pair work

ASK

- Question ❶: *How many parcels are already in the van?*
- Question ❶ a): *How many parcels is Charlie carrying?*
- Question ❶ b): *How many parcels is Mia carrying?*

IN FOCUS This gives a context for adding 10 more, and then adding some 1s in order to find the total. Ensure children recognise that Charlie's parcels can be added on as one whole 10, rather than by counting on. Discuss the place values of 2-digit numbers and observe that first the 10s will change, and then the 1s.

PRACTICAL TIPS Provide place value equipment for children to use to build the numbers in the question and then work through the scenario in stages.

ANSWERS

Question ❶ a): There are 33 parcels in the van.

Question ❶ b): There are now 35 parcels in the van.

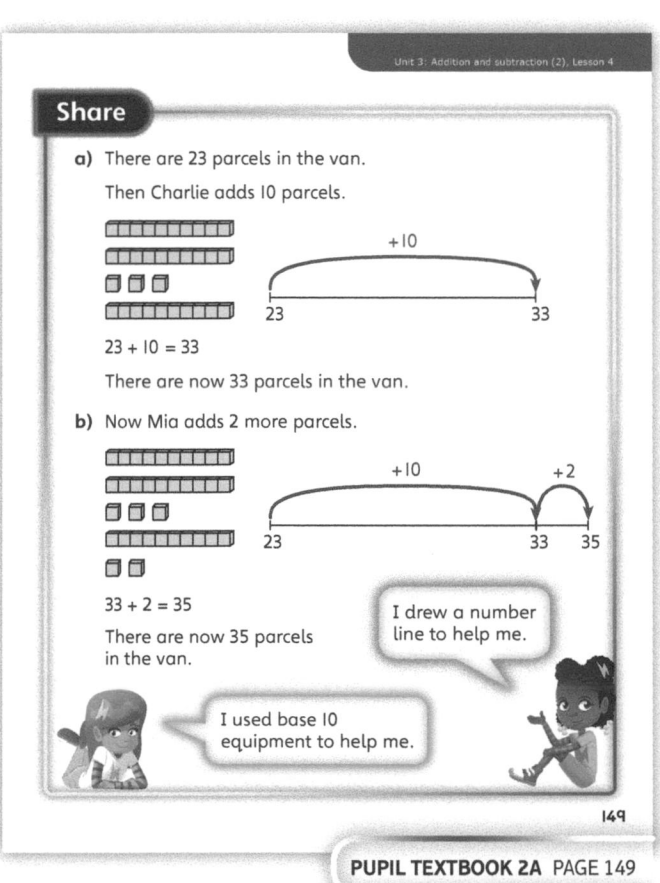

Add two 2-digit numbers – add more 10s then more 1s

Discover

I have 23 parcels in here.

PARCELS

Charlie

Mia

❶ a) Charlie puts his parcels into the van.
How many parcels are now in the van?

b) Next, Mia puts her parcels into the van.
How many parcels are now in the van?

148

PUPIL TEXTBOOK 2A PAGE 148

Share

Whole class teacher led

ASK

- Question ❶ a): *How many 10s and 1s are in the van to begin with?*
- Question ❶ a): *Now 10 more parcels are added. How many 10s now? How many 1s?*
- Question ❶ a): *Can you see how the number line shows this?*
- Question ❶ b): *Can you see the two jumps on the number line? What do they show?*

IN FOCUS Question ❶ a) supports children in adding more 10s to a 2-digit number. In this instance children are only adding one 10 and should notice that the 10s digit increases by 1 and the 1s digit does not change.
In question ❶ b), they then add some more 1s. Discuss how the original number of parcels in the van has now changed.

Share

a) There are 23 parcels in the van.
Then Charlie adds 10 parcels.

+10
23 → 33

23 + 10 = 33

There are now 33 parcels in the van.

b) Now Mia adds 2 more parcels.

+10   +2
23 → 33 → 35

33 + 2 = 35

There are now 35 parcels in the van.

I used base 10 equipment to help me.

I drew a number line to help me.

149

PUPIL TEXTBOOK 2A PAGE 149

Think together

Whole class teacher led (I do, We do, You do)

ASK

• Question **①** b): *What is the first jump? What is the next jump?*
• Question **③**: *What makes the second jump trickier?*

IN FOCUS For question **①**, children are asked to use place value equipment to represent the original number and in question **②**, a number line is used to show the two jumps, of 10s and then 1s. Question **③** asks children to consider what happens when adding the 1s requires them to cross a 10.

STRENGTHEN Ask children to use place value equipment to make the original number and to add the 10s equipment and then the 1s equipment on to it in stages.

DEEPEN Ask children to explore more deeply the link with this method and the method of partitioning and adding the 10s and the 1s separately. Ask: *Which do you prefer? Why?*

ASSESSMENT CHECKPOINT Use question **②** to assess whether children can interpret and use the number line representation to add on some 10s and then some 1s.

ANSWERS

Question **①** a): Children make or draw 35.

Question **①** b): 35 + 13 = 48

Question **②**: 32 + 24 = 56

Question **③** a): 38 + 25 = 63

Question **③** b): 34 + 10 + 8 = 52
56 + 29 = 85
47 + 37 = 84

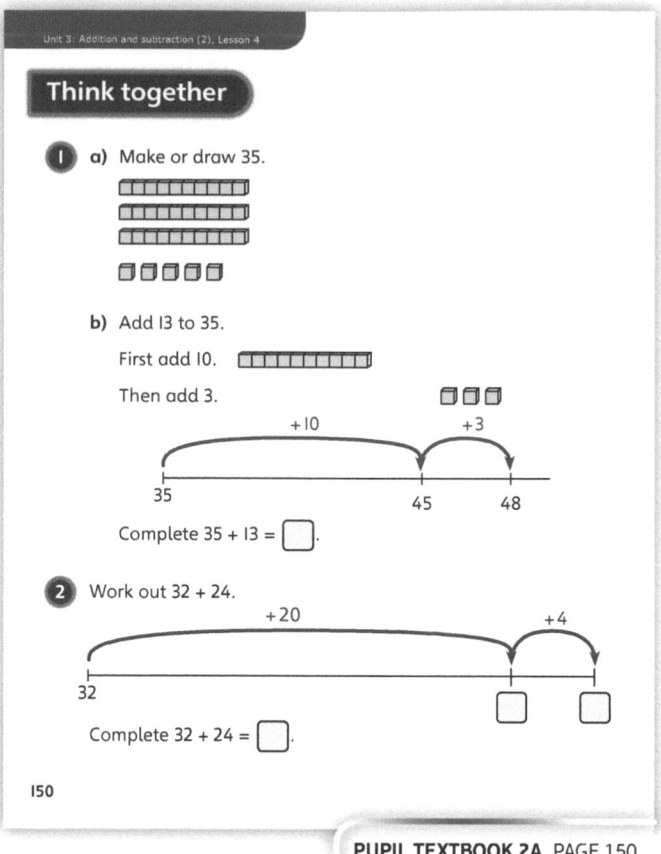

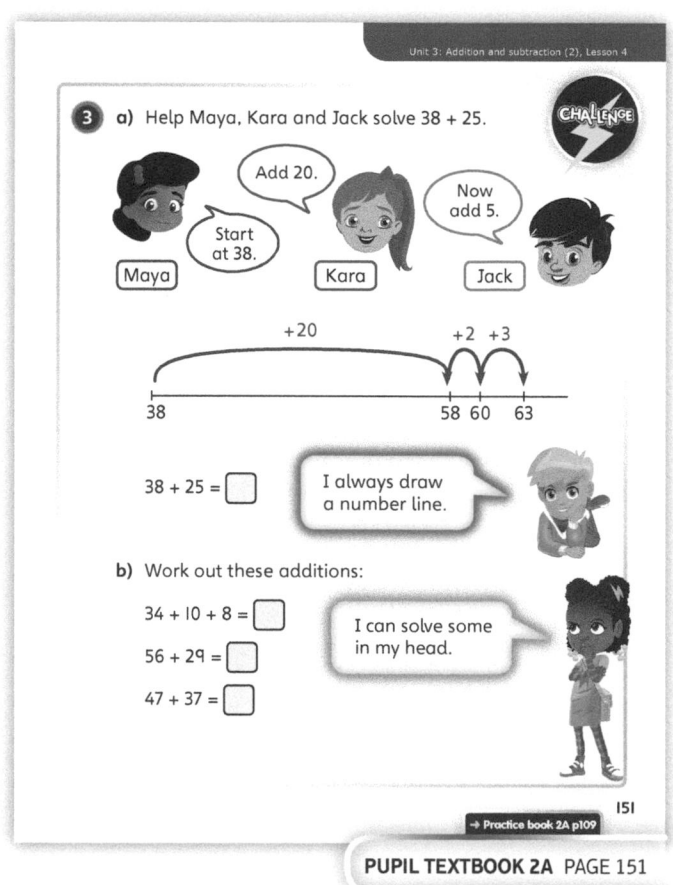

188

Practice

WAYS OF WORKING Independent thinking

IN FOCUS In questions ❶ and ❷, children are using jumps on a number line to add some 10s and then some 1s to a 2-digit number. Question ❸ requires children to add two 2-digit numbers where the 1s cross a 10. Question ❹ does not include a number line representation. Children are encouraged to use mental methods to solve some of these calculations. Ask: *Which can you solve in your head? Can you explain why?*

STRENGTHEN Provide place value equipment for children to build up each of the calculations in stages. Ask them to describe the value of each digit at each stage of their working and to explain what happens to it each time.

DEEPEN Challenge children to explore patterns in groups of calculations such as: 25 + 14, 35 + 14, 45 + 14, 55 + 14 or 32 + 11, 32 + 12, 32 + 13. What do they notice?

ASSESSMENT CHECKPOINT Question ❷ will allow you to assess whether children are confident in adding a 2-digit number in two jumps, the 10s first and then the 1s.

ANSWERS Answers for the **Practice** part of the lesson can be found in the *Power Maths* online subscription.

Reflect

WAYS OF WORKING Pair work

IN FOCUS The **Reflect** part of the lesson will prompt children to think about the best order in which to complete the calculation. They should describe a process that involves two jumps, and will observe that the first number has far fewer 10s but the second number has fewer 1s.

ASSESSMENT CHECKPOINT Assess whether children can decide whether to start with 17 or with 67, and can explain which they feel would be more efficient and why.

ANSWERS Answers for the **Reflect** part of the lesson can be found in the *Power Maths* online subscription.

After the lesson 🔲

- Are children regularly completing 2-digit additions as part of their fluency practice?
- How will you give children the opportunity to discuss the methods that they have chosen rather than just checking that they have found the correct answers?

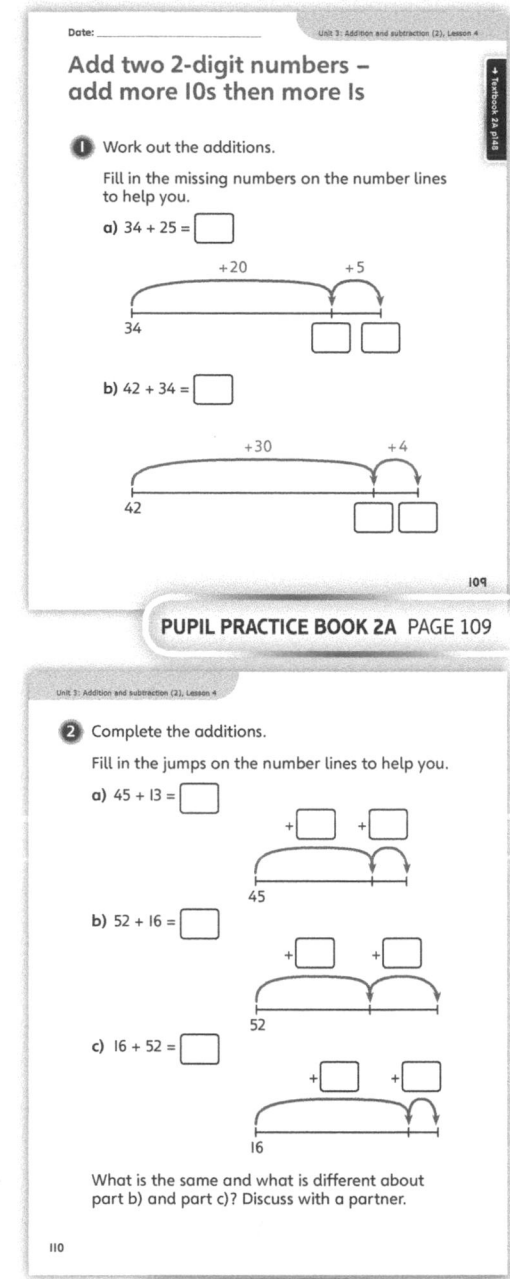

PUPIL PRACTICE BOOK 2A PAGE 109

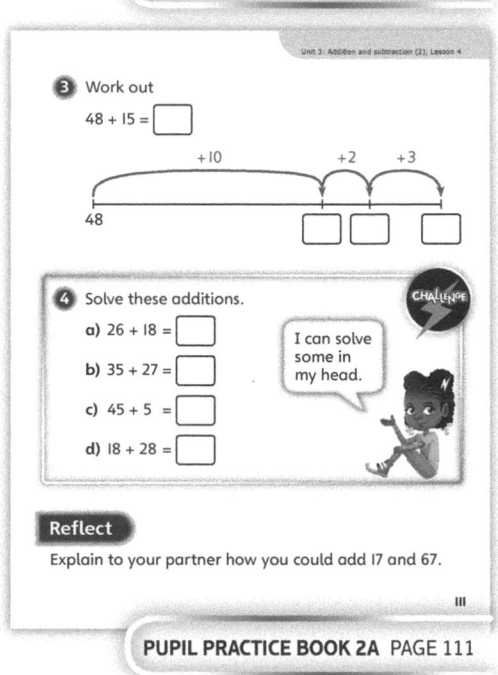

PUPIL PRACTICE BOOK 2A PAGE 110

PUPIL PRACTICE BOOK 2A PAGE 111

189

Subtract a 2-digit number from a 2-digit number – not across 10

Learning focus

In this lesson, children learn to subtract a 2-digit number from a 2-digit number by counting back in 10s, and then counting back in 1s. There is no crossing of 10s in this lesson.

Before you teach

- Can children find 10 less than any given 2-digit number?
- Can children find 1 less than any given 2-digit number?
- Can children recognise a jump of – 10 shown on an open number line?

NATIONAL CURRICULUM LINKS

Year 2 Number – addition and subtraction

Solve problems with addition and subtraction: applying their increasing knowledge of mental and written methods.

Add and subtract numbers using concrete objects, pictorial representations, and mentally, including two 2-digit numbers.

ASSESSING MASTERY

Children can subtract a 2-digit number in two steps, by first subtracting the 10s, and then the 1s.

COMMON MISCONCEPTIONS

Children may not use their number bond knowledge to subtract tens efficiently. Ask:
- *What number bond could help you subtract these 10s?*

STRENGTHENING UNDERSTANDING

Provide place value equipment for children to use alongside the calculation methods to support and check their arithmetic, working and answers.

GOING DEEPER

Challenge children to explore how they will arrive at the same answer by either subtracting the 10s then the 1s, or by subtracting the 1s and then the 10s. What do they notice?

KEY LANGUAGE

In lesson: take away, count back, subtract, tens (10s), ones (1s)

STRUCTURES AND REPRESENTATIONS

Number lines

RESOURCES

Optional: place value equipment

 In the eTextbook of this lesson, you will find interactive links to a selection of teaching tools.

Quick recap

Ask children to count back together in 10s from any given 2-digit number. Repeat several times.

Discover

Pair work

ASK

- Question ❶: *How many eggs are in each group?*
- Question ❶: *How many eggs are there in total?*
- Question ❶: *How many eggs do Seth and Anna take away altogether?*

IN FOCUS This question frames a situation where first 1 ten and then 2 ones are subtracted from a 2-digit number. Discuss the scenario and ensure children recognise that subtraction is the operation that will be needed to find the answer.

PRACTICAL TIPS Encourage children to take the roles of the two chefs and to model the scenario using place value equipment.

ANSWERS

Question ❶ a): There are 15 eggs left.

Question ❶ b): There are 13 eggs left.

PUPIL TEXTBOOK 2A PAGE 152

Share

WAYS OF WORKING Whole class teacher led

ASK

- Question ❶ a): *Can you see the jump back of 10 on the number line? How does this show 'subtract 10'?*
- Question ❶ b): *Why is there a jump back of 10 and then a jump back of 2?*

IN FOCUS In question ❶ a), children subtract 10 from the given 2-digit number. This is modelled with a jump back on the number line. In question ❶ b), they move on to subtract 1 ten and then 2 ones, which is represented as two jumps back on the number line. Help children make the link between taking away and subtraction.

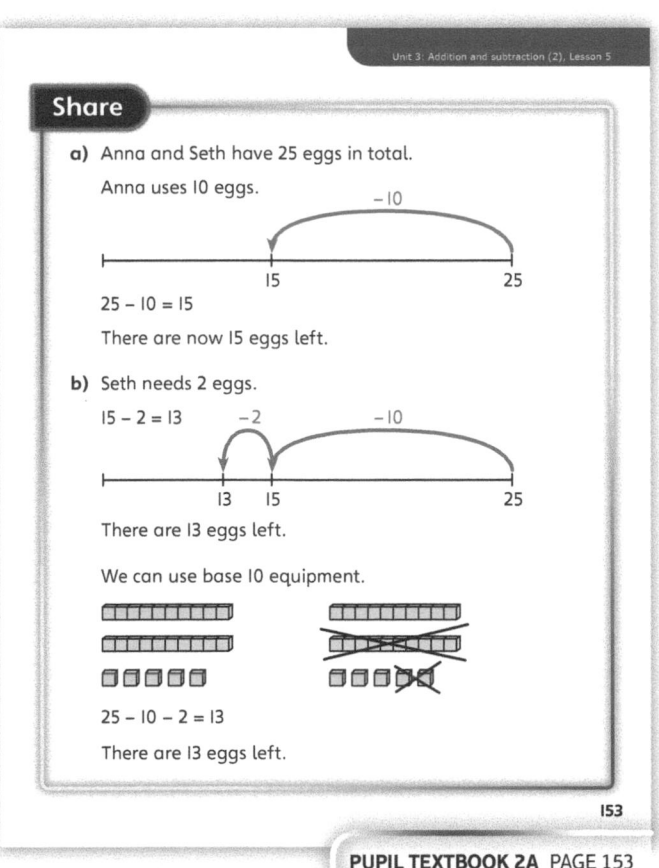

PUPIL TEXTBOOK 2A PAGE 153

Think together

ASK

- Question **1**: *What is the first step? What is the next step?*
- Question **2**: *What do you notice about each calculation?*
- Question **3**: *What is the same and what is different about each method?*

IN FOCUS In questions **1** and **2**, children subtract 10s and then 1s from a 2-digit number. This is shown with jumps on the number line in question **1**. In question **2**, they explore two similar calculations and should notice that the 10s and 1s digits are inverted. Question **3** builds on this by requiring children to compare two methods, subtracting either the 10s or the 1s first. Ask: *Which do you prefer? Why?*

STRENGTHEN Ask children to use place value equipment alongside their working as they complete each of the subtractions in stages.

DEEPEN Challenge children to compare this subtraction method with the addition method of adding more 10s and then more 1s.

ASSESSMENT CHECKPOINT Question **2** will allow you to assess whether children are correctly partitioning and interpreting the value of the 10s and 1s when using this subtraction method.

ANSWERS

Question **1** a): 38 – 10 = 28

Question **1** b): 38 – 15 = 23

Question **2** a): 44 – 13 = 31
44 – 31 = 11

Question **2** b): The starting number is the same. The digits in the second number are the same, just the other way round. This gives a different answer to the calculation.

Question **3**: 64 – 13 = 51
48 – 35 = 13

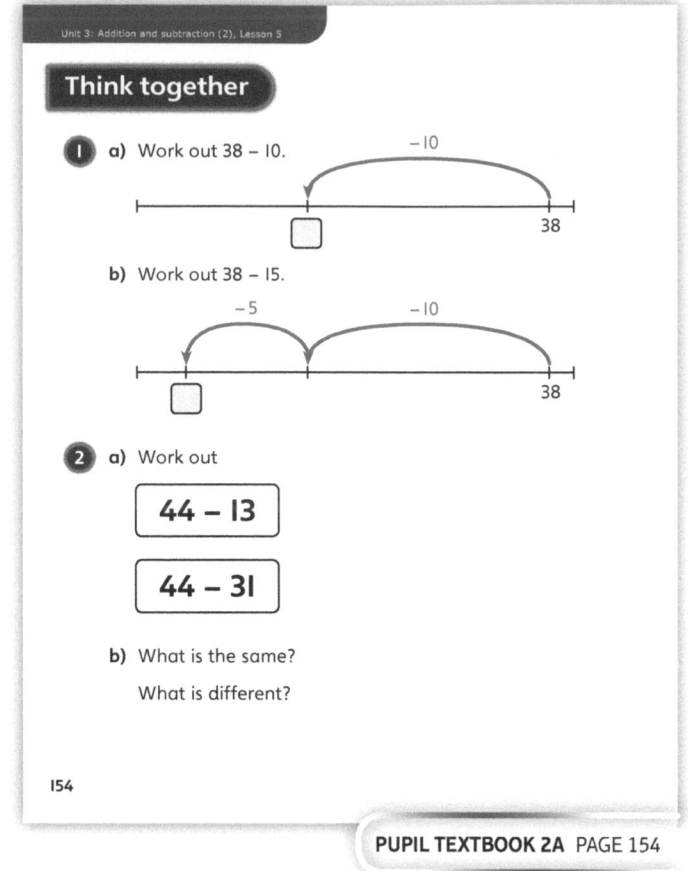

Practice

WAYS OF WORKING Independent thinking

IN FOCUS In question ❶, children complete subtractions on a given number line. Question ❷ requires them to fill in their own number lines to show the calculation steps. In Question ❸, children compare two methods, subtracting either the 10s or the 1s first, and prove that both reach the same answer. Question ❹ requires children to interpret a representation of this method and identify the calculation that it shows.

STRENGTHEN Ensure that children are using place value equipment to build up each of the calculations in stages.

DEEPEN Ask children to investigate how they could use each of the methods shown in question ❸ as a checking strategy for other calculations.

ASSESSMENT CHECKPOINT Use question ❷ to assess whether children are able to represent this subtraction method as jumps back of 10s and 1s on a number line.

ANSWERS Answers for the **Practice** part of the lesson can be found in the *Power Maths* online subscription.

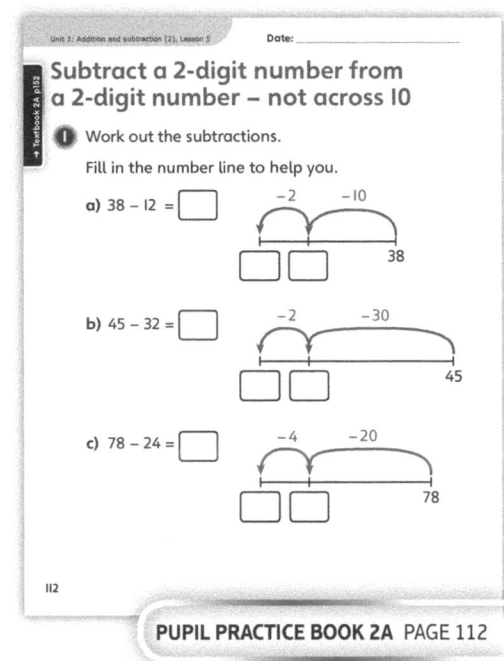

PUPIL PRACTICE BOOK 2A PAGE 112

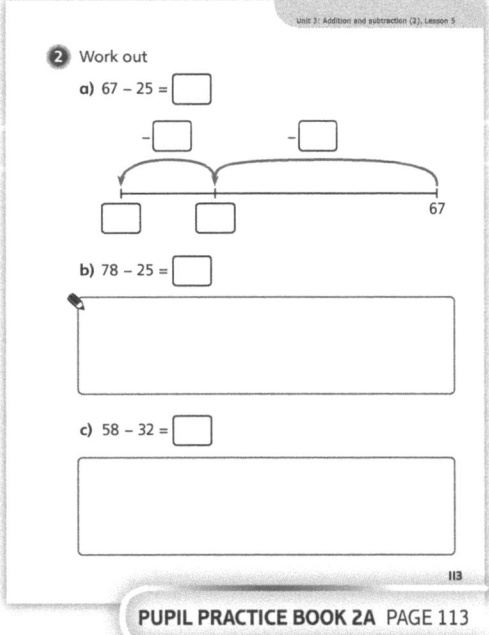

PUPIL PRACTICE BOOK 2A PAGE 113

Reflect

WAYS OF WORKING Pair work

IN FOCUS The **Reflect** part of the lesson is a prompt for children to discuss the steps that they will carry out as they work through this method.

ASSESSMENT CHECKPOINT Assess whether children can explain how to subtract a 2-digit number by first partitioning, and then subtracting first the 10s and then the 1s.

ANSWERS Answers for the **Reflect** part of the lesson can be found in the *Power Maths* online subscription.

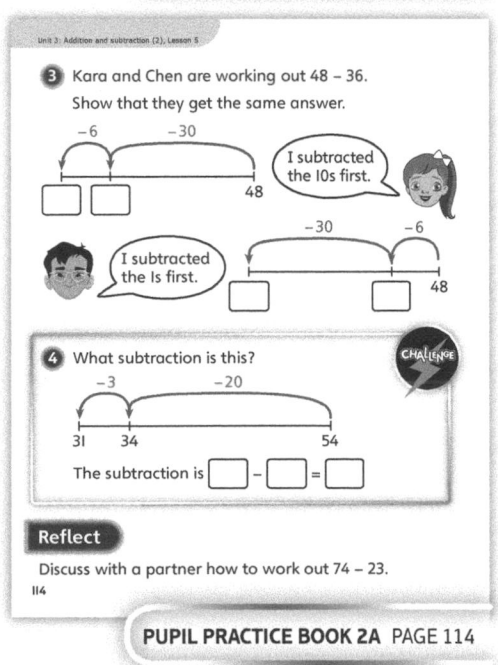

PUPIL PRACTICE BOOK 2A PAGE 114

After the lesson ⏸

- How will you support children in rehearsing this skill regularly as part of their fluency practice?
- Have children had the opportunity to discuss the importance of using this method accurately, rather than just looking for correct or speedy answers?

Subtract a 2-digit number from a 2-digit number – across 10

Learning focus

In this lesson, children build on the method from the previous lesson and begin to subtract 2-digit numbers with calculations that include crossing 10s.

Before you teach

- Can children subtract 10 from a given number?
- Can children subtract in 1s from a given number?
- Can children subtract a 2-digit number with no crossing of 10s?

NATIONAL CURRICULUM LINKS

Year 2 Number – addition and subtraction

Solve problems with addition and subtraction: applying their increasing knowledge of mental and written methods.

Add and subtract numbers using concrete objects, pictorial representations, and mentally, including two 2-digit numbers.

ASSESSING MASTERY

Children can subtract any 2-digit number from another 2-digit number, including those that involve crossing a 10.

COMMON MISCONCEPTIONS

Children may struggle to find an efficient or accurate way to subtract across a ten. Ask:
- *What steps can help you subtract to the previous 10? How many are left?*
- *How could a simple number line drawing help to show the steps?*

STRENGTHENING UNDERSTANDING

Focus on developing this skill by working with children to first subtract across 10 for calculations within 20. Go back and look together at the lessons in Unit 2 to rehearse this skill more fully.

GOING DEEPER

Challenge children to explore how they can use known bonds to help them with calculations within 20 that require them to cross a 10.

KEY LANGUAGE

In lesson: subtract, tens (10s), ones (1s)

STRUCTURES AND REPRESENTATIONS

Number lines

RESOURCES

Mandatory: number lines

Optional: rekenreks

 In the eTextbook of this lesson, you will find interactive links to a selection of teaching tools.

Quick recap

Ask children to count back together in 1s from any given 2-digit number, including counting across 10s. Repeat with different 2-digit numbers.

Discover

WAYS OF WORKING Pair work

ASK

- Question ① a): *Can you see that there are 10 strawberries in each of Kara's full punnets?*
- Question ① a): *Can you see the 4 strawberries that are loose?*
- Question ① a): *How would you subtract 10?*
- Question ① b): *What is tricky about subtracting 5 strawberries?*

IN FOCUS Children see a visual representation of a 2-digit number shown as some 10s and some 1s. In this context, a 10 is taken away, followed by some 1s. This will generate a calculation that requires children to first subtract a 10 and then subtract some 1s, which will include crossing 10. Ensure children understand that a take-away problem will need a subtraction calculation.

PRACTICAL TIPS Ask children to use ten frames to model the problem. This will support them with the crossing of a ten.

ANSWERS

Question ① a): There are 14 strawberries left in Kara's tray.

Question ① b): There are 9 strawberries left in Kara's tray.

Subtract a 2-digit number from a 2-digit number – across 10

Discover

① a) Kara has 24 strawberries. Will takes 10 strawberries.
How many strawberries are left in Kara's tray?

b) Asha takes another 5 strawberries.
How many strawberries are left in Kara's tray?

156

PUPIL TEXTBOOK 2A PAGE 156

Share

WAYS OF WORKING Whole class teacher led

ASK

- Question ① a): *How does the number line show subtract 10?*
- Question ① b): *Which part of this shows that we are subtracting 5?*
- Question ① b): *Why is subtract 5 split up into two jumps?*

IN FOCUS Question ① a) shows a subtraction of 10 as one jump back of 10 on the number line. In question ① b), children then subtract some 1s. This is shown as two jumps on the number line in order to model the crossing of the 10. Discuss with children how their knowledge of number bonds to 10 can be helpful in calculations like these.

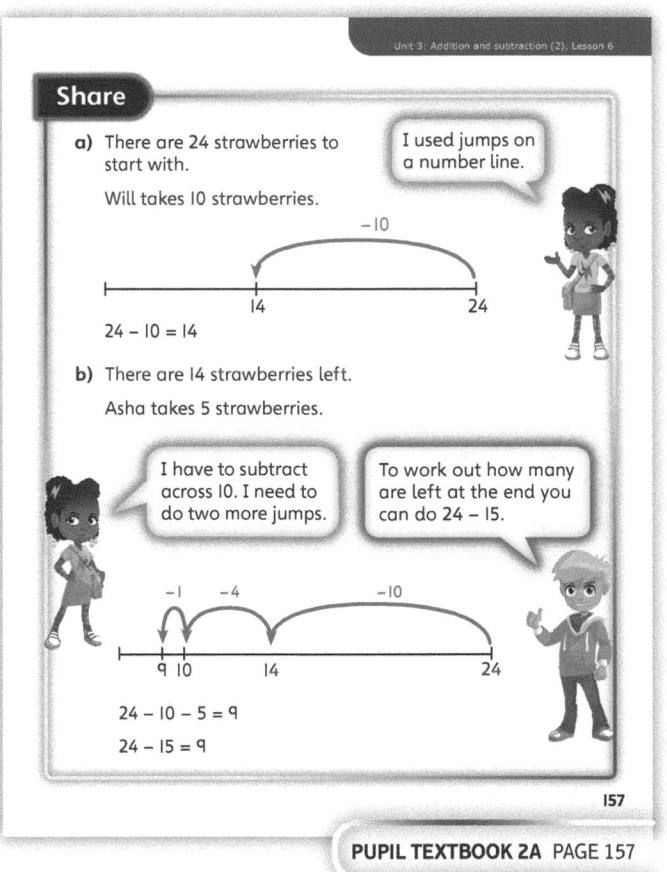

Share

a) There are 24 strawberries to start with.

Will takes 10 strawberries.

I used jumps on a number line.

$$24 - 10 = 14$$

b) There are 14 strawberries left.

Asha takes 5 strawberries.

I have to subtract across 10. I need to do two more jumps.

To work out how many are left at the end you can do 24 – 15.

$$24 - 10 - 5 = 9$$
$$24 - 15 = 9$$

157

PUPIL TEXTBOOK 2A PAGE 157

Think together

Whole class teacher led (I do, We do, You do)

ASK
- Question ① b): *What is the first step? Why are there three jumps?*
- Question ②: *Which jump will you do first? What jumps will help subtract the 1s?*
- Question ③: *What patterns do you see?*

IN FOCUS In question ①, children are subtracting one 10 and then some 1s. This is shown with jumps on a number line. In question ②, children are subtracting one or more 10s and then some 1s. The number line model is provided, but the jumps are not shown. Question ③ requires children to start using known patterns of number bonds within 20 in order to help them to find answers.

STRENGTHEN Ensure children are using ten frames to support them when working through the crossing of a 10 in these subtractions.

DEEPEN Ask children to further explore their use of known bonds within 20 when working through calculations that cross 10. Ask: *Can you explain which bonds will be helpful in each calculation? Why are these bonds helpful?*

ASSESSMENT CHECKPOINT Question ② will allow you to assess whether children are able to use the number line in supporting them with using jumps to subtract some 10s and then some 1s, where a 10 will be crossed.

ANSWERS

Question ① a): 32 − 10 = 22

Question ① b): 32 − 13 = 19

Question ② a): 52 − 15 = 37

Question ② b): 52 − 25 = 27

Children may come up with multiple answers to this question. One thing that is the same is that both jumps cross a ten. A difference is that for b), children are likely to have to make four jumps instead of three to take into account the extra 10.

Question ③ a): 21 − 7 = 14
51 − 7 = 44
31 − 7 = 24
61 − 7 = 54
41 − 7 = 34

Question ③ b): 51 − 17 = 34
61 − 17 = 44
41 − 17 = 24
21 − 17 = 4
81 − 17 = 64

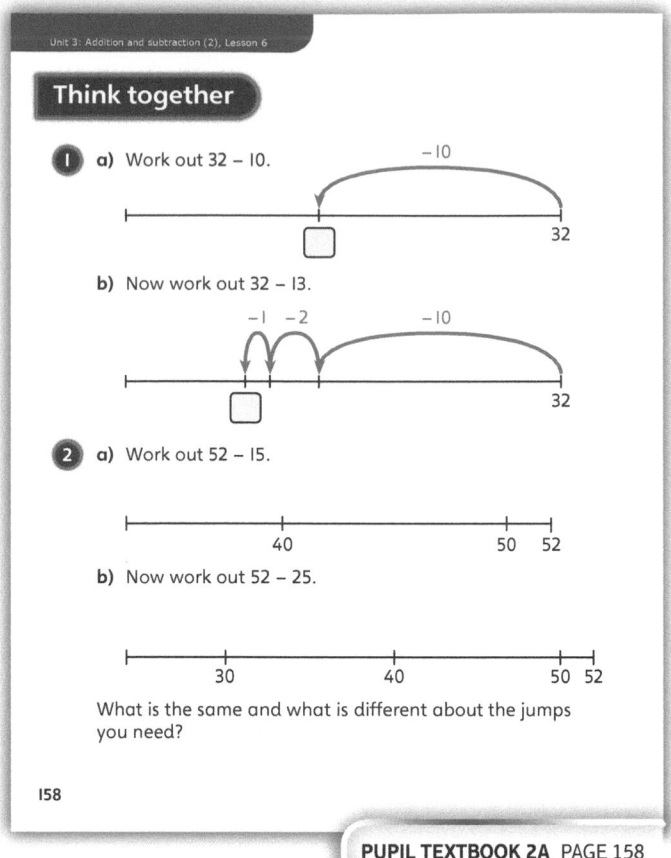

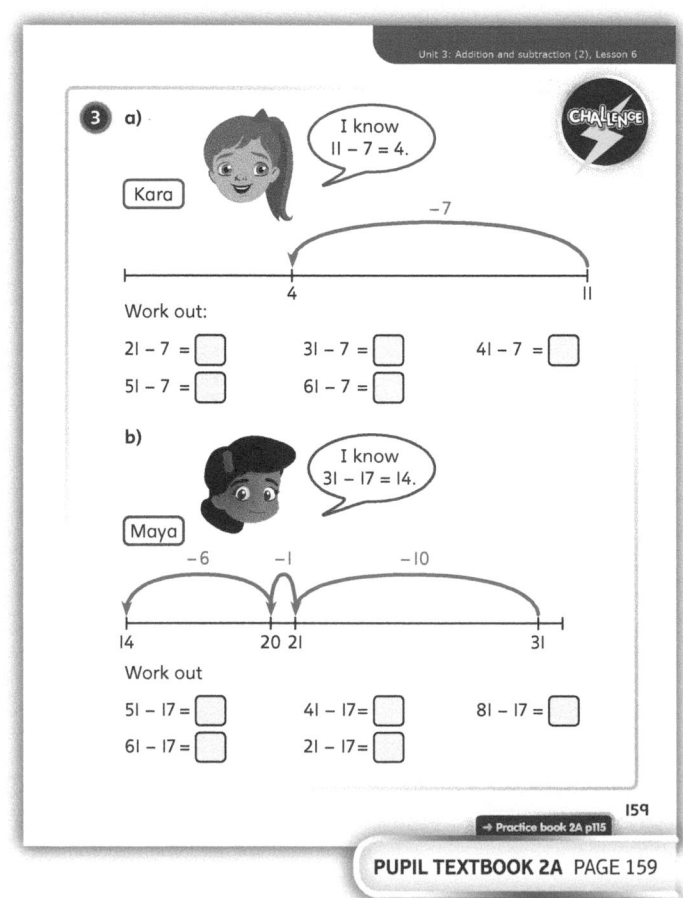

Practice

WAYS OF WORKING Independent thinking

IN FOCUS In questions ① and ②, children will subtract 2-digit numbers in calculations that involve crossing 10s, with the number line model provided. However, in question ②, children are required to decide which steps are needed for the calculation and to label the jumps on the number line themselves. Blank number lines are given for question ③ so that children can draw on their own representations. Question ④ is a puzzle in a problem-solving context.

STRENGTHEN Continue to encourage children to use ten frames or rekenreks to support them when crossing 10s.

DEEPEN Ask children to explore how they can change the order of the steps in calculations and still reach the same answer, for example, 43 – 10 – 5 = ▢ and 43 – 5 – 10 = ▢.

ASSESSMENT CHECKPOINT Use question ② to assess whether children can identify the jumps on a number line that can be used to subtract 10s and 1s, crossing a 10.

ANSWERS Answers for the **Practice** part of the lesson can be found in the *Power Maths* online subscription.

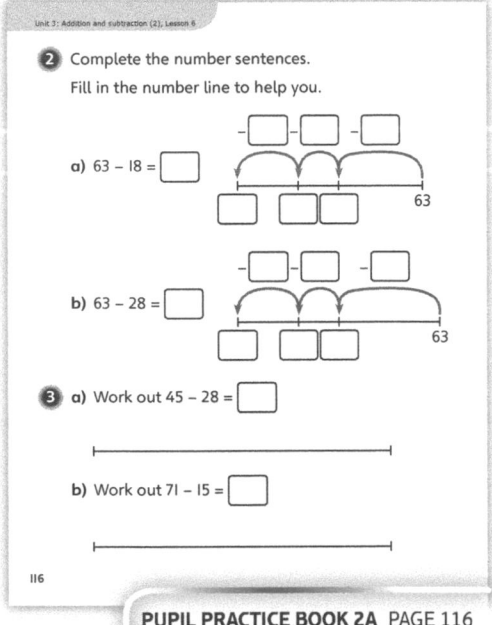

Date: _____

Unit 3: Addition and subtraction (2), Lesson 6

Subtract a 2-digit number from a 2-digit number – across 10

→ Textbook 2A p136

① Complete the number sentences.
Fill in the number line to help you.

a) 52 – 25 = ▢

-3 -2 -20
▢ ▢ ▢ 52

b) 41 – 16 = ▢

-5 -1 -10
▢ ▢ 41

c) 74 – 26 = ▢

-2 -4 -20
▢ ▢ ▢ 74

115

PUPIL PRACTICE BOOK 2A PAGE 115

Unit 3: Addition and subtraction (2), Lesson 6

② Complete the number sentences.
Fill in the number line to help you.

a) 63 – 18 = ▢

-▢ -▢ -▢
▢ ▢ 63

b) 63 – 28 = ▢

-▢ -▢ -▢
▢ ▢ 63

③ a) Work out 45 – 28 = ▢

b) Work out 71 – 15 = ▢

116

PUPIL PRACTICE BOOK 2A PAGE 116

Reflect

WAYS OF WORKING Pair work

IN FOCUS The **Reflect** part of the lesson prompts children to make a comparison of methods for adding and subtracting 2-digit numbers, with some examples that require crossing a 10.

ASSESSMENT CHECKPOINT Assess whether children can identify instances where it will be necessary to cross a 10 and can recognise and describe the similarities and differences in the methods used.

ANSWERS Answers for the **Reflect** part of the lesson can be found in the *Power Maths* online subscription.

④ Here are six digit cards.

CHALLENGE

1 2 3 4 6 8

a) Use each card once to complete the subtraction.

▢▢ - ▢▢ = ▢▢

b) Find another way.

▢▢ - ▢▢ = ▢▢

Reflect

Discuss with a partner how you work out each of these.

43 + 15   45 + 18   45 – 12   45 – 17

117

PUPIL PRACTICE BOOK 2A PAGE 117

After the lesson ⏸

• How will you make sure that children can regularly rehearse 2-digit subtractions as part of their fluency practice?

How many more? How many fewer?

Learning focus

In this lesson, children will answer questions worded 'how many more?' and 'how many fewer?'. They will compare quantities of objects to find the difference and represent this on a number line or bar model.

Before you teach

- Are all children secure with counting on or back from different starting points?
- How familiar are children with the words 'more' and 'fewer'?
- How could you link 'finding more' and 'finding fewer' to subtraction facts?

NATIONAL CURRICULUM LINKS

Year 2 Number – addition and subtraction

Solve problems with addition and subtraction: applying their increasing knowledge of mental and written methods.

Add and subtract numbers using concrete objects, pictorial representations, and mentally, including two 2-digit numbers.

ASSESSING MASTERY

Children can compare different quantities of objects in terms of how many more or fewer there are. Children can understand that the difference is the same when comparing two groups in terms of more and fewer, and express this correctly in a sentence (for example, 'there are 2 more in the first group' and 'there are 2 fewer in the second group', or '6 is 2 more than 4' and '4 is 2 less than 6').

COMMON MISCONCEPTIONS

Children may line up groups of objects inaccurately when comparing them, meaning that objects from one group do not correspond with those from the other. Demonstrate how to line up objects consistently, and explain that anything left over (anything that can't be matched) is the difference between the groups. Ask:

- *How are you going to line up the objects? Which objects are the ones left over? Which group has more? Which group has fewer?*

STRENGTHENING UNDERSTANDING

Strengthen understanding by asking two unequal groups of children to sit in rows on the carpet, with the smaller group spread out to take up the same amount of space. Then ask children to pair up with someone from the other group. Explain that the children with no partners make up the difference between these two groups: there are more children in one group and fewer in the other.

GOING DEEPER

Use classroom stationery objects to deepen understanding by asking children to go on a comparison hunt. Choose a number for comparison, and ask them to find out how many more or fewer pens, pencils, rubbers, sharpeners etc. there are in the room. Provide blank sentence scaffolds and number sentences to help children record their thinking.

KEY LANGUAGE

In lesson: how many more, how many fewer, compare, **difference**, count on, count back

Other language used by the teacher: comparison model

STRUCTURES AND REPRESENTATIONS

Number line, bar model

RESOURCES

Mandatory: blank number lines, cubes and/or counters

Optional: classroom stationery objects, blank sentence scaffolds and number sentences

 In the eTextbook of this lesson, you will find interactive links to a selection of teaching tools.

Quick recap

Ask children to order given sets of 2-digit numbers from least to greatest. Repeat several times.

Discover

Unit 3: Addition and subtraction (2), Lesson 7

WAYS OF WORKING   Pair work

ASK

- Question ❶: In response to the picture: *How many children are in the first row? How many are there in the second row?*
- Question ❶: *What do 'how many more?' and 'how many fewer?' mean?*
- After question ❶ b): *Why are both answers the same?*

IN FOCUS   Question ❶ requires children to understand the concept of more and fewer. Children may count how many children are in each row of the picture, and then not know what to do with the results, or mistakenly say there are '8 more' because 8 is more than 6.

PRACTICAL TIPS   Children can use concrete resources such as cubes to model the number of people in each group. Discuss how arranging the cubes carefully makes it easier to compare the two groups.

ANSWERS

Question ❶ a): There are 2 more children in the back row than there are in the front row.

Question ❶ b): There are 2 fewer children in the front row than there are in the back row.

How many more? How many fewer?

Discover

❶ a) How many more children are there in the back row?

  b) How many fewer children are there in the front row?

160

PUPIL TEXTBOOK 2A PAGE 160

Share

WAYS OF WORKING   Whole class teacher led

ASK

- Question ❶ a): *Why are the children in two rows like this? How does that help us?*
- Question ❶: *Why are the arrows going in different directions on the number lines?*

IN FOCUS   This section uses key terms that you have the opportunity to discuss. Discuss the word 'difference', explaining that it indicates how many more or fewer things there are when you compare them. Highlight what Sparks says: 'You can count on or count back to find the difference.'

Share

a) This is a problem about 'finding the difference'.

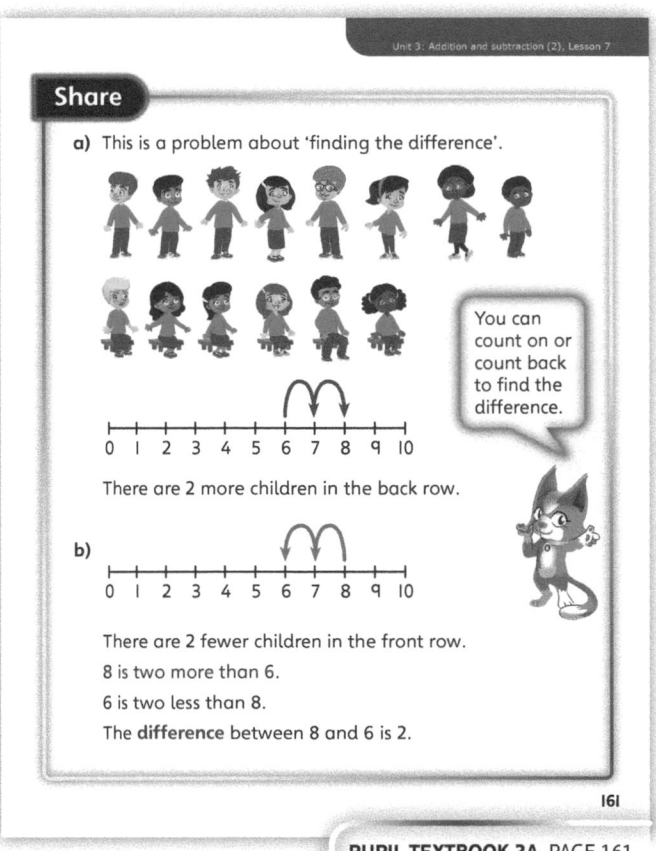

You can count on or count back to find the difference.

There are 2 more children in the back row.

b)

There are 2 fewer children in the front row.

8 is two more than 6.

6 is two less than 8.

The **difference** between 8 and 6 is 2.

161

PUPIL TEXTBOOK 2A PAGE 161

Think together

Whole class teacher led (I do, We do, You do)

ASK

- Question **1**: *Which row has more? How many more? Which row has less? How many less?*
- Question **2**: *Which part of the diagram shows the difference between 7 and 17?*
- Question **3**: *What is the same and what is different? What do you notice?*

IN FOCUS In question **1**, children are finding the difference between two quantities, where the number is represented by beads that are countable. In question **2**, children find the difference between two quantities, but this time it is not countable. Question **3** requires children to investigate pairs of numbers with a given difference.

STRENGTHEN Together think of pairs of numbers that all have a difference of 10. What do children notice?

DEEPEN Challenge children to investigate pairs of numbers with a given difference, such as 9. What do they notice?.

ASSESSMENT CHECKPOINT Assess whether children can explain how to find the difference between two given numbers.

ANSWERS

Question **1**: 11 is 6 more than 5; 5 is 6 less than 11.

Question **2**: 7 is 10 less than 17; 17 is 10 more than 7.

Question **3**: Any two numbers with a difference of 5. From Dexter's number line, the pairs are: 0 and 5; 1 and 6; 2 and 7; 3 and 8; 4 and 9; 5 and 10; 7 and 12; 8 and 13; 9 and 14; 10 and 15.

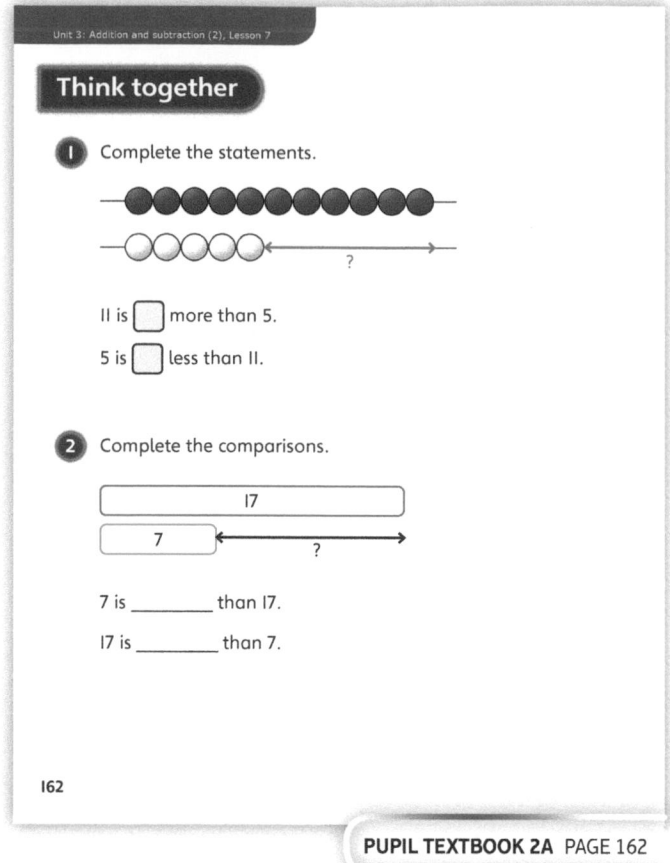

PUPIL TEXTBOOK 2A PAGE 162

PUPIL TEXTBOOK 2A PAGE 163

Practice

IN FOCUS In question **1**, the difference between the numbers is countable in each of the representations, whereas in question **2**, the differences are represented on comparison bar models. In question **3**, children find the difference by choosing their own models or methods.

Question **4** provides no pictorial representations to support thinking. Children could draw the problem or represent it on a number line to help them work it out.

Question **5** requires children to appreciate the difference as an addition or a subtraction.

STRENGTHEN Encourage children to use countable items such as counters or rekenreks to explore the concept fully.

DEEPEN Ask children to come up with their own sentences to match the number cards given in question **4**. As 0 is included, they could come up with a range of sentences that all have '0' as their answer: for example, '9 less than 9 is 0'.

ASSESSMENT CHECKPOINT Use question **3** to assess whether children can identify and represent a suitable method for comparing two numbers and finding the difference between them.

ANSWERS Answers for the **Practice** part of the lesson can be found in the *Power Maths* online subscription.

Reflect

IN FOCUS The **Reflect** part of the lesson prompts children to express their thoughts about finding pairs of numbers with a given difference.

ASSESSMENT CHECKPOINT Assess whether children can accurately find several different examples that satisfy the criteria, and whether they can reason about the patterns that they notice.

ANSWERS Answers for the **Reflect** part of the lesson can be found in the *Power Maths* online subscription.

After the lesson ⏸

- Do children understand that the word 'difference' can mean 'more than' or 'fewer than'?
- Do children understand the importance of matching up objects in different groups in order to compare them accurately?
- Were children able to represent difference on a number line, as a bar model and in a subtraction fact?

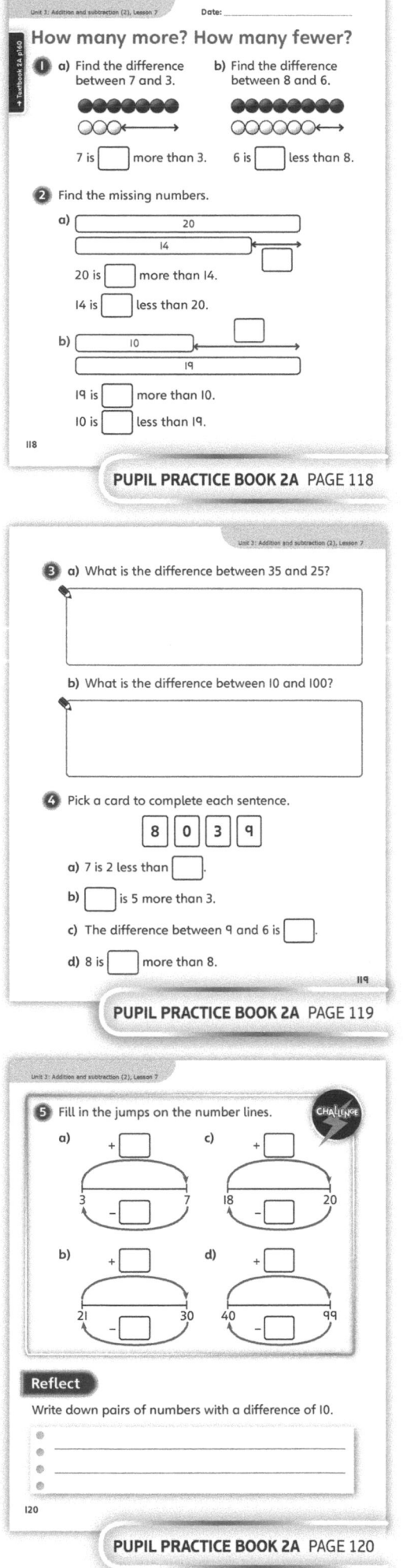

PUPIL PRACTICE BOOK 2A PAGE 118

PUPIL PRACTICE BOOK 2A PAGE 119

PUPIL PRACTICE BOOK 2A PAGE 120

Subtraction – find the difference

Learning focus

In this lesson, children more formally link 'finding the difference' with subtraction, building on their work from the previous lesson about finding 'how many more' or 'how many fewer.'

Before you teach

- Can children say how many more 9 is than 6?
- Can children say how many less 6 is than 9?
- Can children interpret a 'find the difference' question as a comparison between two quantities, such as bead strings, bar models or between two numbers on a number line?

NATIONAL CURRICULUM LINKS

Year 2 Number – addition and subtraction

Solve problems with addition and subtraction: using concrete objects and pictorial representations, including those involving numbers, quantities and measures.

ASSESSING MASTERY

Children can use subtraction to find the difference between two numbers.

COMMON MISCONCEPTIONS

Children may find it difficult to understand the link between subtraction and questions that include the word more, such as 'How many more is 9 than 6?'. Ask:
- *What could you draw to show the difference between these numbers?*

STRENGTHENING UNDERSTANDING

Build understanding of the number line representation by asking children to work initially within 10, then 20, then building up to 100.

GOING DEEPER

Ask children to fully explore how finding the difference between two numbers that are close together can be a good method for completing a subtraction. For example, to complete 98 – 96, it is more efficient to work out that the difference is 2, rather than to subtract 9 tens then 6 ones from 98.

KEY LANGUAGE

In lesson: subtract, difference, find the difference

STRUCTURES AND REPRESENTATIONS

Place value equipment, number lines

RESOURCES

Mandatory: number lines

Optional: base 10 equipment

 In the eTextbook of this lesson, you will find interactive links to a selection of teaching tools.

Quick recap

Ask children to locate given 2-digit numbers on a number line that runs from 0 to 100. Repeat several times.

Discover

WAYS OF WORKING Pair work

ASK

- Question ❶ a): *Can you represent each person's age using place value equipment?*
- Question ❶ a): *Which age is older? Which is younger?*
- Question ❶ b): *Which number is less than the other number?*

IN FOCUS Children can use this context to explore the relationship between subtraction and finding the difference. They should notice that the two answers are the same, even though the questions are presented differently and use different language.

PRACTICAL TIPS Ask children to use place value equipment to model the numbers in the calculation. This should help them to see why a subtraction is required.

ANSWERS

Question ❶ a): John is 13 years older than Sofia.

Question ❶ b): 43 − 30 = 13

PUPIL TEXTBOOK 2A PAGE 164

Share

WAYS OF WORKING Whole class teacher led

ASK

- Question ❶ a): *Which part shows 30? Which part shows 43? Which part shows how many more?*
- Question ❶ b): *What has been taken away? How does finding the difference solve the subtraction?*

IN FOCUS In question ❶ a), children are finding the difference between two numbers. In question ❶ b), they are required to find the difference in order to solve a subtraction. They should notice that the two answers are the same.

PUPIL TEXTBOOK 2A PAGE 165

Think together

Whole class teacher led (I do, We do, You do)

ASK

- Question **1**: *Which part of the diagram shows the difference between 20 and 11?*
- Question **2**: *How is the difference between 21 and 25 related to the subtraction 25 – 21?*
- Question **3**: *Do you notice how easy it is to solve 81 – 79 by finding the difference? What would you do if you solved it by subtracting 10s and 1s?*

IN FOCUS In questions **1** and **2**, children find the difference between two numbers in order to solve a subtraction. Question **1** requires them to consider when finding the difference is a good method for solving a subtraction. They should notice that it works well with two numbers that are close together.

STRENGTHEN Build children's confidence by asking them to solve subtractions where they are finding the difference within 10, then within 20.

DEEPEN Challenge children to find as many subtractions as they can with an answer of 50. They should use finding the difference to prove that their answers are correct.

ASSESSMENT CHECKPOINT Assess whether children can explain how to use finding the difference to solve subtractions.

ANSWERS

Question **1**: 11 is 9 less than 20; 20 is 9 more than 11.
The difference between 11 and 20 is 9.

Question **2**: 25 – 21 = 4; 25 is 4 more than 21;
21 is 4 less than 25.
The difference between 21 and 25 is 4.

Question **3**: 35 – 32 = 3
12 – 9 = 3
60 – 57 = 3
65 – 11 = 54
81 – 79 = 2

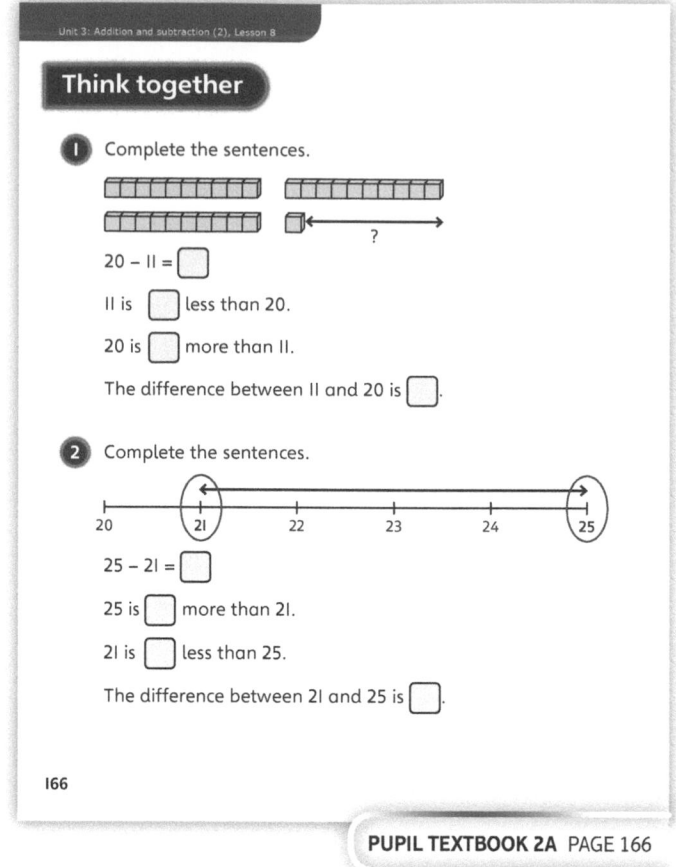

PUPIL TEXTBOOK 2A PAGE 166

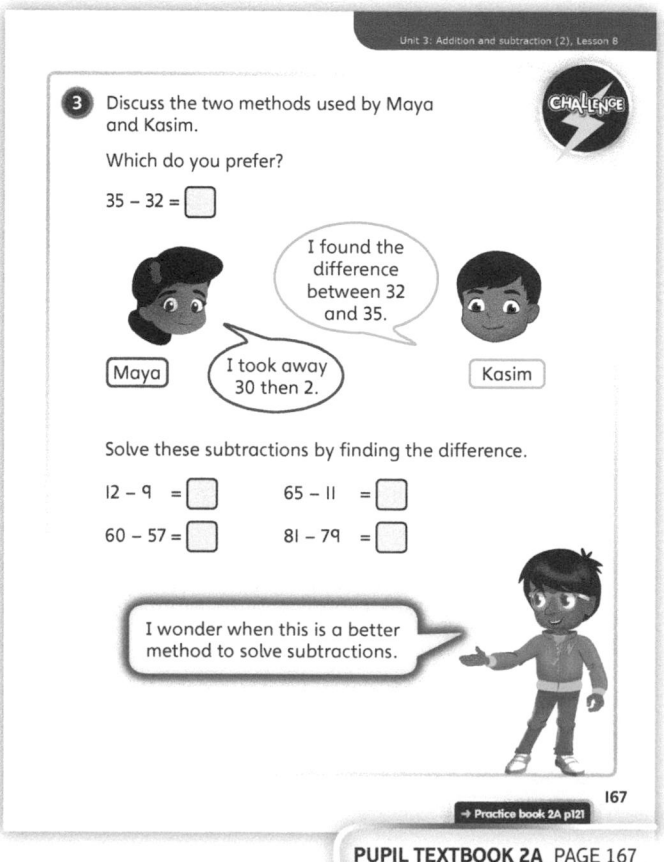

PUPIL TEXTBOOK 2A PAGE 167

Practice

WAYS OF WORKING Independent thinking

IN FOCUS In questions ① and ②, children are finding the difference between numbers that are represented with place value equipment. In question ③, they find the difference between two numbers on a number line. Question ④ involves finding the difference in the context of a word problem. Question ⑤ is an investigation into pairs of numbers that have a constant difference.

STRENGTHEN Give children difference questions that will require them to work first within 10, and then within 20, in order to build their confidence.

DEEPEN Challenge children to compare and contrast different subtraction methods, including 'finding the difference'.

ASSESSMENT CHECKPOINT Question ④ will allow you to assess whether children can recognise and solve a 'finding the difference' question that is presented as a word problem.

ANSWERS Answers for the **Practice** part of the lesson can be found in the *Power Maths* online subscription.

Reflect

WAYS OF WORKING Pair work

IN FOCUS The **Reflect** part of the lesson prompts children to demonstrate their understanding of the link between subtraction and finding the difference.

ASSESSMENT CHECKPOINT Assess whether children can solve a subtraction using the 'find the difference' method.

ANSWERS Answers for the **Reflect** part of the lesson can be found in the *Power Maths* online subscription.

After the lesson

- Can children solve problems involving 'find the difference' by using subtraction?
- Do children know how to solve subtractions by finding the difference between two numbers?

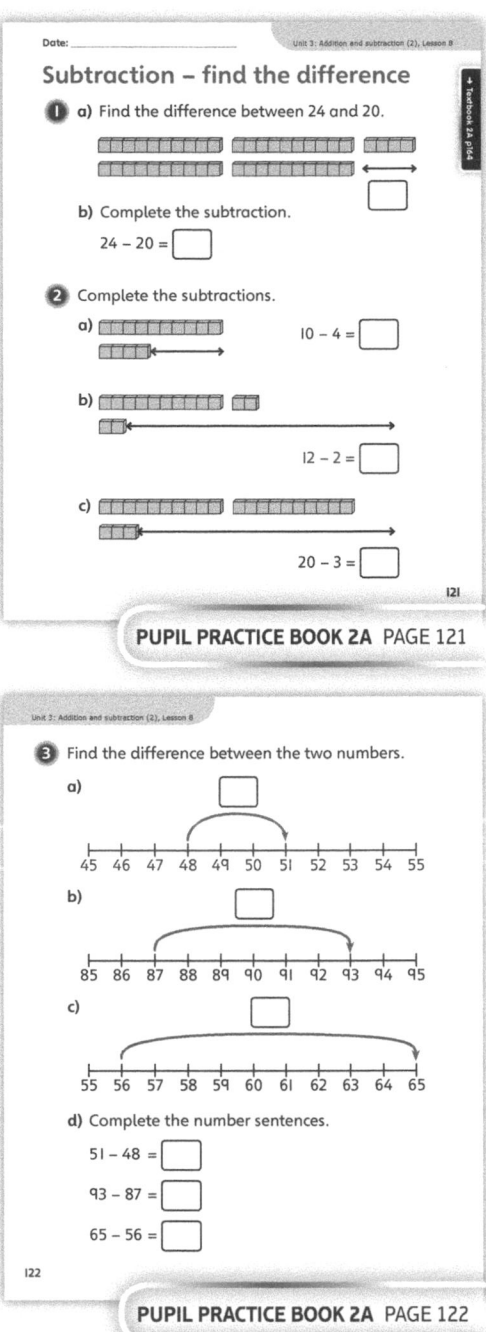

PUPIL PRACTICE BOOK 2A PAGE 121

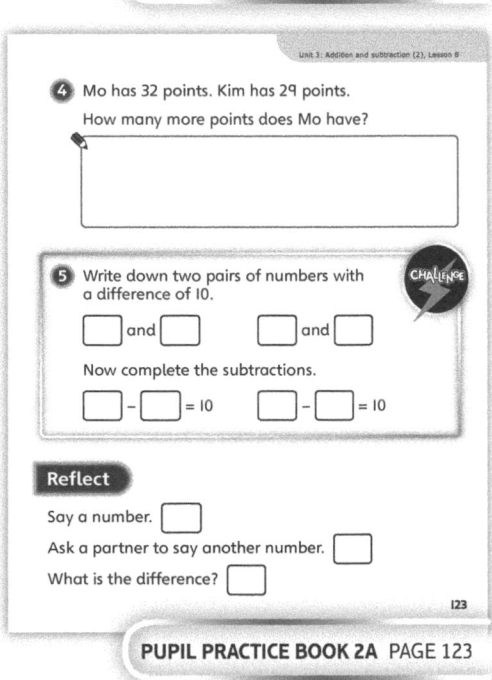

PUPIL PRACTICE BOOK 2A PAGE 122

PUPIL PRACTICE BOOK 2A PAGE 123

Compare number sentences

Learning focus

In this lesson, children recognise that two calculations can be compared using the symbols =, < and >.

Before you teach ⏸

- Can children add confidently within 20?
- Can children subtract confidently within 20?
- Can children use the < and > symbols correctly?

NATIONAL CURRICULUM LINKS

Year 2 Number – addition and subtraction

Solve problems with addition and subtraction: using concrete objects and pictorial representations, including those involving numbers, quantities and measures.

Recall and use addition and subtraction facts to 20 fluently, and derive and use related facts up to 100.

ASSESSING MASTERY

Children can use the symbols =, < and > to compare calculations.

COMMON MISCONCEPTIONS

Children may work straight through the calculations on a left-to-right basis, looking only at the number that is immediately after the symbol, for example: 12 + 3 = 15 + 2 = 17. Ask:
- *What is on this side of the symbol? What is on that side?*

STRENGTHENING UNDERSTANDING

Help children to focus on thinking about the calculations on each side of the symbol as two separate entities, rather than assuming that the equals sign means that an answer will immediately follow.

GOING DEEPER

Challenge children to fully examine and explore the pattern that is implied in **Think together** question ❸ b). Ask: *What do you notice about how each of the numbers changes each time?*

KEY LANGUAGE

In lesson: compare, equal, greater than, less than

Other language to be used by the teacher: addition, number sentence, smallest

STRUCTURES AND REPRESENTATIONS

< and > signs

RESOURCES

Mandatory: cubes or counters

 In the eTextbook of this lesson, you will find interactive links to a selection of teaching tools.

Quick recap

Ask children to use the signs < and > to compare two given 2-digit numbers. Repeat several times.

Discover

Unit 3: Addition and subtraction (2), Lesson 9

WAYS OF WORKING Pair work

ASK

- Question ❶ a): *Each child has 11 cubes. What parts have they made?*
- Question ❶ a): *Could you make any different parts with 11 cubes?*

IN FOCUS Children see how the same number of cubes can be arranged in a variety of ways to show several different parts. They can use this to help them recognise that there can be different partitions of a constant total.

PRACTICAL TIPS Ask children to make towers of 11 cubes and to then try and break their towers into several parts. Did everybody make different parts? Collect the results and support children in recording them as additions.

ANSWERS

Question ❶ a): 7 + 1 + 3; 1 + 2 + 8; 8 + 3

Question ❶ b): 8 + 3 = 10 + 1

Compare number sentences

Discover

Make a tower of 11 cubes. Break it into parts.

❶ a) Write an addition for each partition of 11.

b) Complete the number sentence.

8 + 3 = 10 + ☐

168

PUPIL TEXTBOOK 2A PAGE 168

Share

WAYS OF WORKING Whole class teacher led

ASK

- Question ❶ a): *Can you see how these show the different parts of the towers?*
- Question ❶ a): *What is the same and what is different each time?*
- Question ❶ a): *Why do the number sentences all equal 11?*
- Question ❶ b): *Can you see that 8 + 3 is equal to 10 + 1? What does this mean?*

IN FOCUS Question ❶ a) shows the different partitions of a given total, represented with groups of cubes and as number sentences. In question ❶ b), the equals sign is introduced in a context where it means 'is the same total as' rather than 'the answer is'.

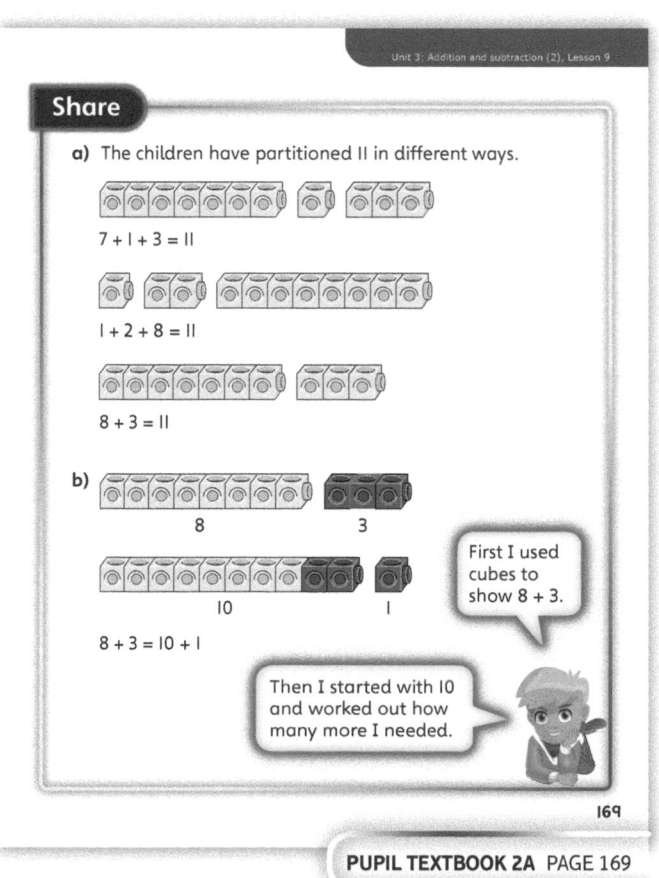

Share

a) The children have partitioned 11 in different ways.

7 + 1 + 3 = 11

1 + 2 + 8 = 11

8 + 3 = 11

b)

8   3

10   1

8 + 3 = 10 + 1

First I used cubes to show 8 + 3.

Then I started with 10 and worked out how many more I needed.

169

PUPIL TEXTBOOK 2A PAGE 169

Think together

WAYS OF WORKING Whole class teacher led (I do, We do, You do)

ASK

• Question **1**: *What is the same, what is different?*
• Question **2**: *Can you work out the correct symbol without calculating?*
• Question **3**: *What patterns do you spot?*

IN FOCUS In question **1**, cubes are used to represent an equation with two additions. Question **2** requires children to recognise and reason about inequalities of calculations using the < and > symbols. In question **3**, children explore and reason about equalities and inequalities in order to find possible missing numbers where more than one answer is possible.

STRENGTHEN Ask children to use cubes or counters to model the calculations and support their reasoning throughout

DEEPEN Challenge children to continue the pattern in question **3** b) as far as they can. Can they find every answer up to 39 + 16 = 40 + ☐ and describe the pattern as they go?

ASSESSMENT CHECKPOINT Assess whether children can justify their decisions about missing numbers or symbols using reasoning.

ANSWERS

Question **1** a): 6 + 2 = 5 + 3

Question **1** b): 6 + 2 = 4 + 4

Question **2** a): 4 + 3 > 4 + 2

Question **2** b): 8 − 2 > 8 − 3

Question **3** a): 1st box: any number between 0 and 7 (smallest number is 0);
2nd box: 0, 1 or 2 (smallest number is 0)

Question **3** b): 5 + 1 = 6 + 0
8 + 3 = 9 + 2
9 + 9 = 10 + 8
39 + 16 = 40 + 15

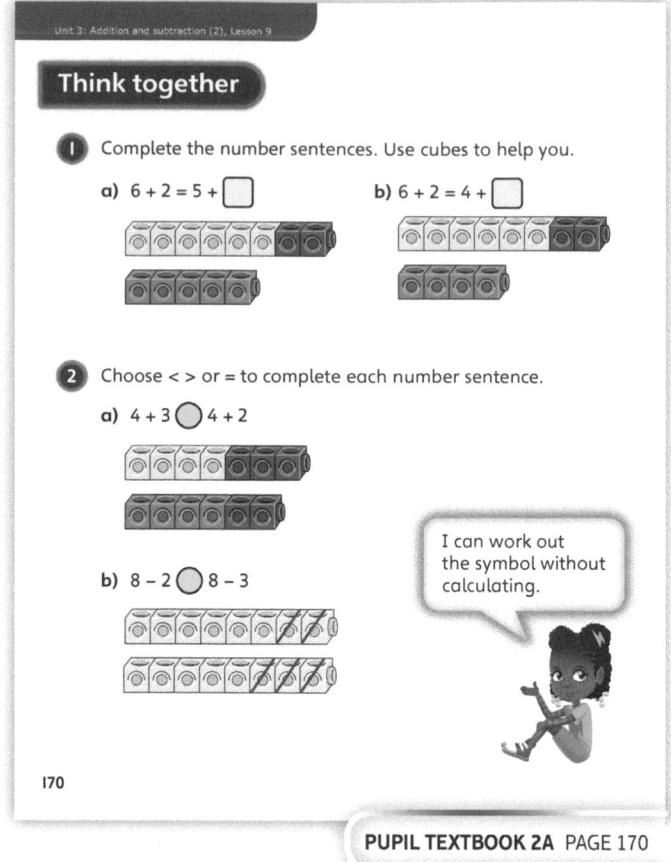

PUPIL TEXTBOOK 2A PAGE 170

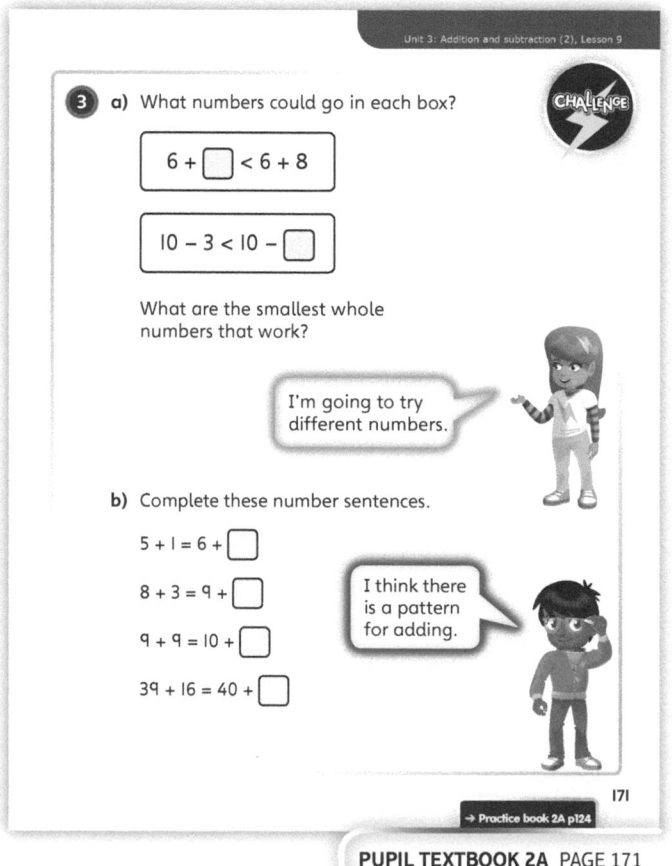

PUPIL TEXTBOOK 2A PAGE 171

Practice

WAYS OF WORKING Independent thinking

IN FOCUS In question ①, children complete number sentences with equivalent additions. Question ② requires children to compare calculations. This is also represented visually. In question ③, children use the symbols <, > or = to compare calculations. In question ④, the comparison symbols are given, and children use them to solve missing number problems. They should notice that there can be more than one possible answer.

STRENGTHEN Encourage children to use cubes or counters to model the numbers and make comparisons.

DEEPEN Challenge children to find all the possible missing numbers that would work to make this equality statement correct:

$20 - \boxed{} = 10 + \boxed{}$

THINK DIFFERENTLY Question ③ encourages children to reason more deeply, as the similarities between the two sides of the calculations are less obvious.

Do children find that they can reason the correct symbol each time, rather than having to calculate?

ASSESSMENT CHECKPOINT Use question ③ to assess children's flexibility with numbers. How many number sentences can they complete without working out the answers on both sides? For example, did they reason that 8 + 8 > 7 + 7 because 8 > 7?

ANSWERS Answers for the **Practice** part of the lesson can be found in the *Power Maths* online subscription.

Reflect

WAYS OF WORKING Pair work

IN FOCUS In this **Reflect** question, the numbers 2, 3, 4 and 5 have been given to fill in the number sentence scaffolds. In the addition scaffold, children can put in the numbers in any order. They can then work out the total of each side and add the correct comparative symbol. Once they become more familiar with the numbers they are using, they can consider how to construct their subtraction calculation more carefully.

ASSESSMENT CHECKPOINT Assess whether children can justify their choices using accurate reasoning. Do they recognise that there are more constraints when trying to make two different subtraction facts, as the bigger number has to come first on each side?

ANSWERS Answers for the **Reflect** part of the lesson can be found in the *Power Maths* online subscription.

After the lesson

- Are children able to identify this pattern and complete these calculations:

  $10 + 2 = 9 + \boxed{}$ $10 + 2 = 7 + \boxed{}$

  $10 + 2 = 8 + \boxed{}$ $10 + 2 = 6 + \boxed{}$

- Can they explain what they notice?

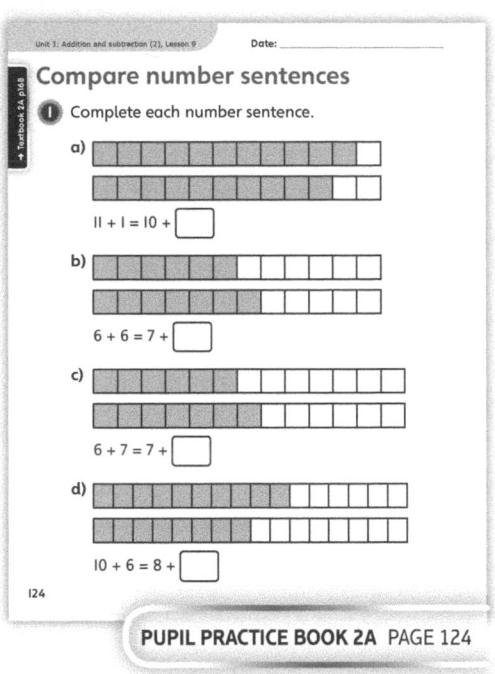

PUPIL PRACTICE BOOK 2A PAGE 124

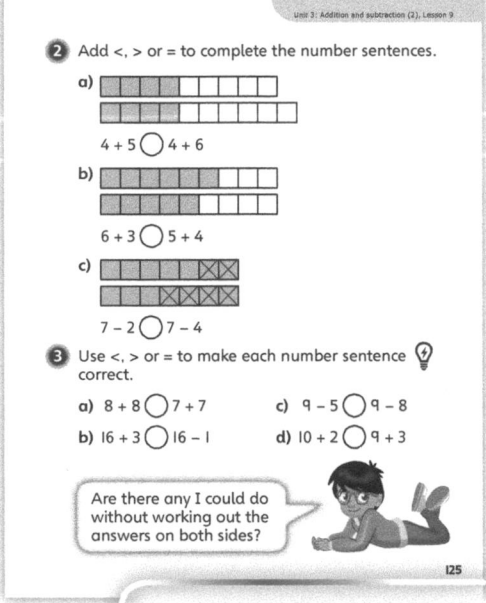

PUPIL PRACTICE BOOK 2A PAGE 125

Unit 3: Addition and subtraction (2), Lesson 9

④ Find numbers that make these number sentences correct.

a) $3 + 7 > 3 + \boxed{}$ c) $5 + 14 > \boxed{} + 14$

b) $6 - 2 < 6 - \boxed{}$ d) $12 + \boxed{} = 6 + 12$

⑤ Find **one** number that makes **both** of these number sentences correct.   **CHALLENGE**

$7 + 6 < 6 + \boxed{}$ $14 - \boxed{} > 14 - 10$

Reflect

Make number sentences using the numbers 2, 3, 4 and 5.

$\boxed{} + \boxed{} \bigcirc \boxed{} + \boxed{}$

$\boxed{} + \boxed{} \bigcirc \boxed{} + \boxed{}$

Ask a partner to check your number sentences.

126

PUPIL PRACTICE BOOK 2A PAGE 126

Missing number problems

Learning focus

In this lesson, children use their calculation skills to solve missing number problems such as:

☐ + 12 = 25.

Before you teach

- Can children add and subtract confidently?
- Are children able to explain the methods that they use?

NATIONAL CURRICULUM LINKS

Year 2 Number – addition and subtraction

Solve problems with addition and subtraction: using concrete objects and pictorial representations, including those involving numbers, quantities and measures.

Recall and use addition and subtraction facts to 20 fluently, and derive and use related facts up to 100.

ASSESSING MASTERY

Children can use addition or subtraction to find a missing number where either a part or the whole is not known.

COMMON MISCONCEPTIONS

Children may not understand how the numbers they know will help them to work out the missing number. Ask:
- *What would be a good way to show this calculation? Is that a part or a whole? What do we know already and what do we need to find out?*

STRENGTHENING UNDERSTANDING

Children can use base ten equipment to represent the numbers in each calculation. Encourage them to use this approach to identify whether an addition or a subtraction is needed.

GOING DEEPER

Challenge children to write their own missing number questions for a partner to solve. They may choose to represent their questions however they like, for example on a part-whole model or bar model, or with shapes, symbols or pictures in place of missing numbers.

KEY LANGUAGE

In lesson: missing number, part, whole

Other language to be used by the teacher: add, subtract, calculation, number sentence, represent

STRUCTURES AND REPRESENTATIONS

Part-whole models, bar models, number lines, place value equipment

RESOURCES

Optional: number lines, base 10 equipment, sticky notes, counters

 In the eTextbook of this lesson, you will find interactive links to a selection of teaching tools.

Quick recap

Present all the number bonds to 10 as missing number problems; 1 + ? = 10, 2 + ? = 10, 3 + ? = 10, and so on. Ask children to fill in the missing numbers.

Repeat with all the bonds to 20.

Discover

WAYS OF WORKING Pair work

ASK

- Question **1**: *What can you see?*
- Question **1** a): *Can you guess what is covered with ink?*
- Question **1** a): *What sort of work was Fred doing?*
- Question **1** b): *What is missing, a part or the whole?*

IN FOCUS Children can see an addition calculation where one of the parts is obscured and cannot be read. They will need to recognise whether a part or a whole is missing from the calculation and identify which operation to use to find the missing number.

PRACTICAL TIPS Ask children to try at random several different possible answers for the missing number, such as 1, 40, 3, and so on. What do they notice?

ANSWERS

Question **1** a):

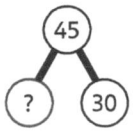

Question **1** b): 45 – 30 = 15.
  Fred's missing number is 15.

Missing number problems

Discover

Maya   Fred

> I will draw a part-whole model to help work out the missing number.

1 a) Draw a part-whole model for this calculation.

  b) What is Fred's missing number?

172

PUPIL TEXTBOOK 2A PAGE 172

Share

WAYS OF WORKING Whole class teacher led

ASK

- Question **1** a): *What is missing, the whole or a part? How do the diagrams show this?*
- Question **1** a): *What calculation could you use to find a missing part?*

IN FOCUS To answer question **1** a), children will need to recognise whether it is a part or the whole that is the missing element of the calculation. A part-whole model and bar model can be used to show that, in this instance, it is a part that is missing. In question **1** b), children then carry out a subtraction to calculate the missing number, using their knowledge of parts and wholes.

Share

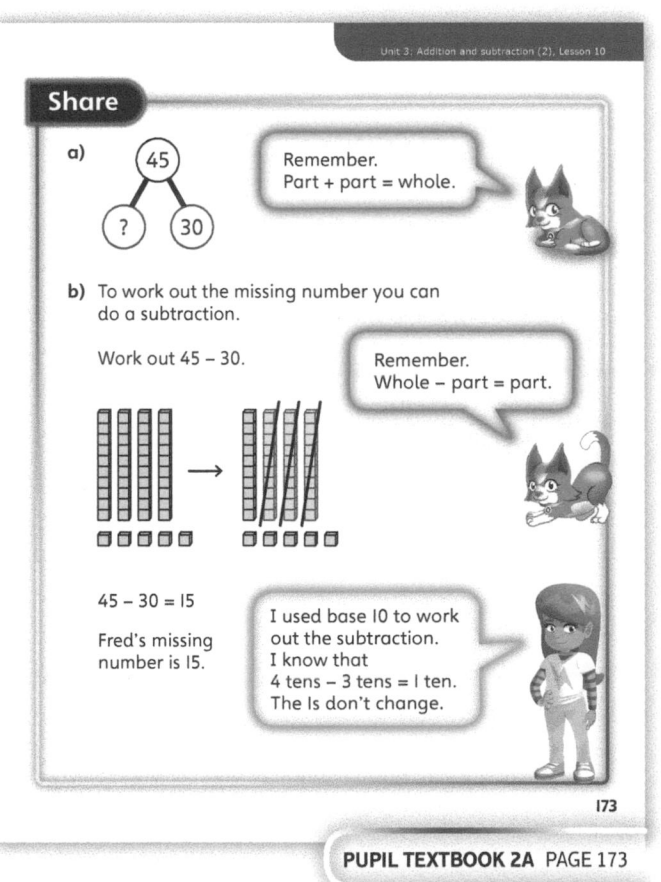

a)

> Remember.
> Part + part = whole.

b) To work out the missing number you can do a subtraction.

Work out 45 – 30.

> Remember.
> Whole – part = part.

45 – 30 = 15

Fred's missing number is 15.

> I used base 10 to work out the subtraction. I know that 4 tens – 3 tens = 1 ten. The 1s don't change.

173

PUPIL TEXTBOOK 2A PAGE 173

Think together

WAYS OF WORKING Whole class teacher led (I do, We do, You do)

ASK

• Question ❶: *What does the bar model show you?*
• Question ❷: *What is missing, the whole or a part?*
• Question ❸: *What is the same, and what is different about each puzzle?*

IN FOCUS In question ❶, children will use a bar model to support them in finding the missing part in a subtraction. Question ❷ has a variety of addition and subtraction calculations where children are finding a missing part each time. In question ❸, children will find both missing parts and missing wholes.

STRENGTHEN Ask children to draw their own part-whole models throughout to represent each calculation and to support their understanding

DEEPEN Challenge children to create their own missing-number puzzles by writing a calculation on mini-whiteboards or on paper and then covering one aspect of it with a sticky note, a small flat shape or a counter.

ASSESSMENT CHECKPOINT Use question ❸ to assess whether children can identify whether it is a part or the whole that is hidden in a calculation, and whether they can use this understanding to find the missing number.

ANSWERS

Question ❶ a): 4

Question ❶ b): 19 − 4 = 15

Question ❷: ☆ = 20; ☁ = 1; △ = 5

Question ❸ a): ◇ = 2; △ = 16; ☁ = 16; ♥ = 2

Question ❸ b): ● = 5; △ = 9; ♥ = 7.
The first two columns do not add up to 20.

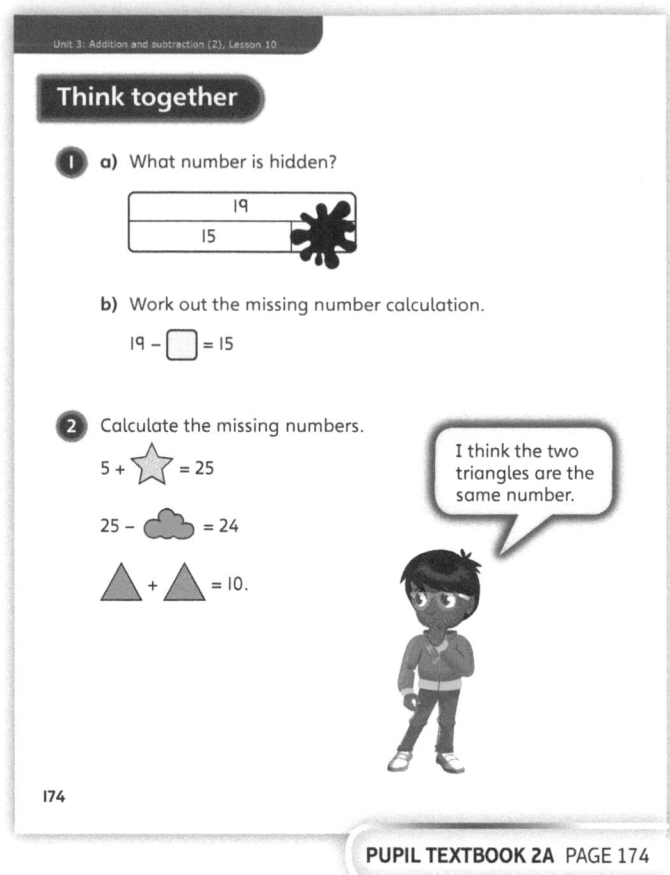

Think together

❶ a) What number is hidden?

	19
15	

b) Work out the missing number calculation.

$19 - \square = 15$

❷ Calculate the missing numbers.

$5 + ☆ = 25$

$25 - ☁ = 24$

$△ + △ = 10.$

I think the two triangles are the same number.

174

PUPIL TEXTBOOK 2A PAGE 174

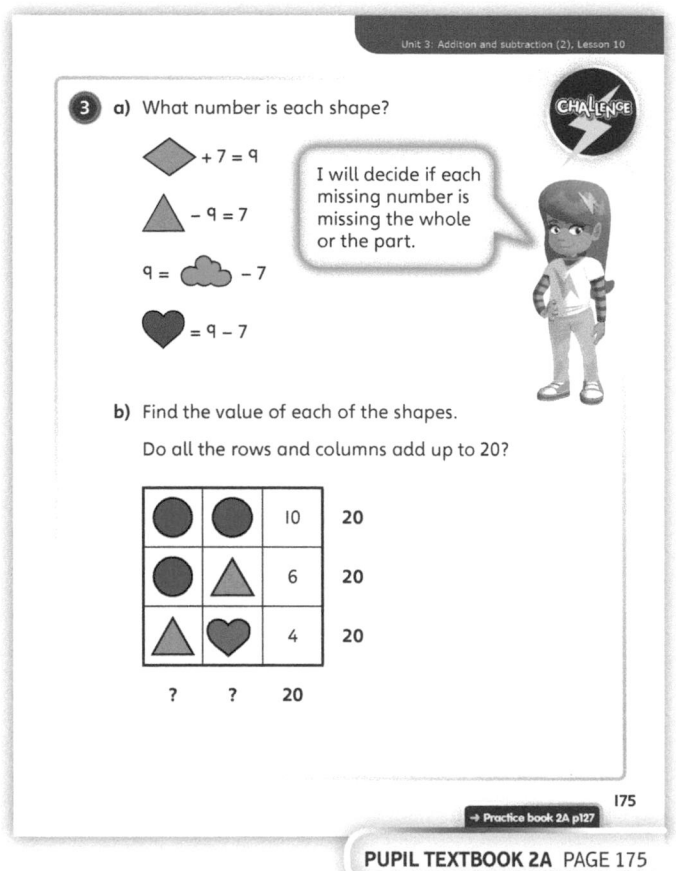

❸ a) What number is each shape? **CHALLENGE**

$◇ + 7 = 9$

$△ - 9 = 7$

$9 = ☁ - 7$

$♥ = 9 - 7$

I will decide if each missing number is missing the whole or the part.

b) Find the value of each of the shapes.

Do all the rows and columns add up to 20?

●	●	10	20
●	△	6	20
△	♥	4	20
?	?	20	

175

→ Practice book 2A p127

PUPIL TEXTBOOK 2A PAGE 175

Practice

WAYS OF WORKING Independent thinking

IN FOCUS In question ①, children are finding the missing part in additions. In question ②, the missing part in each addition is represented as a jump on a number line. In questions ③ and ④, children are finding the missing parts in subtractions. Question ⑤ requires children to find the missing wholes in subtractions. Question ⑥ is a problem-solving question presented as a puzzle where shapes are used to represent missing numbers. Children use what they know about each shape to solve a further calculation.

STRENGTHEN Ask children to draw their own part-whole models to represent the calculations throughout. If necessary, start by supporting them to work with numbers in smaller ranges in order to build confidence.

DEEPEN Challenge children to create more missing-number shape puzzles, like the one in question ⑥, for a partner to solve.

THINK DIFFERENTLY When children reach question ⑤, they should recognise that up to this point they have been finding a missing part, but now they are being asked to find the missing whole. Do they recognise that an addition is needed to do this?

ASSESSMENT CHECKPOINT Assess whether children can identify if the element that is missing from a given calculation is a part or is the whole.

ANSWERS Answers for the **Practice** part of the lesson can be found in the *Power Maths* online subscription.

Reflect

WAYS OF WORKING Independent thinking

IN FOCUS The **Reflect** part of the lesson prompts children to consider whether they are finding a missing part or a missing whole.

ASSESSMENT CHECKPOINT Assess whether children can identify whether they need to find the whole or a part and whether they can use the correct operation to find the answer.

ANSWERS Answers for the **Reflect** part of the lesson can be found in the *Power Maths* online subscription.

After the lesson ⏸

- Are there any children who are still not able to distinguish between a part and a whole?
- How will you support these children in clarifying this point?
- How can you give children regular opportunities to solve simple missing-number problems within 20 as part of their fluency practice?

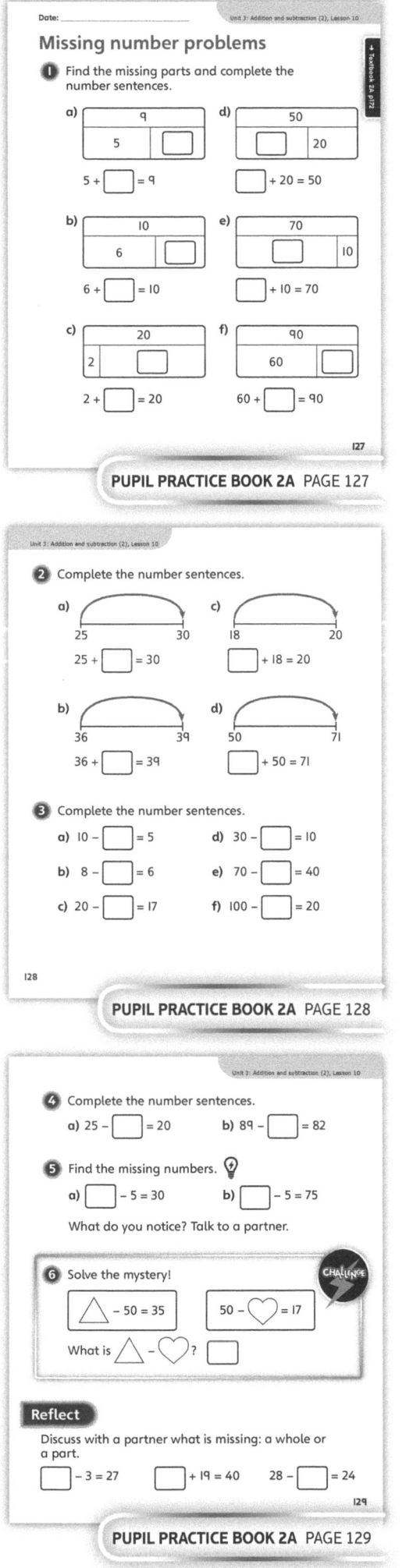

PUPIL PRACTICE BOOK 2A PAGE 127

PUPIL PRACTICE BOOK 2A PAGE 128

PUPIL PRACTICE BOOK 2A PAGE 129

Mixed addition and subtraction

Learning focus

In this lesson, children will represent word problems using single bar models. They will use the words 'part' and 'whole' to help them identify whether the calculation is addition or subtraction.

Before you teach

- How might you scaffold questioning to help children reflect on their assumptions regarding the size or choice of the bar for each number?
- How can you continue to promote the importance of mental calculation strategies within this lesson?

NATIONAL CURRICULUM LINKS

Year 2 Number – addition and subtraction

Solve problems with addition and subtraction: using concrete objects and pictorial representations, including those involving numbers, quantities and measures.

Solve problems with addition and subtraction: applying their increasing knowledge of mental and written methods.

ASSESSING MASTERY

Children can confidently use bar models to represent addition and subtraction. They can choose whether it is addition or subtraction that is needed, and use this to find missing numbers.

COMMON MISCONCEPTIONS

The size of the bars should be proportional to the size of the numbers that they represent. Children may initially find this difficult but the focus is on deciding which number should go in each bar; encourage them to consider the size of the bar when they make their choice. Ask:

- *Is the number with the greatest value represented by the largest bar?*

Children may confuse addition and subtraction until they become familiar with the bar model. To help them decide whether a word problem requires addition or subtraction, encourage them to explain the problem in their own words, using the terms 'part' and 'whole'. Ask:

- *Do you need to add or subtract to answer the question? Can you explain why?*

STRENGTHENING UNDERSTANDING

To help children become familiar with the bar model, give them strips of paper of fixed length to represent the bars within a bar model. Ask children to manipulate the strips to represent addition or subtraction.

GOING DEEPER

When children have completed a bar model, ask them if they agree with the relative size of each bar. Although it is not the focus of the lesson, doing this will deepen children's sense of numbers and help them be more accurate in future lessons.

KEY LANGUAGE

In lesson: +, –, =, bar model, represent, altogether, left, part, whole, total

Other language to be used by the teacher: proportion, addition, subtraction, relative, accurate, difference, remaining

STRUCTURES AND REPRESENTATIONS

Bar model

RESOURCES

Mandatory: base 10 equipment

Optional: strips of paper to represent bars, place value grid, counters

 In the eTextbook of this lesson, you will find interactive links to a selection of teaching tools.

Quick recap

As a class, discuss the parts and the whole in given number sentences. For example

$2 + 3 = 5$ or $5 – 3 = 2$.

Ask children to represent the parts and the whole on a diagram.

Discover

Unit 3: Addition and subtraction (2), Lesson 11

WAYS OF WORKING Pair work

ASK

- Question ①: *What key words could you look for to help you determine if a question involves addition or subtraction?*

IN FOCUS Question ① helps children distinguish between addition and subtraction. Discuss how they know which operation they need to use.

PRACTICAL TIPS Encourage children to use base 10 equipment to represent the stickers in the question. Can they arrange the equipment to represent the parts and the whole on a bar model?

ANSWERS

Question ① a): Mr Dean has 87 stickers.

Question ① b): Mr Dean has 42 stickers left.

Mixed addition and subtraction

Discover

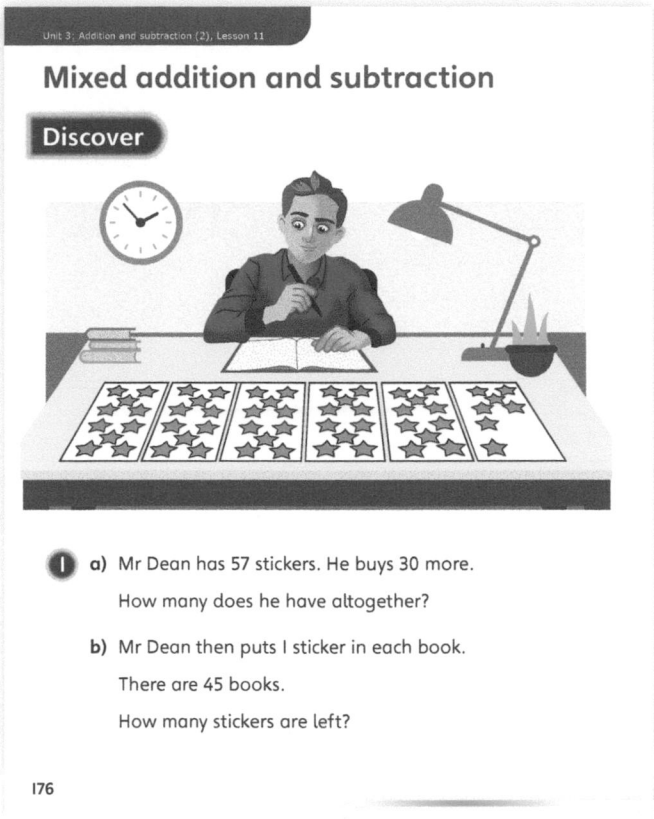

① a) Mr Dean has 57 stickers. He buys 30 more.
   How many does he have altogether?

b) Mr Dean then puts 1 sticker in each book.
   There are 45 books.
   How many stickers are left?

176

PUPIL TEXTBOOK 2A PAGE 176

Share

WAYS OF WORKING Whole class teacher led

ASK

- Question ①: *Which question does Astrid use 'part + part = whole' for? Which does she use 'whole – part = part' for?*
- Question ① a): *What is the same and what is different about the different representations?*
- Question ① b): *Could you swap the parts of the bar model around? Would it change the whole?*

IN FOCUS In question ① a), two different representations of the calculation are given. Identifying what is the same in them will help children to see that they can both be used to find the answer.

Share

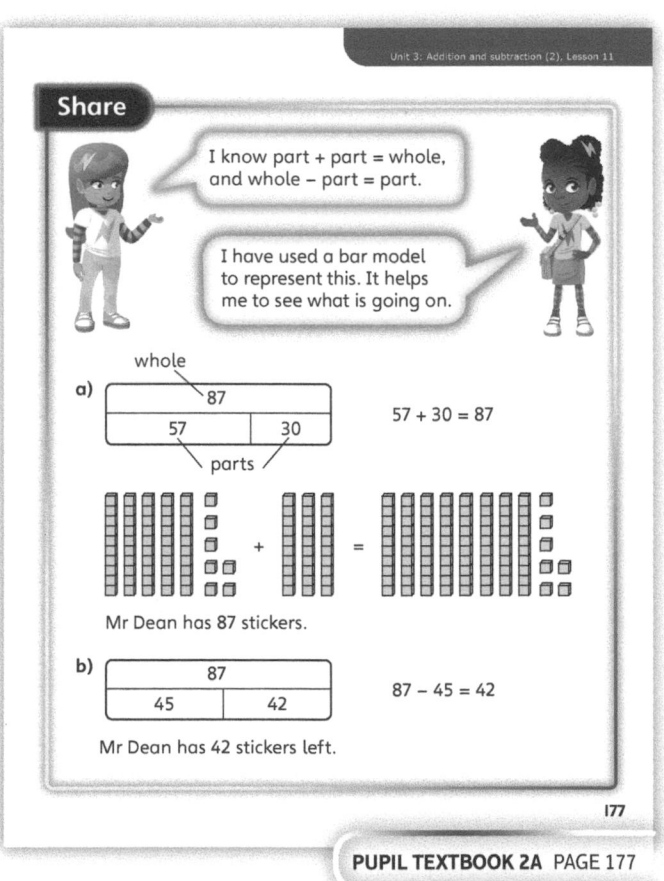

177

PUPIL TEXTBOOK 2A PAGE 177

Think together

Whole class teacher led (I do, We do, You do)

ASK

- Question ❶: *Can the size of the bars help you estimate the size of the parts and/or whole?*
- Question ❷: *Can you think of anything you learned in a previous lesson that will help you find the answers to these problems?*

IN FOCUS Use question ❶ to emphasise that the bar model is used only to represent the problem; it is not a way to calculate the answer. Prompt children to recognise that to answer question ❶ they need to find the difference between 27 and 45, using a method practised in previous lessons.

Question ❸ introduces children to a bar model with three parts. Encourage them to make links to the previous lesson to consolidate their understanding of this representation.

STRENGTHEN Using strips of paper to represent the bars in questions ❶ and ❷ will help children distinguish whether they need to add or subtract to find the solution.

DEEPEN From the three parts shown on the bar model in question ❸, how many different ways can the addition equation be written? How many different ways could be used to make the whole (the total number of stickers) if the quantities changed, but the number of parts stayed the same?

ASSESSMENT CHECKPOINT Assess whether children can justify why they have decided to add or subtract, explaining the question in their own words and with reference to the bar model.

ANSWERS

Question ❶: 45 – 27 = 18. 18 pupils have a packed lunch.

Question ❷: 35 + 16 = 51. There are 51 stickers altogether.

Question ❸: 7 + 5 + 9 = 21. Mrs Bell uses 21 stickers altogether.

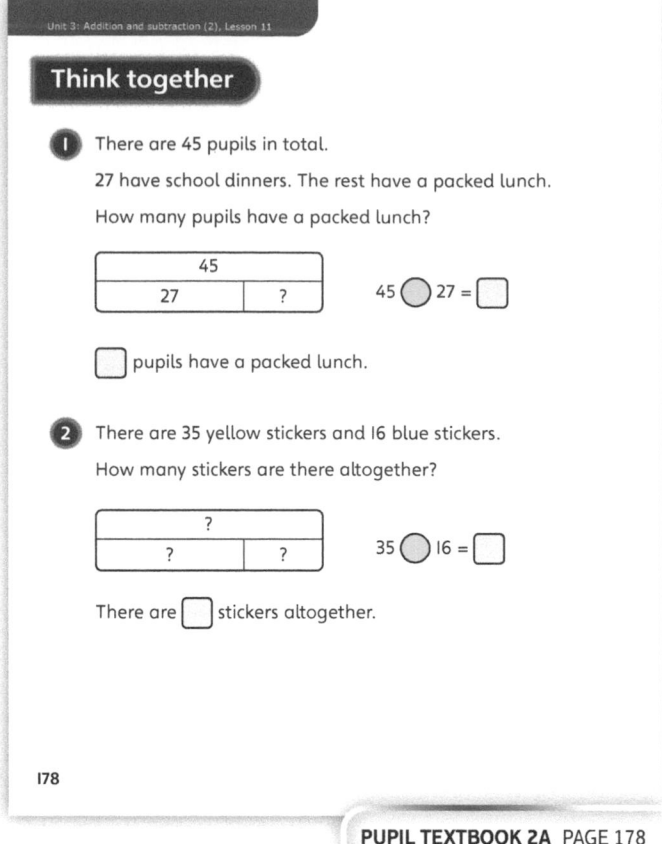

PUPIL TEXTBOOK 2A PAGE 178

PUPIL TEXTBOOK 2A PAGE 179

Practice

WAYS OF WORKING Independent thinking

IN FOCUS In question ❶, children fill in the missing numbers on bar models in order to solve addition or subtraction calculations. In questions ❷ and ❸, real-life contexts are applied to the problems. Children should see how the word problems relate to the position of each number in the bar model. The different parts, and therefore the bars, are significantly different in size. This will help children see where each number should be placed in the bar model.

STRENGTHEN If children make calculation errors that prevent them understanding the concept of the bar model, give them some completed calculations and ask them to draw them in different ways on an empty bar model. Similarly, you could ask children to create their own word problems based on a complete bar model.

DEEPEN Challenge children to estimate the value of a missing part or whole based on the relative size of the bars in a bar model. After they have calculated the answer, they can assess the accuracy of their estimations.

THINK DIFFERENTLY In question ❹, children are using the information in a word problem and a bar model to find the missing third part when two parts and the whole are given. Encourage them to explain the method and operation(s) they will use to work out the missing number.

ASSESSMENT CHECKPOINT Assess children's understanding by asking them to justify their placement of the numbers in a bar model.

ANSWERS Answers for the **Practice** part of the lesson can be found in the *Power Maths* online subscription.

Reflect

WAYS OF WORKING Whole class teacher led

IN FOCUS The **Reflect** part of the lesson requires children to recognise that the problem must involve subtraction, since the whole is known but one of the parts is unknown.

DEEPEN Challenge children to write two different subtraction problems, one using take away and the other using difference.

ASSESSMENT CHECKPOINT Assess whether children are using the vocabulary 'part' and 'whole'. Do they recognise that this bar model represents a subtraction?

ANSWERS Answers for the **Reflect** part of the lesson can be found in the *Power Maths* online subscription.

After the lesson ⏸

- Are children confident identifying whether a bar model represents addition or subtraction?
- Can children create their own word problems that involve addition and subtraction?

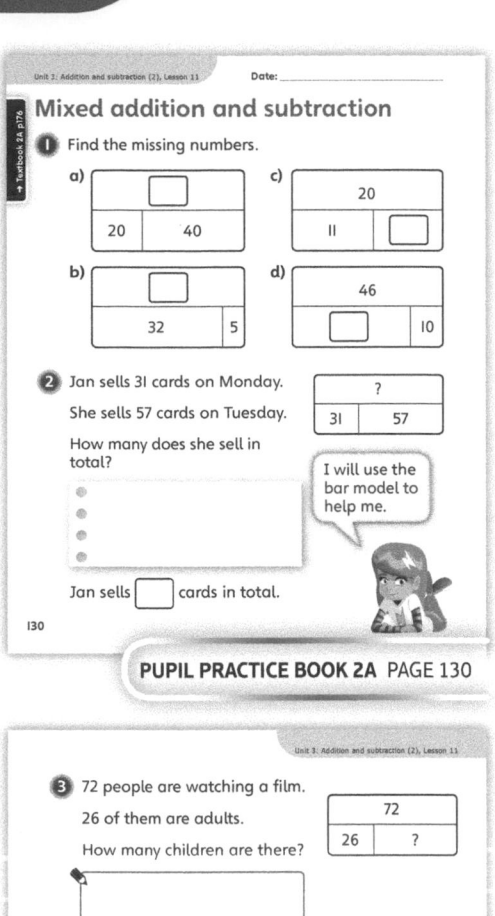

PUPIL PRACTICE BOOK 2A PAGE 130

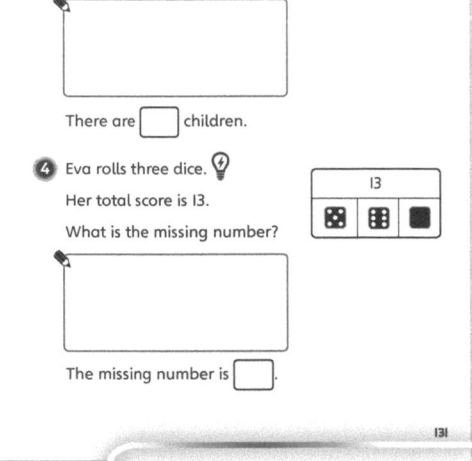

PUPIL PRACTICE BOOK 2A PAGE 131

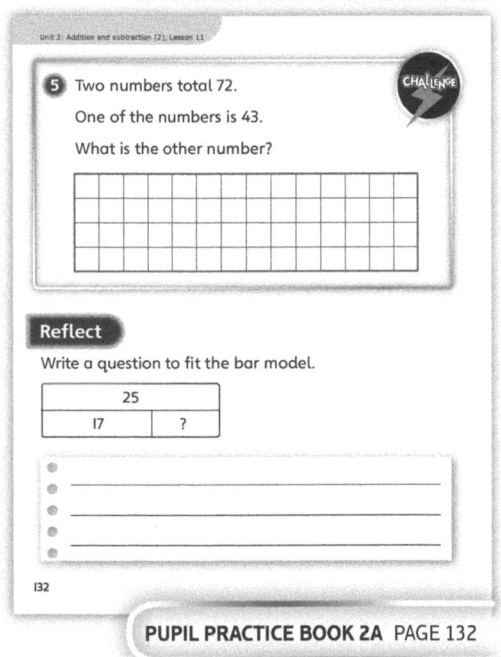

PUPIL PRACTICE BOOK 2A PAGE 132

Two-step problems

Learning focus

In this lesson, children will represent word problems, including two-step problems, using single and comparison bar models. They will use the words 'part' and 'whole' to help them identify whether the calculations are addition or subtraction.

Before you teach

- Based on the previous lesson, what will be the key barriers to overcome during this lesson?
- How can you refine children's representations of the bar model throughout the lesson?

NATIONAL CURRICULUM LINKS

Year 2 Number – addition and subtraction

Solve problems with addition and subtraction: using concrete objects and pictorial representations, including those involving numbers, quantities and measures.

Solve problems with addition and subtraction: applying their increasing knowledge of mental and written methods.

ASSESSING MASTERY

Children can use bar models to help them see which operation or operations to use when completing word problems, leading to increased conceptual understanding.

COMMON MISCONCEPTIONS

Children may confuse the single and comparison bar models and may struggle to distinguish which is appropriate for a particular question type. Ask:
- *Does the bar model that you have chosen to draw represent the information presented in the question? Can you identify where on the bar model the answer is found?*

Children may continue to make calculation errors if they do not use the appropriate methods from previous lessons. Ask:
- *Have you used methods that you learned in previous lessons to calculate the answer? Can you explain how you know your answer is correct?*

STRENGTHENING UNDERSTANDING

Encourage children to simulate word problems using real-life objects. This will give them a greater understanding of what they are being asked to calculate.

GOING DEEPER

Challenge children to create their own word problems to match given bar models, or to create a combination of addition and subtraction problems (including difference) using a given set of numbers.

KEY LANGUAGE

In lesson: +, −, =, bar model, altogether, more than, added

Other language to be used by the teacher: part, whole, comparison, addition, subtraction, difference, total, remaining, fewer

STRUCTURES AND REPRESENTATIONS

Bar model

RESOURCES

Optional: resources to represent real-life objects (e.g. base 10 equipment, counters), bean bags, hoops, strips of paper

 In the eTextbook of this lesson, you will find interactive links to a selection of teaching tools.

Quick recap 🔁

Ask children to solve given additions of two 2-digit numbers in two steps, by first adding the 10s and then the 1s. They should show the two steps on open number lines.

Repeat with subtractions, subtracting the 10s and then the 1s.

Discover

PUPIL TEXTBOOK 2A PAGE 180

WAYS OF WORKING Pair work

ASK

- Question ❶ a): *What do you think you need to do to find the solution?*
- Question ❶ b): *How could you represent difference using bar models?*
- Question ❶: *What additional questions could you ask about this picture?*

IN FOCUS Question ❶ b) is the first time that the single bar model taught in the previous lesson cannot be used to represent the question. Establish that the question asks for the difference, not for a part or whole, and agree that a different representation is therefore needed.

PRACTICAL TIPS Provide children with strips of paper to make each of the bar models. Ensure they see the difference between the single bar model in question ❶ a) and the comparison bar model in question ❶ b).

ANSWERS

Question ❶ a): There are 61 marbles in Amy's pot and Kasim's pot altogether.

Question ❶ b): There are 25 more marbles in Kat's pot than in Ben's pot.

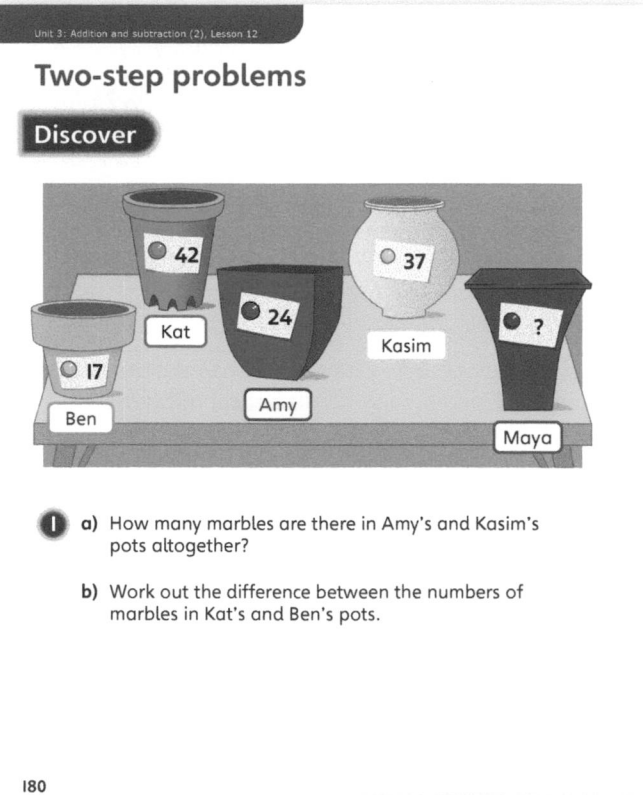

Two-step problems

Discover

❶ a) How many marbles are there in Amy's and Kasim's pots altogether?

b) Work out the difference between the numbers of marbles in Kat's and Ben's pots.

180

Share

PUPIL TEXTBOOK 2A PAGE 181

WAYS OF WORKING Whole class teacher led

ASK

- Question ❶: *How are these bar models different from the ones you used in the previous lesson?*
- Question ❶: *Why is the question mark in a different place in the two representations?*

IN FOCUS Draw children's attention to the location of the question mark in the two parts of question ❶. This shows where the answer is located within the model. It is important that children understand what the question mark and the bar boundaries represent in each case.

STRENGTHEN If children find the concept of difference represented in this way difficult, they could play a game, attempting to throw bean bags into a hoop laid on the ground. They could then compare how many bags landed inside and outside the hoop, and find the difference. They could line the bean bags up in in rows to emulate the bar model.

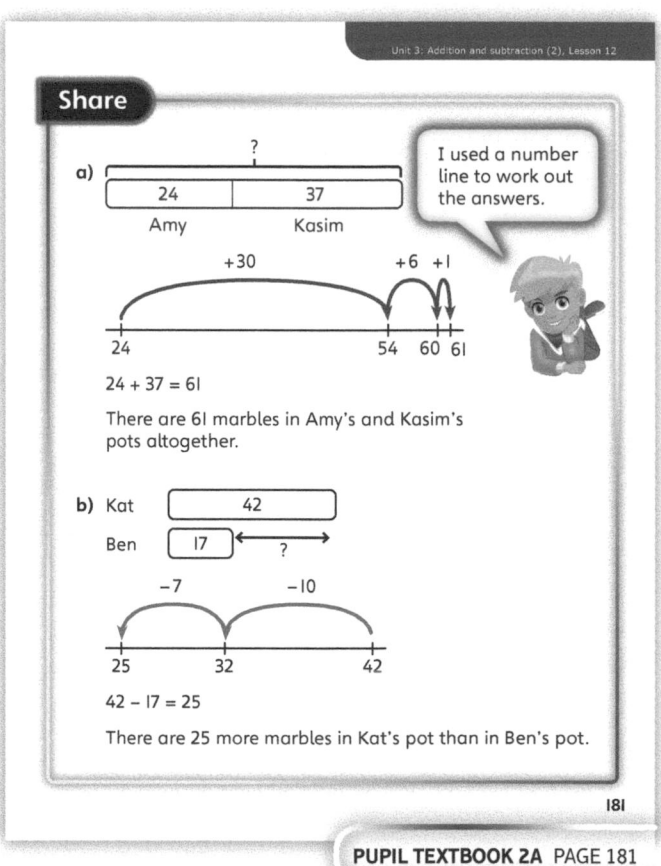

Share

a)

$$24 + 37 = 61$$

There are 61 marbles in Amy's and Kasim's pots altogether.

b) Kat 42
Ben 17

$$42 - 17 = 25$$

There are 25 more marbles in Kat's pot than in Ben's pot.

I used a number line to work out the answers.

181

Think together

Whole class teacher led (I do, We do, You do)

ASK

- Question **1**: *Where is the value you need to calculate shown on the bar model?*
- Question **2**: *Which of the representations could you also draw as a single bar model?*

IN FOCUS Question **2** includes the first instance of a bar model used to find a quantity that is a known amount more than a known quantity. Ensure children understand which part of the bar model they need to calculate.

Question **3** is a two-step question. Encourage children to show clearly the steps of their working, following Flo's advice, rather than trying to work mentally. Can they explain why they have done what they have?

STRENGTHEN Using concrete manipulatives to represent the numbers within the bars takes children a step back and may enable them to understand what they are calculating more easily.

DEEPEN Challenge children to answer questions relating to question **3**, such as: *What is the smallest total of marbles you could make by combining two of the pots? What is the second smallest total you could make? What is the largest number of marbles you could make by combining two of the pots?*

ASSESSMENT CHECKPOINT Assess whether children can explain the position of the bars in the representation for a specific question, and identify or explain where the answer is found.

ANSWERS

Question **1**: 17 + 24 = 41
    There are 41 marbles altogether.

Question **2**: There are 57 marbles in Maya's pot.

Question **3**: 17 + 19 = 36
    There are 36 marbles in Ben's pot.
    36 − 24 = 12
    There are now 12 more marbles in Ben's pot than in Amy's pot.

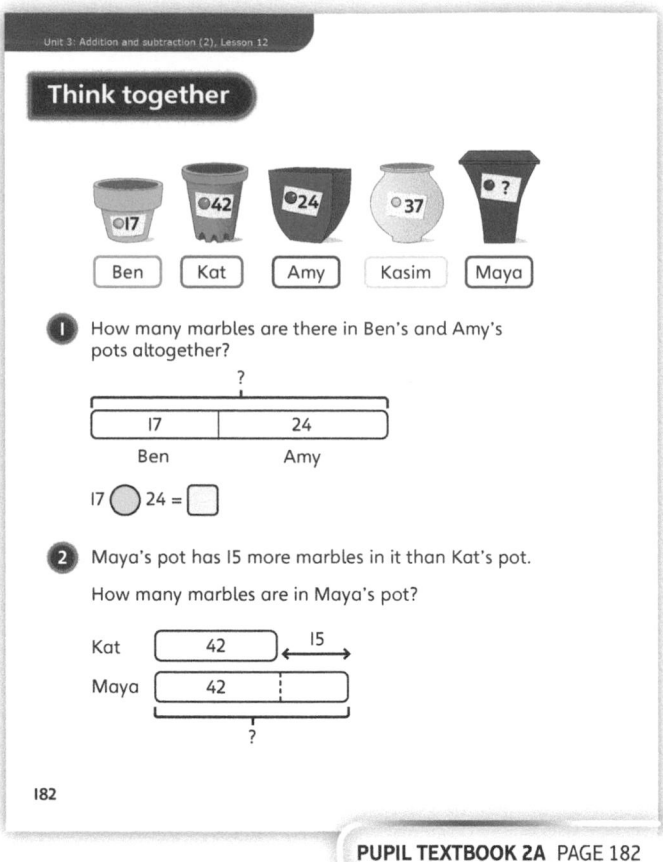

PUPIL TEXTBOOK 2A PAGE 182

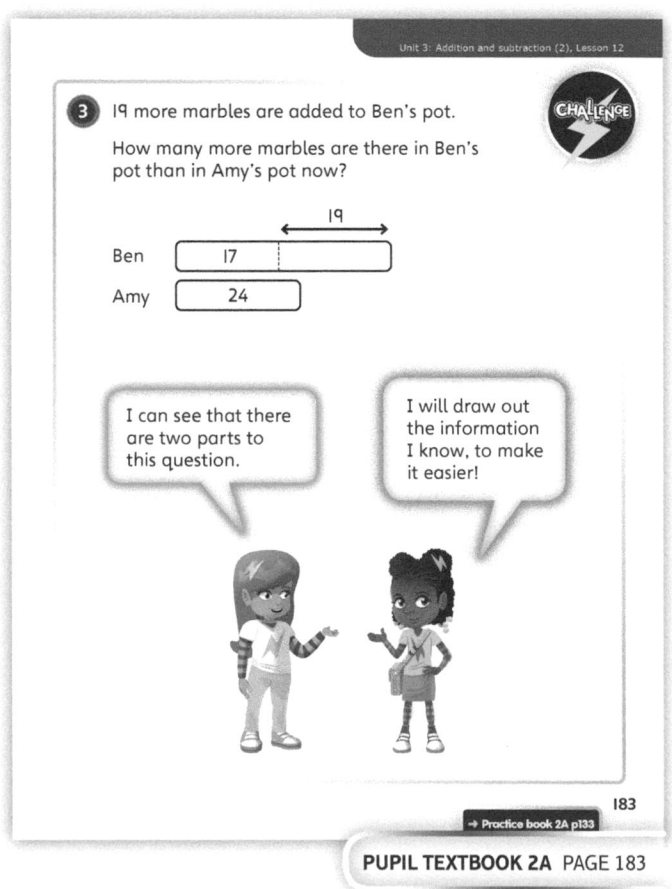

PUPIL TEXTBOOK 2A PAGE 183

Practice

WAYS OF WORKING Independent thinking

IN FOCUS The questions in the **Practice** section have progressively less scaffolding. Questions ❸ and ❹ both require children to draw their own bar model. Children should notice that questions ❹ and ❺ are both two-step problems.

STRENGTHEN Giving children the calculations needed for questions ❸ (75 – 48 = ☐) and ❹ (25 + 25 + 16 = ☐) will allow children to focus on correctly drawing the bar models.

DEEPEN Challenge children to change one (or more) of the questions so that it uses the other operation (i.e. question ❶ becomes an addition question), while keeping the context and numbers used the same.

ASSESSMENT CHECKPOINT Ask children to explain why they have chosen to draw a single or a comparison bar model for each of questions ❸ and ❹. A confident explanation indicates a secure understanding of the concept.

ANSWERS Answers for the **Practice** part of the lesson can be found in the *Power Maths* online subscription.

Reflect

WAYS OF WORKING Independent thinking

IN FOCUS The **Reflect** part of the lesson requires children to reflect on the different problems they have encountered in the unit. Encourage them to explain why they found that question easier than the others. Encourage them to think carefully about the type of question they are asking a partner to solve; you could ask them to provide a scaffold of the bar model to assist their partner.

ASSESSMENT CHECKPOINT Assess whether the language they use is clear. If not, it is likely that the child has not fully understood the concept of addition or subtraction in the context of bar models.

ANSWERS Answers for the **Reflect** part of the lesson can be found in the *Power Maths* online subscription.

After the lesson ⏸

- Have children demonstrated that they have mastered addition and ways to calculate and represent addition problems?
- Have children demonstrated that they have mastered subtraction and ways to calculate and represent subtraction problems?
- Are there any common misconceptions that the class are still finding difficult?

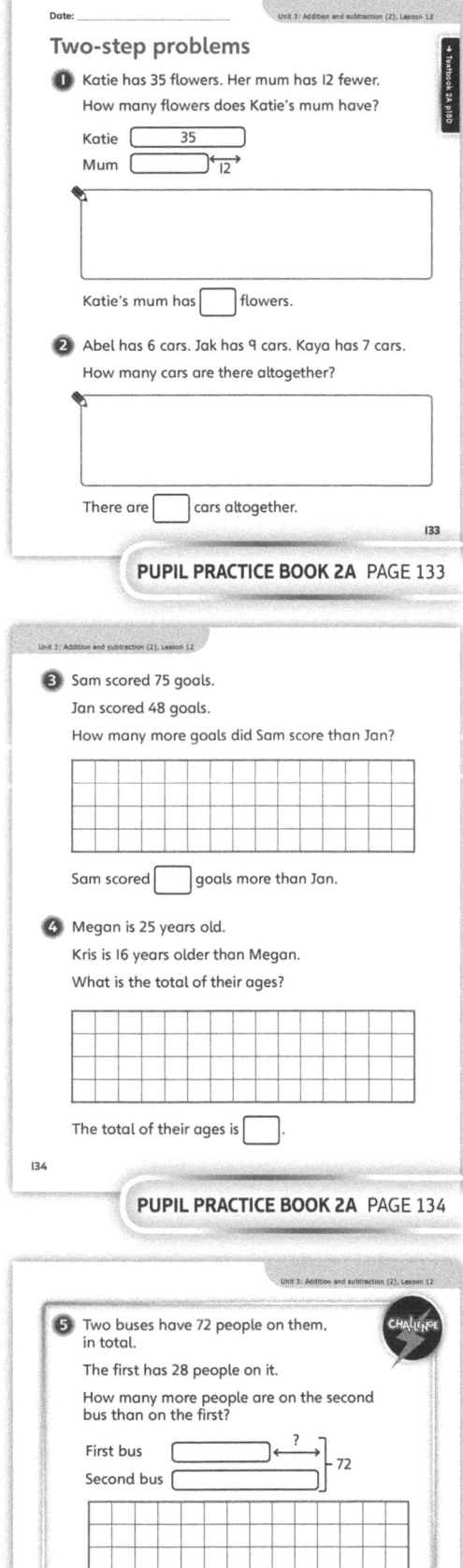

PUPIL PRACTICE BOOK 2A PAGE 133

PUPIL PRACTICE BOOK 2A PAGE 134

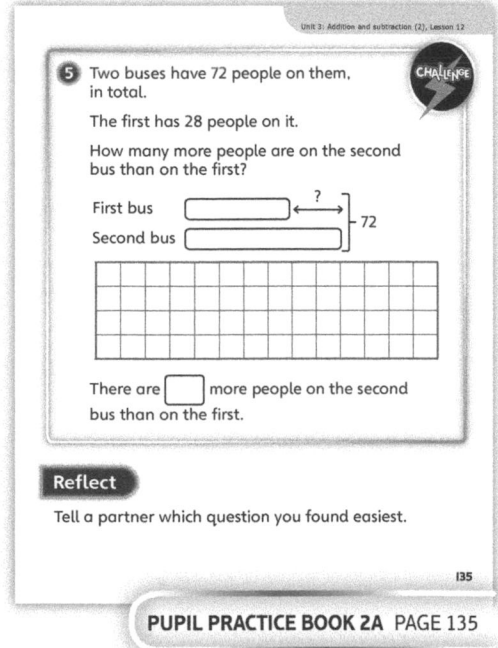

PUPIL PRACTICE BOOK 2A PAGE 135

End of unit check

> **Don't forget the unit assessment grid in your *Power Maths* online subscription.**

WAYS OF WORKING Group work teacher led

IN FOCUS These questions assess children's understanding of partitioning numbers in different ways and how this supports different mental calculations. The questions also assess children's understanding of the links between addition and subtraction and how to move between the operations.

Think!

WAYS OF WORKING Pairs or small groups

IN FOCUS

- This question requires children to understand that all questions require addition to calculate the unknown whole.
- Children must then be able to understand, as a result of the numbers that they are presented with, that the 1s in two of the calculations have a total that is more than 9, whereas in the other calculation the total of the 1s is less than 9.
- If children are finding this difficult, ask: *Which is the easiest calculation?*
- Children should be encouraged to use all of the vocabulary given to them to explain their answer.

ANSWERS AND COMMENTARY Children who have mastered this unit will be able to differentiate between addition and subtraction problems. They will be able to understand how to represent the numbers provided within the context described in different ways, using different resources. Children should be flexible with the methods they can use to calculate different problems as a result of each problem's complexity and should be working efficiently as a result of using known number facts within mental calculations.

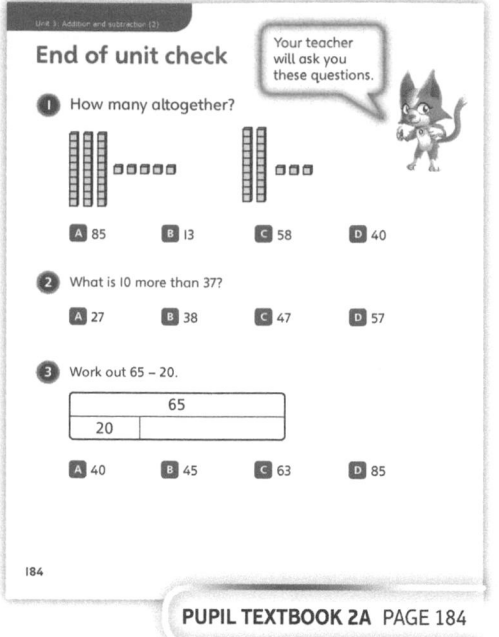

PUPIL TEXTBOOK 2A PAGE 184

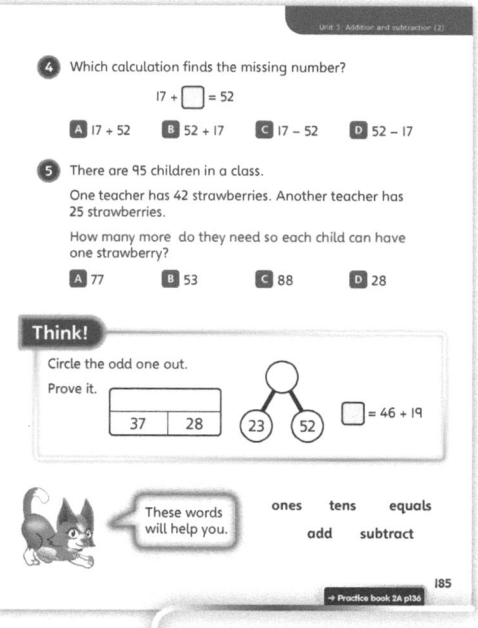

PUPIL TEXTBOOK 2A PAGE 185

Q	A	WRONG ANSWERS AND MISCONCEPTIONS	STRENGTHENING UNDERSTANDING
1	C	B suggests that children have simply added up the total number of blocks without considering the different values of the 10s and the 1s.	Children should have the opportunity to work with base 10 equipment as much as possible to correctly understand how to partition the same number in different ways. Numbers should be partitioned into two and more parts.
2	C	A suggests children have found 10 less than 37 (mistaken addition for subtraction). B suggests children have found 1 more than 37.	
3	B	D suggests that children have not interpreted the bar model correctly and have found the total of the two numbers given rather than the difference.	Children should also practise writing all possible calculations from a complete part-whole model, understanding that the equals sign can occur in different places and means 'is the same as'.
4	D	C suggests the child thinks that subtraction is also commutative and the order of the numbers does not matter.	Creating number stories that require addition, subtraction or a combination of the two operations will increase children's understanding of the differences between the operations.
5	D	B suggests that the child has calculated how many more 95 is than the number of strawberries that the first teacher has.	

My journal

WAYS OF WORKING Independent thinking

ANSWERS AND COMMENTARY The part-whole model is the odd one out because the total of the 1s is less than 10; 3 ones plus 2 ones equals 5 ones, 2 tens and 5 tens equals 7 tens, so the answer is 75. The other questions are also addition as we know both of the parts and not the whole but in those, the total of the 1s is more than 10; 6 ones plus 9 ones is 15 ones, and 7 ones and 8 ones is 15 ones, and 15 ones is greater than 10 ones.

Power check

WAYS OF WORKING Independent thinking

ASK

- *How confident do you feel about using a number line to show the steps of addition and subtraction?*
- *How confident do you feel about using the bar model to represent additions or subtractions?*
- *Which resource do you find easiest to help you with addition and subtraction?*

Power puzzle

WAYS OF WORKING Pair work or small group

IN FOCUS Use this **Power puzzle** to see if children can work in small groups to find the same totals using different digits. This will assess their fluency of working with numbers within 20 to make different totals. It will also check to see how confident children are at choosing a starting point of their own and working from this point to find different solutions to the problem.

ANSWERS AND COMMENTARY If children can complete the **Power puzzle**, it suggests that they are confident at problem solving and making mistakes in the process of arriving at the correct answer. Children who find the problem difficult may need some help or some hints to start. For example, ask: *What is the total amount of all of the cards? Can you make 15 in three different ways using each card once?*

Unequal piles: 9 + 6, 7 + 5 + 3, 8 + 4 + 2 + 1 or 9 + 6, 8 + 5 + 2, 7 + 4 + 3 + 1

Equal piles: 9 + 1 + 5, 8 + 3 + 4, 7 + 6 + 2 or 9 + 2 + 4, 8 + 6 + 1, 7 + 5 + 3

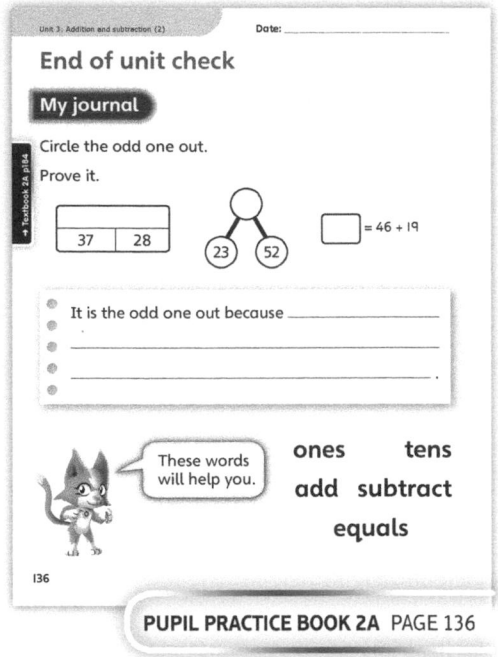

PUPIL PRACTICE BOOK 2A PAGE 136

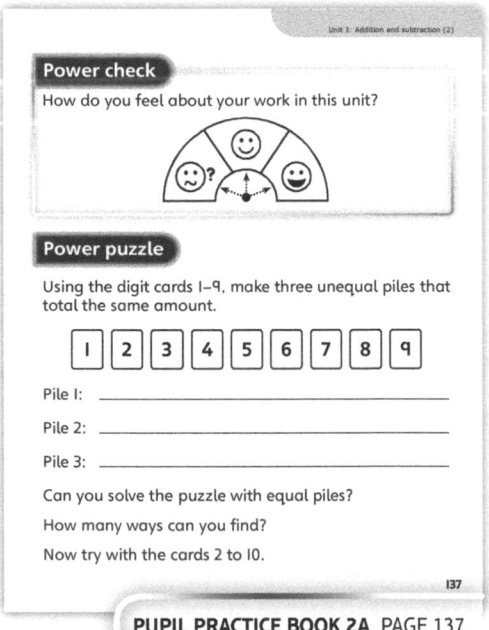

PUPIL PRACTICE BOOK 2A PAGE 137

After the unit ⏸

- Were children more confident at calculating addition or subtraction problems?
- What visual representations did children find most difficult to use accurately?

Strengthen and **Deepen** activities for this unit can be found in the *Power Maths* online subscription.

Unit 4
Properties of shapes

Mastery Expert tip! 'There is a lot of topic-specific key language in this lesson. I found that displaying a list of this language in the classroom and encouraging children to contribute to it helped them to learn and understand the language so that they could use it during whole-class and peer discussions.'

Don't forget to watch the Unit 4 video!

WHY THIS UNIT IS IMPORTANT

This unit focuses on the properties of 2D and 3D shapes. Children will learn to describe and sort shapes based on the shapes' mathematical properties, using the correct terminology. Although this is the first unit covering geometry in Year 2, children have experience of recognising, naming, describing and sorting 2D and 3D shapes from Unit 5 in Year 1.

Children will also draw on their counting skills and their ability to compare and order numbers. In this unit, children will learn to describe and categorise shapes based on their number of sides, vertices, edges and faces.

WHERE THIS UNIT FITS

→ Unit 3: Addition and subtraction (2)
→ **Unit 4: Properties of shapes**
→ Unit 5: Money

Children should already be able to recognise and name familiar 2D and 3D shapes. Children will be familiar with using the word 'face' to describe a flat surface of a 3D shape and they will be able to describe the shape of the faces. Children have also experienced identifying and describing repeating patterns using 2D and 3D shapes.

Before they start this unit, it is expected that children:
- know how to distinguish between 2D and 3D shapes
- understand that shapes are categorised based on specific properties
- know the names of common 2D and 3D shapes and some of their properties.

ASSESSING MASTERY

Children who have mastered this unit will be able to use key language (such as faces, edges and vertices) fluently when describing 2D and 3D shapes. Children will be able to sort shapes in different ways based on different mathematical properties and create both repeating and symmetrical patterns involving shapes of increasing complexity.

COMMON MISCONCEPTIONS	STRENGTHENING UNDERSTANDING	GOING DEEPER
Children may confuse key language, such as identifying edges as vertices.	Give children concrete 2D and 3D shapes to handle and practise counting the different properties. Encourage children to contribute to a classroom display of the definitions of the different terms.	Ask children to explore what happens to the number of faces, vertices, sides or edges when they join shapes together. Prompt children to look for patterns and to suggest explanations for what they observe.
Children may miscount the number of edges, vertices, faces or sides.	Give children dry-wipe markers and concrete 3D shapes so that they can mark off a specific property as they count it. Ask children to record the number of each property and add it to the classroom display, so that they can refer to it in future.	

Unit 4: Properties of shapes

UNIT STARTER PAGES

Use these pages to introduce the unit and to prompt children to recall what they already know about 2D and 3D shapes. Focus on the key vocabulary and discuss the potential meanings of this vocabulary. This will indicate what children already know and expose any misconceptions that they may have about the properties of shapes.

STRUCTURES AND REPRESENTATIONS

Sorting circles: Sorting circles are a useful way to order and compare objects under different headings according to shared properties.

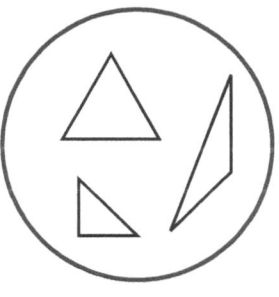 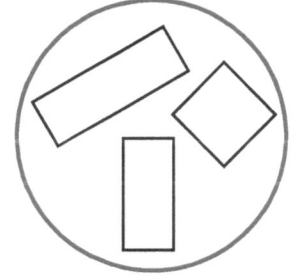

KEY LANGUAGE

There is some key language that children will need to know as part of the learning in this unit:

→ circle, semicircle

→ oval, triangle, square, rectangle, quadrilateral

→ polygon, pentagon, hexagon

→ sphere, hemisphere

→ cone, cylinder

→ triangle-based pyramid, square-based pyramid, pentagon-based pyramid, hexagon-based pyramid

→ cube, cuboid

→ prism, triangular prism, pentagonal prism, hexagonal prism

→ 2D, 3D

→ properties

→ side, vertex, vertices, edge, face

→ pattern

→ symmetry, symmetrical, line of symmetry

→ curved surface

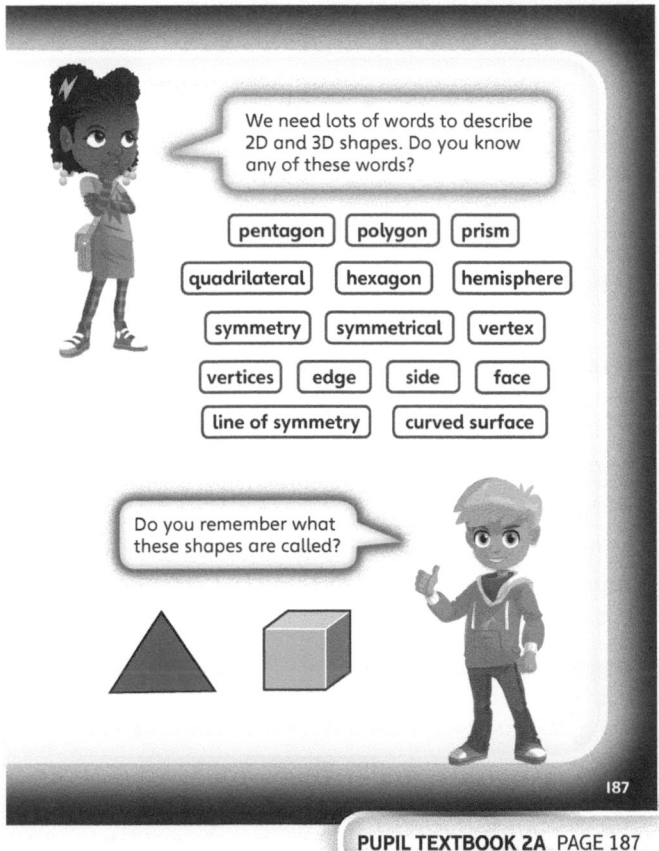

In this unit we will …
⚡ Recognise 2D and 3D shapes
⚡ Count the sides and vertices on 2D shapes
⚡ Learn about symmetry
⚡ Count the faces, edges and vertices on 3D shapes
⚡ Sort 2D and 3D shapes

How are these shapes similar? How are they different?

186

PUPIL TEXTBOOK 2A PAGE 186

We need lots of words to describe 2D and 3D shapes. Do you know any of these words?

pentagon | polygon | prism
quadrilateral | hexagon | hemisphere
symmetry | symmetrical | vertex
vertices | edge | side | face
line of symmetry | curved surface

Do you remember what these shapes are called?

187

PUPIL TEXTBOOK 2A PAGE 187

Recognise 2D and 3D shapes

Learning focus

In this lesson, children will recognise and name 2D and 3D shapes and make links between them. They will begin to identify common features of different types of 2D and 3D shapes.

Before you teach

- What do children already know about 2D and 3D shapes?
- How can you reinforce the vocabulary used in this lesson?
- How will you support children in identifying the key properties for classifying 2D and 3D shapes?

NATIONAL CURRICULUM LINKS

Year 2 Geometry – properties of shape

Compare and sort common 2D and 3D shapes and everyday objects.

ASSESSING MASTERY

Children can name and describe a range of 2D and 3D shapes, identifying what features determine the type of shape and any similarities and differences across different types of shapes. Children will begin to describe the faces of 3D shapes.

COMMON MISCONCEPTIONS

Children may misname a shape when its orientation changes. Hold up different shapes and rotate them. Ask:
- *How has it changed? Is it now a different shape?*

Children may apply names of 2D shapes to 3D shapes. Hold up a cube. Ask:
- *What do you call this shape? What can you tell me about it?*

Hold up a square next to the cube. Ask:
- *Are these the same? How are they different? What do you call these shapes?*

STRENGTHENING UNDERSTANDING

Give children a range of 2D and 3D shapes to explore and manipulate. Provide corresponding name labels that children have to match to the shapes. Children can then use these as a point of reference throughout the lesson.

GOING DEEPER

Challenge children to explore combining 2D or 3D shapes to create new shapes. Discuss how children know what type of shape it is, so that they begin to think carefully about common features.

KEY LANGUAGE

In lesson: rectangle, square, **quadrilaterals**, **sides**, triangle, cuboid, 2D, 3D

Other language to be used by the teacher: right angle, cone, prism, polygons, pyramid, sphere, cube, oblong, cylinder, circle

RESOURCES

Mandatory: a range of 2D and 3D shapes with labels

Optional: materials for printing with 3D shapes

 In the eTextbook of this lesson, you will find interactive links to a selection of teaching tools.

Quick recap

Give children manipulatives of rectangles, triangles, circles and squares, so that they can practise handling and naming 2D shapes. Discuss the name and features of each shape together.

Discover

Pair work

ASK

- Question ❶ a): *What different shapes can you see?*
- Question ❶ a): *How did you know which shapes were squares?*
- Question ❶ a): *How is the square the same as or different from the oblong?*

IN FOCUS This part of the lesson reinforces the concept that size, colour and orientation are not important when naming a 2D shape. Identify misconceptions that children may have when naming a shape that has been rotated. For example, children may see the red square in the middle picture as a 'diamond'.

ANSWERS

Question ❶ a): This is Mia's picture.

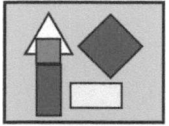

Question ❶ b): This is Sunil's picture.

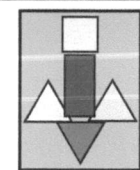

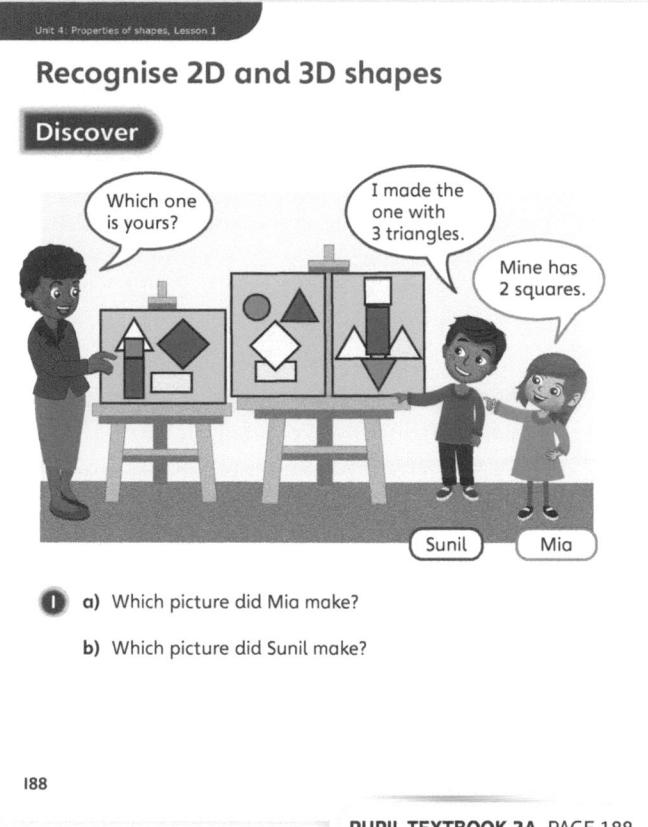

Recognise 2D and 3D shapes

Discover

Which one is yours?

I made the one with 3 triangles.

Mine has 2 squares.

Sunil   Mia

❶ a) Which picture did Mia make?

b) Which picture did Sunil make?

188

PUPIL TEXTBOOK 2A PAGE 188

Share

Whole class teacher led

ASK

- Questions ❶ a) and b): *How do the shapes differ?*
- Questions ❶ a) and b): *What features can we ignore when naming the shapes?*
- Questions ❶ a) and b): *What features are important when naming shapes?*

IN FOCUS Sparks introduces the idea that there is a family of shapes known as quadrilaterals. This may be the first time that children have come across this term, so it is important to ensure that they understand what constitutes a quadrilateral.

Children may have the misconception that a square is not a rectangle. Explain to children that a rectangle has four sides with four right angles and that there are two types of quadrilateral in the pictures: a square rectangle and an oblong rectangle. Discuss how the shapes are different.

Share

a) Mia's picture has 2 squares.
   Which picture has 2 squares?

I think a square looks like this ☐.

This is Mia's picture.

Rectangles and squares are **quadrilaterals** because they have **4 sides**.

b) This is Sunil's picture.

189

PUPIL TEXTBOOK 2A PAGE 189

Think together

WAYS OF WORKING Whole class teacher led (I do, We do, You do)

ASK

- Question ❷: *Is there more than one possibility for printing each shape? How can you be sure you have found them all?*
- Question ❸: *What shapes can you use to draw a rectangle? What shapes can you not use?*

IN FOCUS Questions ❷ and ❸ prompt children to look at the faces of the 3D shapes. Although the term 'face' is not used, children are looking specifically at describing the types of faces they can see. Children will realise that different 3D shapes may share the same shaped face and that, as a result, there is more than one way to answer the questions.

STRENGTHEN Provide 3D shapes for children to draw around or print with in order to help them identify the shapes of different faces.

DEEPEN Ask children if they can create different 2D shapes by combining faces. For example, ask: *Can you create an oblong rectangle using just a cube? Can you use the triangle face of a pyramid to create a four-sided 2D shape?*

ASSESSMENT CHECKPOINT Question ❶ will show whether children can identify rectangles and whether they understand that squares are a special type of rectangle.

ANSWERS

Question ❶: There are 7 rectangles in this picture (2 squares and 5 oblongs).

Question ❷: The 3D shapes are a cylinder, cube, cuboid, triangular prism and a square-based pyramid. The circle can be printed using the cylinder. The square can be printed using the cube, cuboid or square-based pyramid. The oblong can be printed using the cuboid or triangular prism. The triangle can be printed using the triangular prism or square-based pyramid.

Question ❸: The cube can produce a square. The cuboid can produce a square or an oblong. The square-based pyramid can produce a square or a triangle. The triangular prism can produce a triangle or an oblong. The cone can produce a circle. The cylinder can produce a circle. The sphere cannot be used to produce a 2D shape.

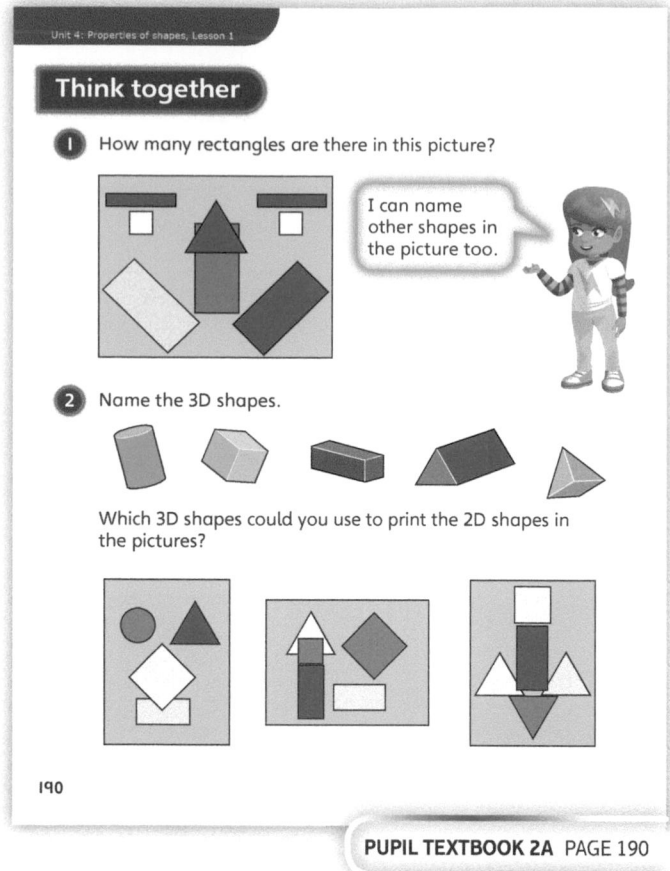

PUPIL TEXTBOOK 2A PAGE 190

PUPIL TEXTBOOK 2A PAGE 191

Practice

WAYS OF WORKING Pair work

IN FOCUS Question ❶ requires children to identify 2D shapes correctly and question ❷ helps children to recognise 3D shapes. Question ❹ requires children to reason about the shape each child will draw based on information that they can see.

STRENGTHEN Provide children with a range of 2D and 3D shapes with name labels as prompts. For question ❹, children could use the 3D shapes to draw around to help identify the shapes of the faces.

DEEPEN Ask children to create their own picture using 2D shapes. Children can describe their picture to a partner, who then has to replicate it behind a screen. Children can then compare pictures to see how accurate they were in giving and following instructions.

ASSESSMENT CHECKPOINT Question ❶ will determine whether children are able to identify triangles and squares correctly, while question ❷ will determine whether children can identify cuboids, pyramids and spheres correctly. Question ❹ will determine whether children can identify the shapes of the faces of a cylinder and a triangle-based pyramid (tetrahedron).

ANSWERS Answers for the **Practice** part of the lesson can be found in the *Power Maths* online subscription.

Reflect

WAYS OF WORKING Pair work

IN FOCUS This activity encourages children to identify 2D and 3D shapes in the real world. In order to do this, children have to focus on the shapes' mathematical properties rather than on their aesthetic or contextual characteristics. This activity could be turned into a game of 'I spy', where one child says the name of the shape they can see and the other child has to point to it.

ASSESSMENT CHECKPOINT This activity will determine whether children are able to identify and name a range of 2D and 3D shapes.

ANSWERS Answers for the **Reflect** part of the lesson can be found in the *Power Maths* online subscription.

After the lesson 🔢

- Are children confident naming 2D and 3D shapes?
- Are children beginning to use the mathematical properties when describing shapes?
- How can you display the shapes and vocabulary in the classroom in the next lesson to reinforce children's understanding?

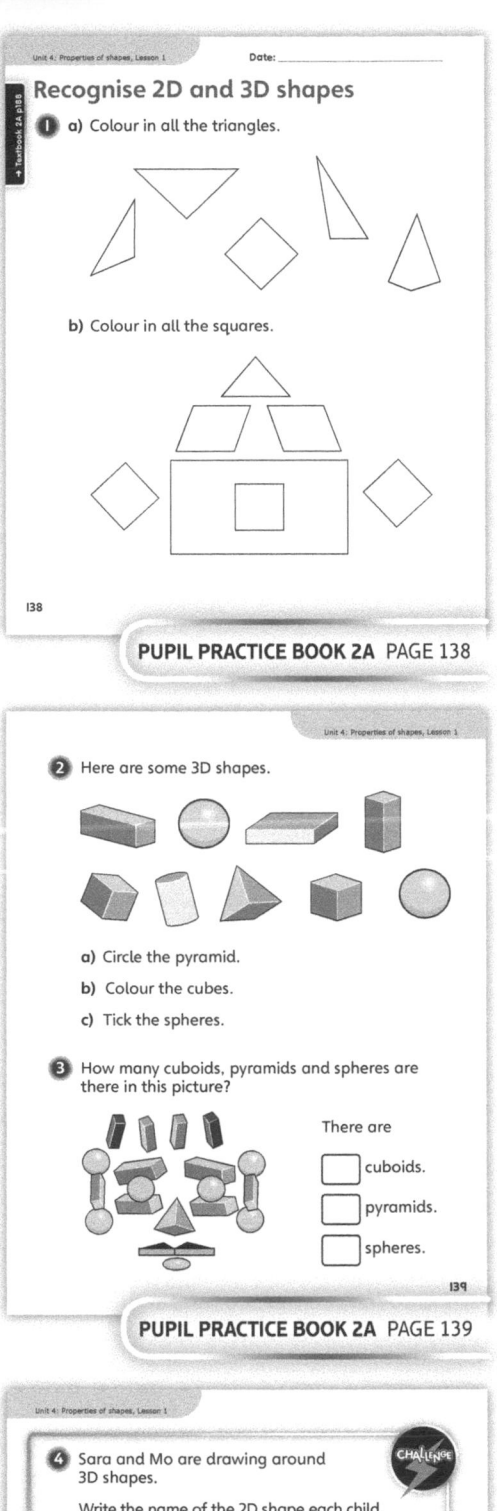

PUPIL PRACTICE BOOK 2A PAGE 138

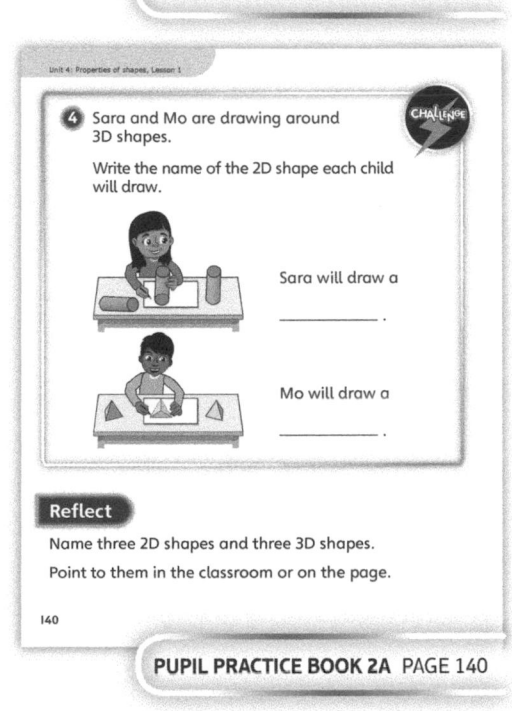

PUPIL PRACTICE BOOK 2A PAGE 139

PUPIL PRACTICE BOOK 2A PAGE 140

Count sides on 2D shapes

Learning focus

In this lesson, children will count the number of sides on 2D shapes and will learn to use this knowledge to categorise different shapes.

Before you teach

- Do children need support to reliably count the sides of 2D shapes?
- Are children secure in recognising different triangles, squares and oblongs?
- Where could you include practical activities in this lesson?

NATIONAL CURRICULUM LINKS

Year 2 Geometry – properties of shape

Identify and describe the properties of 2D shapes, including the number of sides and line symmetry in a vertical line.

ASSESSING MASTERY

Children can identify how many sides a 2D shape has and can use this information to categorise the shape. Children will be able to name and sort irregular polygons by counting the number of sides.

COMMON MISCONCEPTIONS

Children may miscount the number of sides due to either being unsystematic in their approach to counting or losing track of where they started counting. Ask:

- *How will you ensure that you count all the sides only once?*

Children may misname shapes if they are irregular, particularly if one of the internal angles is a reflex angle. Show children some irregular polygons. Ask:

- *What shape is this? How do you know? How many sides does it have?*

STRENGTHENING UNDERSTANDING

Provide a selection of regular and irregular 2D shapes. Work with children to sort them into sorting hoops by counting the number of sides. Children could mark which side they start with using a dry-wipe marker in order to aid their counting. Encourage children to touch or mark each side as they count it. Support children in writing labels for each group.

GOING DEEPER

Using pattern shapes, children explore what happens to the number of sides when they combine two of the same shape. Do they notice a pattern? Why do they think the number of sides changes the way that it does? Is it the same pattern for all the shapes? What happens if they combine two different shapes?

KEY LANGUAGE

In lesson: sides, 2D, corners, **pentagon**, **hexagon**, quadrilateral

Other language to be used by the teacher: triangle, square, oblong, rectangle, polygon

RESOURCES

Mandatory: regular and irregular 2D shapes, sorting hoops, pattern shapes, rulers

Optional: dry-wipe markers, sticks, art straws, sticky notes, screen or bag to hide 2D shapes

 In the eTextbook of this lesson, you will find interactive links to a selection of teaching tools.

Quick recap

Reveal a 2D shape slowly from behind a screen or inside a bag. Prompt children to guess what the shape is before it is fully revealed.

Discover

WAYS OF WORKING Pair work

ASK

- Question ① b): *Can you name any of the shapes?*
- Question ① b): *Can you order the shapes by number of sides? What do you notice?*
- Question ① b): *What shapes do you know that also have four sides?*

IN FOCUS This part of the lesson prompts children to focus on the property of 'number of sides'. Children may immediately recognise certain shapes, particularly regular polygons, without the need to count the sides. However, the number of sides of a 2D shape is a crucial element in determining the type of shape, especially when the shape is not instantly recognisable.

ANSWERS

Question ① a): Kirsty will use five different coloured pens.

Question ① b): No. Kirsty has enough pens for four of the shapes, but she does not have enough pens to draw the hexagon.

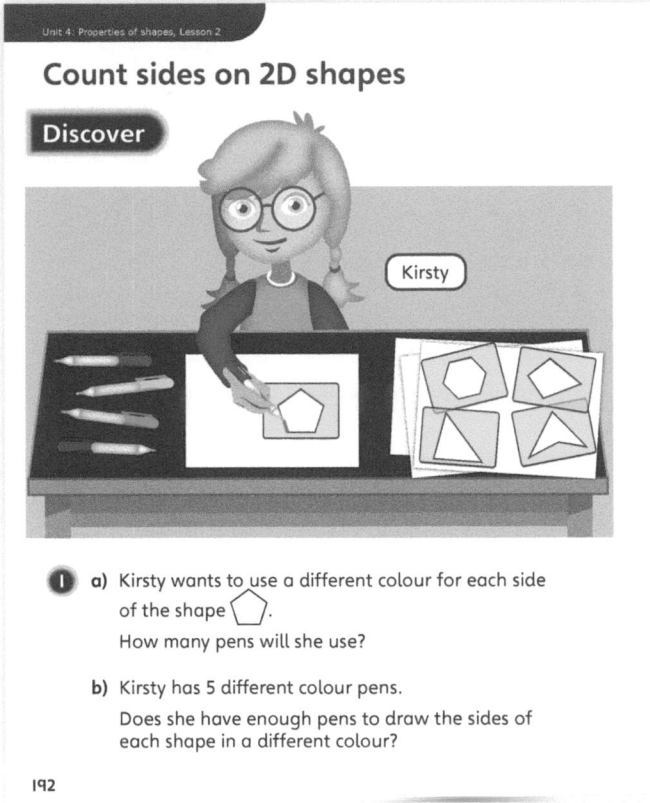

Count sides on 2D shapes

Discover

Kirsty

① a) Kirsty wants to use a different colour for each side of the shape ⬠.
How many pens will she use?

b) Kirsty has 5 different colour pens.
Does she have enough pens to draw the sides of each shape in a different colour?

192

PUPIL TEXTBOOK 2A PAGE 192

Share

WAYS OF WORKING Whole class teacher led

ASK

- Question ① a): *Did you count the sides in the same way as Astrid? Why do you think she counted the sides in the way that she did?*
- Question ① b): *Look at the two shapes with four sides. How are they the same? How are they different?*

IN FOCUS This part of the lesson illustrates that the number of sides a shape has determines what type of shape it is. The illustrated shapes with numbered sides highlight how we can reliably count the number of sides by having a systematic approach. This section also introduces children to pentagons and hexagons.

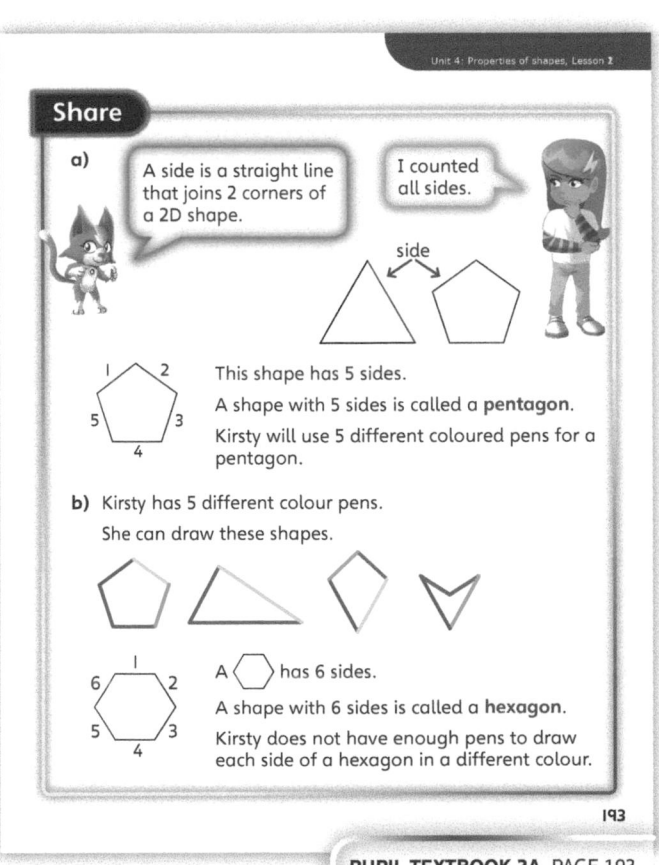

Share

a) A side is a straight line that joins 2 corners of a 2D shape.

I counted all sides.

side

This shape has 5 sides.
A shape with 5 sides is called a **pentagon**.
Kirsty will use 5 different coloured pens for a pentagon.

b) Kirsty has 5 different colour pens.
She can draw these shapes.

A ⬡ has 6 sides.
A shape with 6 sides is called a **hexagon**.
Kirsty does not have enough pens to draw each side of a hexagon in a different colour.

193

PUPIL TEXTBOOK 2A PAGE 193

Think together

Think together

WAYS OF WORKING Whole class teacher led (I do, We do, You do)

ASK

- Question **1**: *What different shapes can you see? How do you know what the shapes are?*
- Question **2**: *How would you sort the shapes into those with more than four sides and those with fewer than four sides?*
- Question **3**: *Did you find any shapes difficult to identify?*

IN FOCUS Question **3** is a collection of irregular polygons. Children will not be able to immediately identify the shapes and will, therefore, have to count the sides in order to name them. The hexagon and the reflex kite both have reflex internal angles. Some children may find it challenging to identify all the sides in these shapes, as they are used to shapes having only acute, right angle and obtuse internal angles.

STRENGTHEN Ask children to use sticks or straws to replicate the shapes in question **2**. Encourage children to make different triangles, quadrilaterals, pentagons and hexagons using the sticks or straws.

DEEPEN Ask children to walk around the classroom or school and label the different polygons they can see using sticky notes. See if they can find examples of triangles, quadrilaterals, pentagons and hexagons.

ASSESSMENT CHECKPOINT Question **1** will determine whether children can identify both regular and irregular quadrilaterals. Question **2** will determine whether children can accurately count the number of sides of different polygons. Question **3** will determine whether children are able to correctly identify the type of polygon based on the number of sides.

ANSWERS

Question **1**: The quadrilaterals are square, oblong, kite and reflex kite.

Question **2**: The shapes have 3 sides, 4 sides, 5 sides and 5 sides respectively.

Question **3**: From left to right, top to bottom, the number of sides are: 4, 3, 3, 5, 4, 6, 4, 4.
There are four quadrilaterals.
There is one pentagon.
There is one hexagon.

PUPIL TEXTBOOK 2A PAGE 194

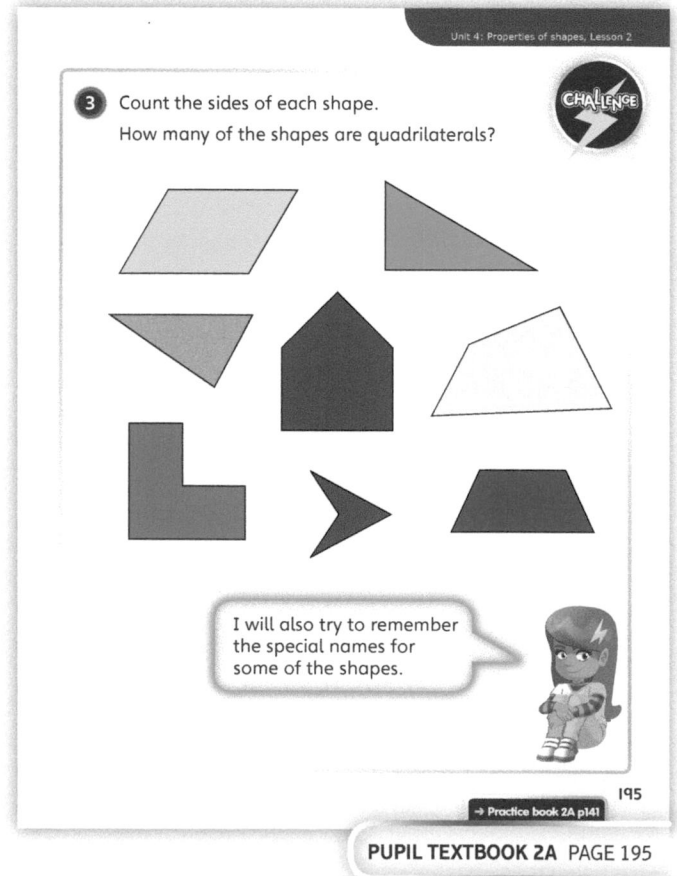

PUPIL TEXTBOOK 2A PAGE 195

Practice

WAYS OF WORKING Independent thinking

IN FOCUS Question ④ uses word problems to relate properties of shapes to multiplication and division. The main challenge here is identifying what operation is required in order to solve the problem. Using practical materials, such as straws, may help support children in understanding what they need to do in order to find the solution.

STRENGTHEN For question ②, suggest that children mark each side as they count them and write the number of sides next to each shape.

Give children a laminated version of question ③. They can then use dry-wipe pens to practise joining the vertices before completing the drawings in their books. For question ④, provide straws so that children can carry out the task practically.

DEEPEN Give children an A4 piece of paper. Ask them to draw a number of lines randomly across the paper so that they intersect to make different types of polygons. They then have to identify all the triangles, the quadrilaterals, the pentagons and the hexagons by giving each type of shape a different pattern, such as spots or stripes.

ASSESSMENT CHECKPOINT Question ① will determine whether children can correctly count the number of sides of different polygons with a familiar representation and name them. Question ② will determine whether children can count the sides and reason about unfamiliar irregular polygons. Question ③ will determine whether children can draw sides to join vertices and count the number of sides. Question ④ will determine whether children can apply their knowledge of 2D shapes, multiplication and division to solve problems.

ANSWERS Answers for the **Practice** part of the lesson can be found in the *Power Maths* online subscription.

Reflect

WAYS OF WORKING Pair work

IN FOCUS This section asks children to group the shapes by counting the number of sides and then justify their decisions. The irregular nature of the polygons forces children to physically count the sides. They have to be systematic in their counting to ensure it is reliable.

ASSESSMENT CHECKPOINT This section will determine whether children are able to count the number of sides accurately and whether they understand that polygons are categorised by how many sides they have.

ANSWERS Answers for the **Reflect** part of the lesson can be found in the *Power Maths* online subscription.

After the lesson ⏸

- Did children find it challenging to identify irregular and unfamiliar polygons?
- How successful were the practical opportunities that you provided for children?
- Are children secure in understanding that polygons are categorised by the number of sides?

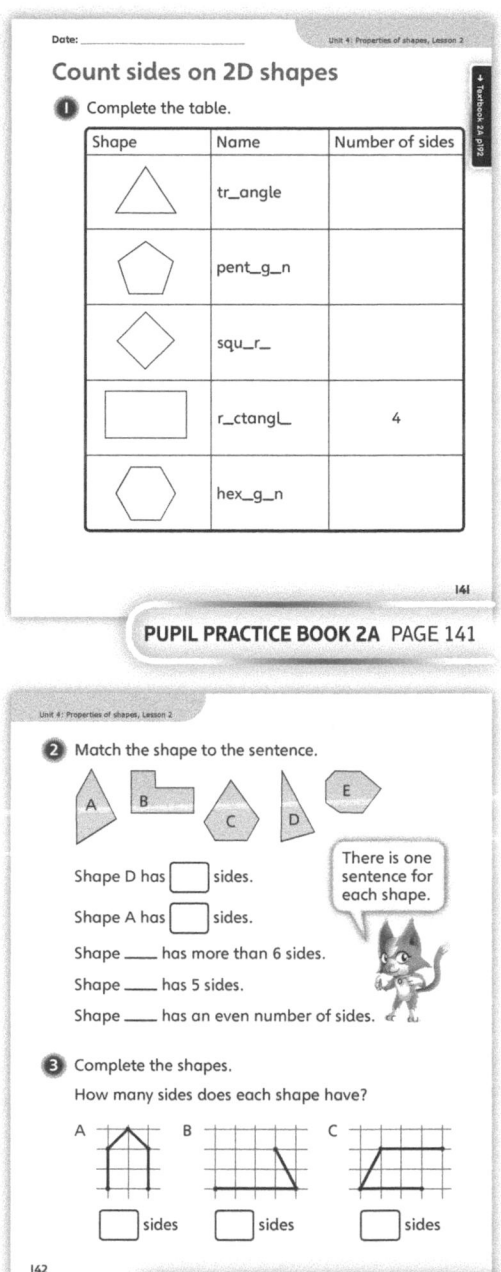

PUPIL PRACTICE BOOK 2A PAGE 141

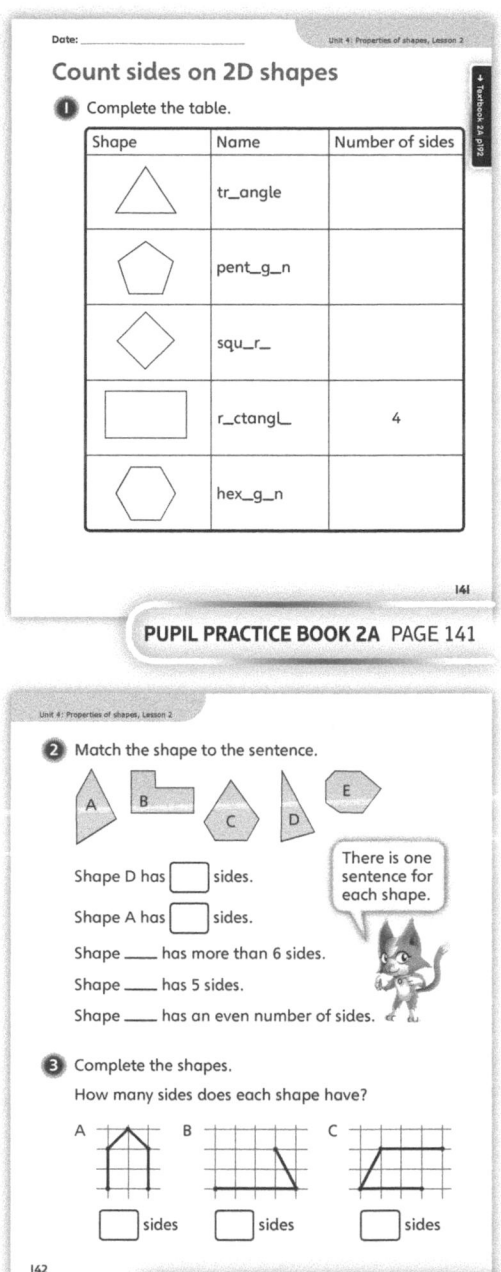

PUPIL PRACTICE BOOK 2A PAGE 142

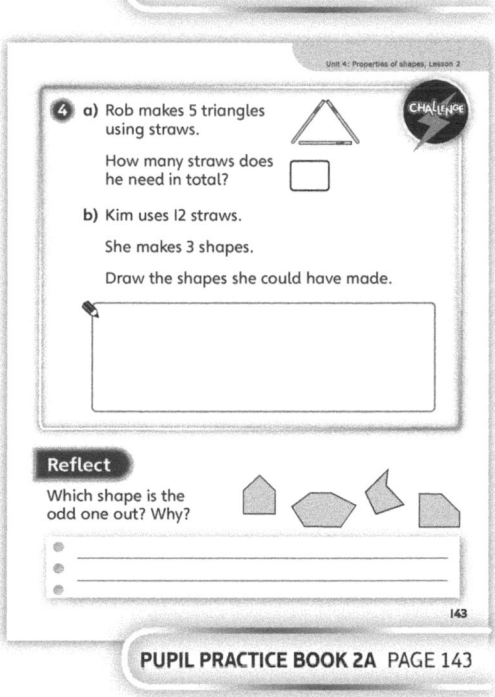

PUPIL PRACTICE BOOK 2A PAGE 143

Count vertices on 2D shapes

Learning focus

In this lesson, children will learn that vertices are points where two or more sides of a polygon meet. Children will learn that they can classify shapes by the number of vertices.

Before you teach

- Are children secure in identifying shapes by the number of sides?
- How can you display the vocabulary in this lesson for children to refer to?
- Where can you provide practical opportunities for children in this lesson?

NATIONAL CURRICULUM LINKS

Year 2 Geometry – properties of shape

Identify and describe the properties of 2D shapes, including the number of sides and line symmetry in a vertical line.

ASSESSING MASTERY

Children can identify and count the vertices of a variety of regular and irregular polygons. Children understand that the corner where two sides meet is called a vertex and that they can categorise shapes by the number of vertices.

COMMON MISCONCEPTIONS

Children may not identify a vertex if it is a reflex internal angle. Show children a quadrilateral with a reflex internal angle on a peg board. Ask:
- *How many sides does this shape have? What is a vertex? How many vertices does this shape have?*

STRENGTHENING UNDERSTANDING

Take children outside with a large loop of string. Ask children to use the string round their ankles to create vertices. Can they make a triangle, quadrilateral, pentagon and hexagon? Ask: *How many ankles are needed for each shape? How many sides does each shape have? How many vertices?*

GOING DEEPER

Using pattern shapes, ask children to explore what happens to the number of vertices when they combine two of the same shape. If children did this activity in the last lesson exploring the number of sides, can they predict what will happen to the number of vertices and justify their predictions? Do they notice a pattern? Why do they think the number of vertices changes the way that it does? Is it the same pattern for all the shapes? What happens if they combine two different shapes?

KEY LANGUAGE

In lesson: square, corners, **vertex**, **vertices**, triangle, hexagon, sides, pentagon

Other language to be used by the teacher: oblong, rectangle, quadrilateral, polygon

RESOURCES

Mandatory: string, pattern shapes, straws, a selection of regular and irregular 2D shapes

Optional: peg board, dry-wipe pen, isometric paper

 In the eTextbook of this lesson, you will find interactive links to a selection of teaching tools.

Quick recap

Play 'Guess my shape'. Say: *I am thinking of a shape. It has four sides. Draw a picture of what you think my shape looks like.*

Discover

WAYS OF WORKING Pair work

ASK

- Question ❶ a): *What do you notice about the part of the shape where the children's fingers are?*
- Question ❶ a): *What shapes can you make using four fingers?*

IN FOCUS This part of the lesson draws the focus from sides to vertices of polygons. By using their fingers, children create 'corners', or vertices. Discuss how two sides meet at each finger. Some children may begin to see the relationship between number of sides and number of vertices.

ANSWERS

Question ❶ a): Four fingers are needed to create a square.
Five fingers are needed to create a pentagon.

Question ❶ b): The only shape you can make with three fingers is a triangle.

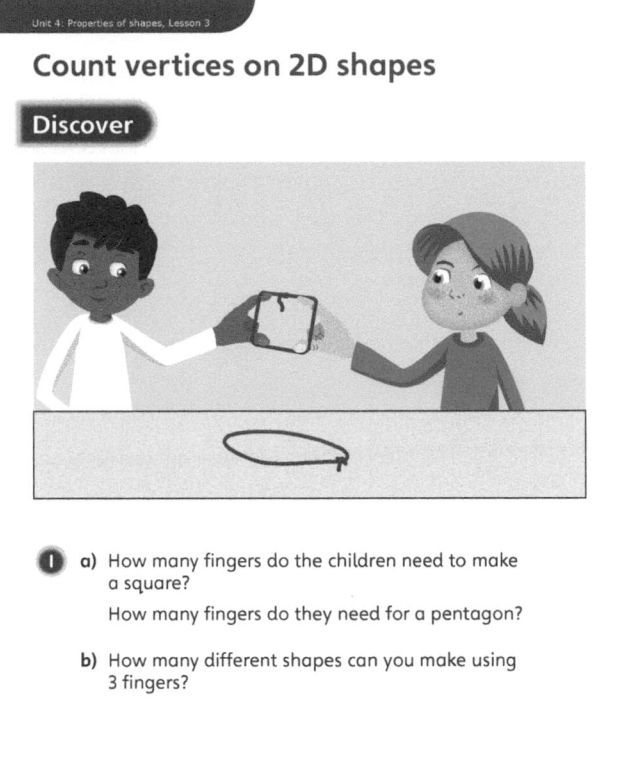

Count vertices on 2D shapes

Discover

❶ a) How many fingers do the children need to make a square?
How many fingers do they need for a pentagon?

b) How many different shapes can you make using 3 fingers?

196

PUPIL TEXTBOOK 2A PAGE 196

Share

WAYS OF WORKING Whole class teacher led

ASK

- Question ❶ a): *How can you make sure that you count all the vertices only once?*
- Question ❶ a): *How is this similar to what you did in the last lesson?*

IN FOCUS Sparks's comment introduces children to the terms 'vertex' and 'vertices'. Previously, children may have referred to vertices as corners. Emphasise that a vertex is created where two sides of a polygon meet.

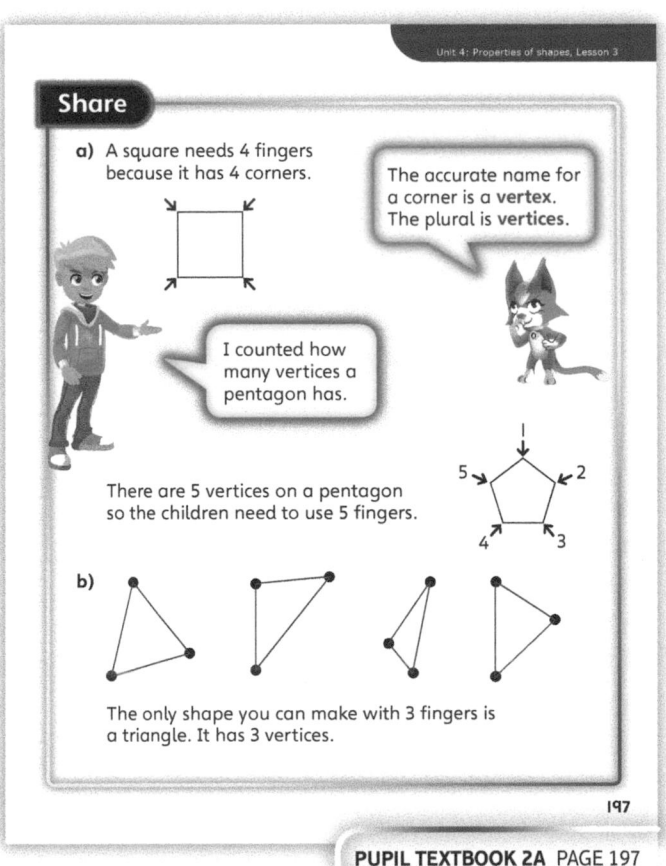

Share

a) A square needs 4 fingers because it has 4 corners.

The accurate name for a corner is a **vertex**. The plural is **vertices**.

I counted how many vertices a pentagon has.

There are 5 vertices on a pentagon so the children need to use 5 fingers.

b)

The only shape you can make with 3 fingers is a triangle. It has 3 vertices.

197

PUPIL TEXTBOOK 2A PAGE 197

Think together

WAYS OF WORKING Whole class teacher led (I do, We do, You do)

ASK

- Question ❸: *What do you notice about the number of sides and the number of vertices of a shape? Is this always true? Why do you think that is?*
- Question ❸: *Which way do you find easier to identify a polygon? Counting sides or counting vertices?*

IN FOCUS Question ❷ has a number of shapes with at least one internal reflex angle. Some children may not recognise these as vertices. To help children identify all vertices, remind them that a vertex is created where two or more sides meet.

STRENGTHEN Ask children to carry out questions ❶ and ❷ practically. This will help them to identify the vertices. When counting vertices in a pictorial representation of a polygon, encourage children to mark off the vertices as they count them.

DEEPEN Ask children to explore what polygons they can create by cutting a straight line from one vertex to another to divide a number of shapes into two new polygons. What is the total number of vertices of the shapes created? How does it relate to the number of vertices children started with? Why do they think that is? Can children predict the total number of vertices after cutting one line, based on the starting number of vertices? For example, ask: *A square has four vertices. After cutting one line from one vertex to another, you create two triangles with a total of six vertices. This is two more than the original square. If I start with a hexagon and cut a line from a vertex to vertex to create two new polygons, will I have a total of eight vertices?*

ASSESSMENT CHECKPOINT Question ❶ will determine whether children can count the vertices of a variety of shapes. Question ❷ will determine whether children can identify all vertices of both regular and irregular polygons, including those with reflex internal angles. Question ❸ will determine whether children can identify and count the sides and vertices of polygons and identify the relationship between them.

ANSWERS

Question ❶: The oblong and kite have four vertices; the regular and irregular pentagons have five vertices; the hexagon has six vertices.

Question ❷: From left to right and top to bottom, the numbers of vertices are: 6, 6, 6, 6, 5, 6.

Question ❸: Shape A has 6 sides and 6 vertices.
Shape B has 4 sides and 4 vertices.
Shape C has 8 sides and 8 vertices.

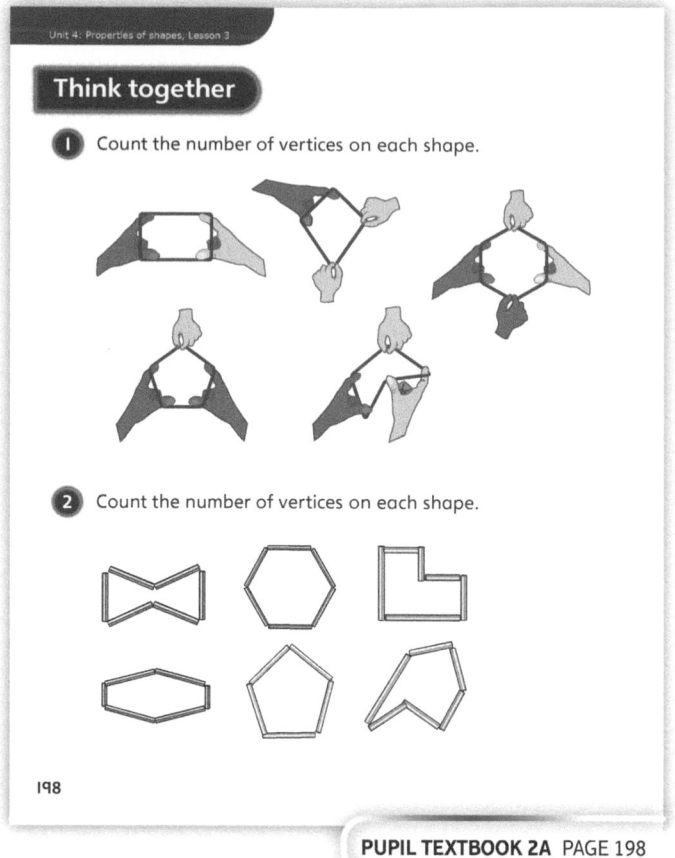

PUPIL TEXTBOOK 2A PAGE 198

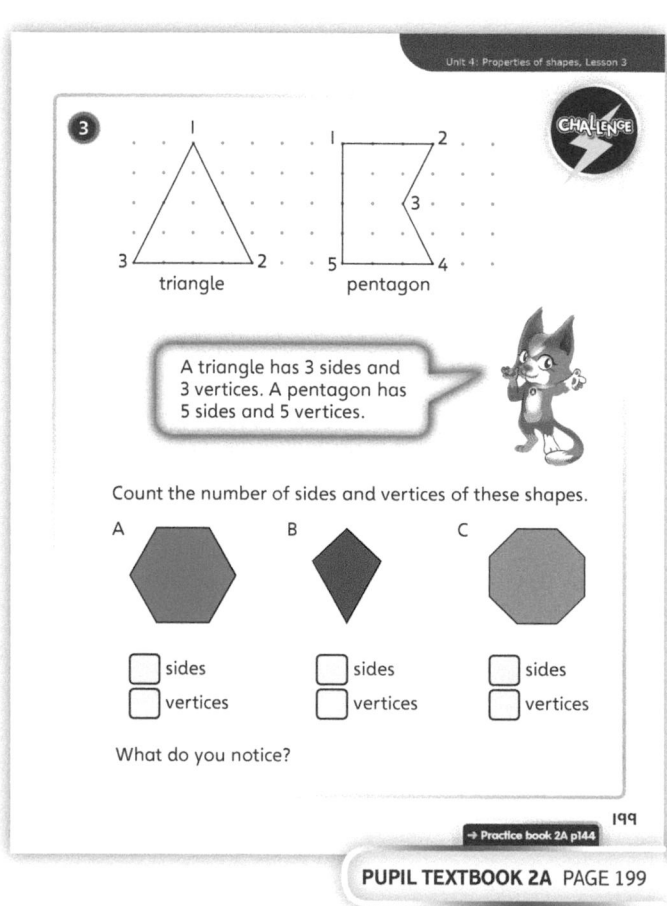

PUPIL TEXTBOOK 2A PAGE 199

Practice

IN FOCUS Questions **1** and **2** provide children with pictorial representations of shapes and ask them to identify the number of vertices in each one. Question **3** becomes more abstract, as children are asked to complete sentences with the names of shapes according to how many vertices and sides they have.

STRENGTHEN For questions **1** and **2**, encourage children to mark off each vertex as they count it. For question **3**, have the corresponding 2D shapes available for children to refer to. Provide children with a laminated version of question **6** and a dry-wipe pen so that they can practise, review and correct their responses before drawing them.

DEEPEN Provide children with isometric paper. Encourage them to create different quadrilaterals, pentagons and hexagons. See if children can create polygons with seven, eight and nine vertices.

THINK DIFFERENTLY In question **4**, children explore a misconception about the number of vertices in a shape. Remind them that a vertex is only created when two or more sides meet to help them see that the number of dots on the square dotted paper is not relevant.

ASSESSMENT CHECKPOINT Questions **1** and **2** will determine whether children can correctly count the number of vertices in a polygon. Question **3** will determine whether children can match the names of polygons to the correct number of vertices and sides. Question **4** will determine whether children understand that a vertex is created where two sides of a polygon meet. Question **6** will determine whether children can accurately draw polygons with a given number of vertices.

ANSWERS Answers for the **Practice** part of the lesson can be found in the *Power Maths* online subscription.

Reflect

WAYS OF WORKING Pair work

IN FOCUS This section highlights the misconception that a vertex is not present where an internal angle in a polygon is reflex. Children may struggle to articulate the differences between the two shapes. Support children by suggesting they think about the angles as turns. Tell children to imagine standing on the vertices facing a side. They then have to turn to face inside the shape and then continue turning until they end up facing the other side. Would they have to make more than a half turn on any of the vertices?

ASSESSMENT CHECKPOINT This section will determine whether children can correctly identify vertices even where there is an internal reflex angle.

ANSWERS Answers for the **Reflect** part of the lesson can be found in the *Power Maths* online subscription.

After the lesson

- Do children understand the definition of a vertex?
- Can children identify vertices even on unfamiliar irregular polygons?
- Do children understand the relationship between the number of vertices and the number of sides of polygons?

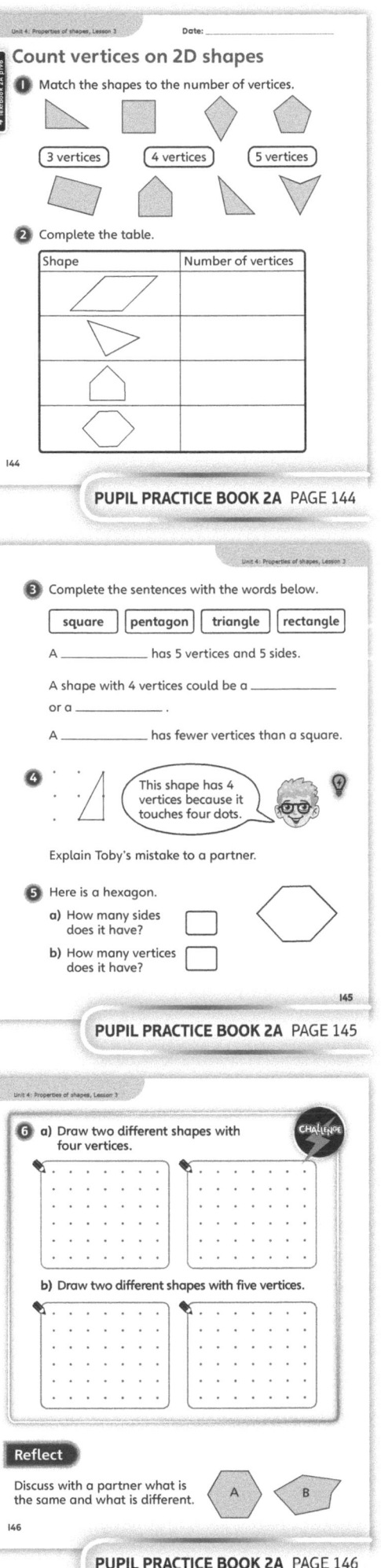

PUPIL PRACTICE BOOK 2A PAGE 144

PUPIL PRACTICE BOOK 2A PAGE 145

PUPIL PRACTICE BOOK 2A PAGE 146

Draw 2D shapes

Learning focus

In this lesson, children will apply what they have learned about the properties of shapes in order to accurately draw 2D shapes.

Before you teach ▮▮

- Are children secure in recognising the properties of rectangles and triangles?
- How can you reinforce understanding of what is required for a square, oblong and rectangle?
- What resources and prompts could you use to support understanding?

NATIONAL CURRICULUM LINKS

Year 2 Geometry – properties of shape

Identify and describe the properties of 2D shapes, including the number of sides and line symmetry in a vertical line.

ASSESSING MASTERY

Children can accurately draw triangles, squares and oblongs, identifying and including the properties of these shapes. They can create different types of triangles, recognising the need for three straight lines only, and different types of rectangles, recognising the need for four straight lines and four right angles.

COMMON MISCONCEPTIONS

When drawing triangles, children may not recognise irregular triangles as triangles. Show children an equilateral and a scalene triangle. Ask:

- *What is the same about these shapes? What is different about them? Are they both triangles?*

When drawing rectangles, children may not realise the need for four right angles. Show children a rectangle and a trapezium. Ask:

- *Are they both rectangles? What is the same about these shapes? What is different about them?*

STRENGTHENING UNDERSTANDING

Children may need to have shapes to draw around first. Ask children to place dots on the vertices so that they can see how the shape is formed. Children could then use art straws to recreate the shapes using sticky tack on the vertices. Not all 'rectangles' created in this way will have accurate right angles, so be on hand to discuss what makes a rectangle a special type of quadrilateral.

GOING DEEPER

Children could combine 2D shapes to create new ones. Ask children to explore joining squares or triangles to create rectangles and other quadrilaterals. Help children to notice that they can only make a true rectangle by joining two matching rectangles or two matching right-angled triangles.

KEY LANGUAGE

In lesson: 2D, square, sides, triangle, rectangle

Other language to be used by the teacher: oblong

RESOURCES

Mandatory: 2D shapes, square dotted paper, squared paper, plain paper, rulers

Optional: art straws, sticky tack, dry-wipe markers, isometric paper, laminated 2D shapes

 In the eTextbook of this lesson, you will find interactive links to a selection of teaching tools.

Quick recap

Play 'Guess my shape'. Say: *I am thinking of a shape. It has three vertices. Draw a picture of what you think my shape looks like.* Repeat with shapes with four or five vertices.

Discover

WAYS OF WORKING Pair work

ASK

- Question ❶ a): *What shapes can you see? How do you know?*
- Question ❶ a): *Why are Meg's shapes not accurate?*
- Question ❶ a): *How would you draw a shape more accurately?*

IN FOCUS This part of the lesson highlights that an approximate representation of a 2D shape is not adequate. The shapes drawn do not have straight sides and they do not all join to create vertices. Children may see the top two shapes as squares, even though the sides are not of equal length. It is important to show that precision is needed in order to draw an accurate representation of a 2D shape.

ANSWERS

Question ❶ a): Children must draw a shape with four straight sides of the same length and with four right angles, using a ruler and squared paper.

Question ❶ b): Children must draw a shape with three straight sides and three vertices, using a ruler.

Share

WAYS OF WORKING Whole class teacher led

ASK

- Question ❶ a): *How do the squared paper and ruler help?*
- Question ❶ a): *How can you measure the sides of the square to ensure they are of equal length? How do you know your shape has four right angles?*
- Question ❶ b): *Is that the only way to draw a triangle?*

IN FOCUS This section focuses on the fact that polygons must have straight sides and should be drawn with precision.

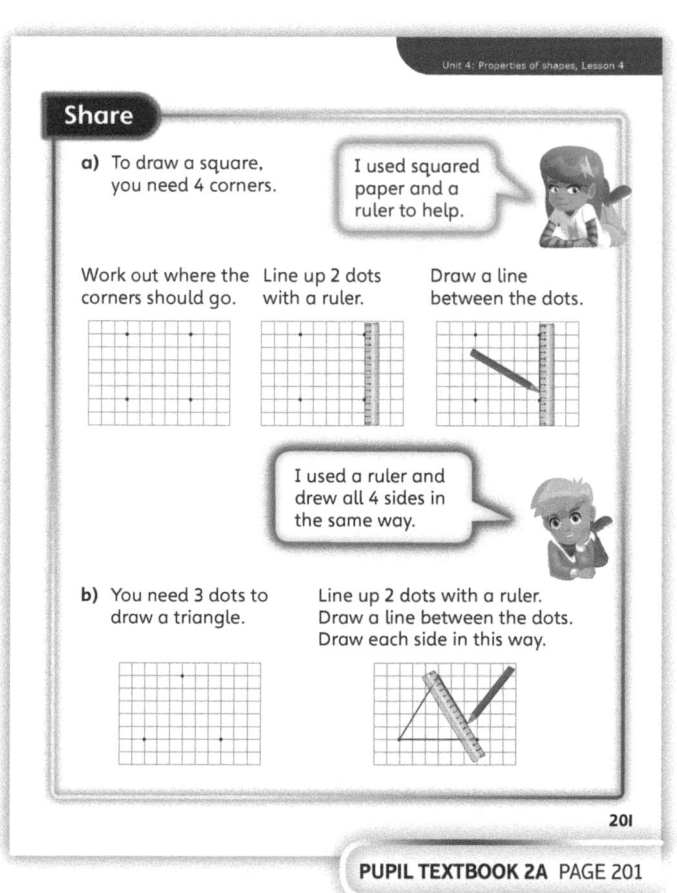

Think together

WAYS OF WORKING Whole class teacher led (I do, We do, You do)

ASK

- Question ❶: *Which paper makes it easier to draw polygons: squared paper or square dotted paper?*
- Question ❶: *Is it possible to draw a different rectangle?*
- Question ❷: *Can you draw any other types of triangles using the square dotted paper?*

IN FOCUS Question ❶ asks children to use their knowledge of drawing squares in **Share** and apply this to oblong rectangles.

Question ❸ b) requires children to draw the shapes without the aid of squared or dotted paper. Be on hand to help children draw perpendicular sides for the rectangles, perhaps using a right-angled object. For the isosceles triangles, help children find the midpoint of the base and create a perpendicular line, so they can then find where the other equal sides meet.

STRENGTHEN Provide children with laminated squared and dotted paper for them to explore creating squares, oblongs and triangles. Encourage children to identify the shapes they have drawn and describe their properties.

DEEPEN Ask children to draw a square. Can they draw a single line to create two new shapes that are identical? Can they create new shapes that are identical by drawing two lines?

ASSESSMENT CHECKPOINT Question ❶ will determine whether children recognise the mathematical properties of a rectangle and can draw them accurately with some vertices already given. Question ❷ will determine whether children can recognise the mathematical properties of triangles and accurately copy them using dotted paper. Question ❸ a) will determine whether children can accurately copy shapes using squared paper. Question ❸ b) will determine whether children can use a ruler to accurately copy squares, rectangles and triangles.

ANSWERS

Question ❶: A rectangle, four squares by two squares; a square, three squares by three squares.

Question ❷: The triangles should accurately match those in the **Textbook**.

Question ❸: The shapes should accurately match those in the **Textbook**.

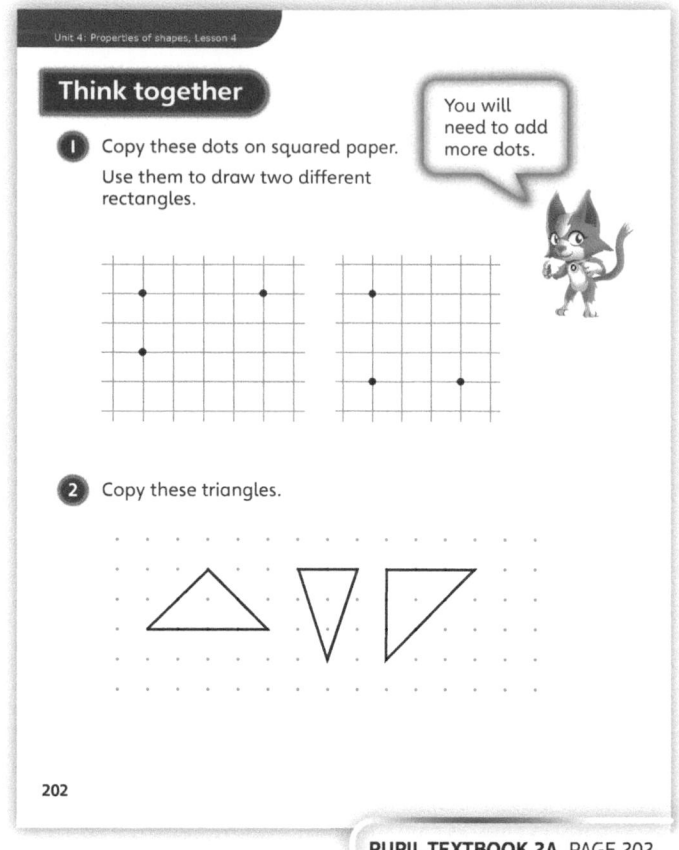

Think together

❶ Copy these dots on squared paper. Use them to draw two different rectangles.

You will need to add more dots.

❷ Copy these triangles.

202

PUPIL TEXTBOOK 2A PAGE 202

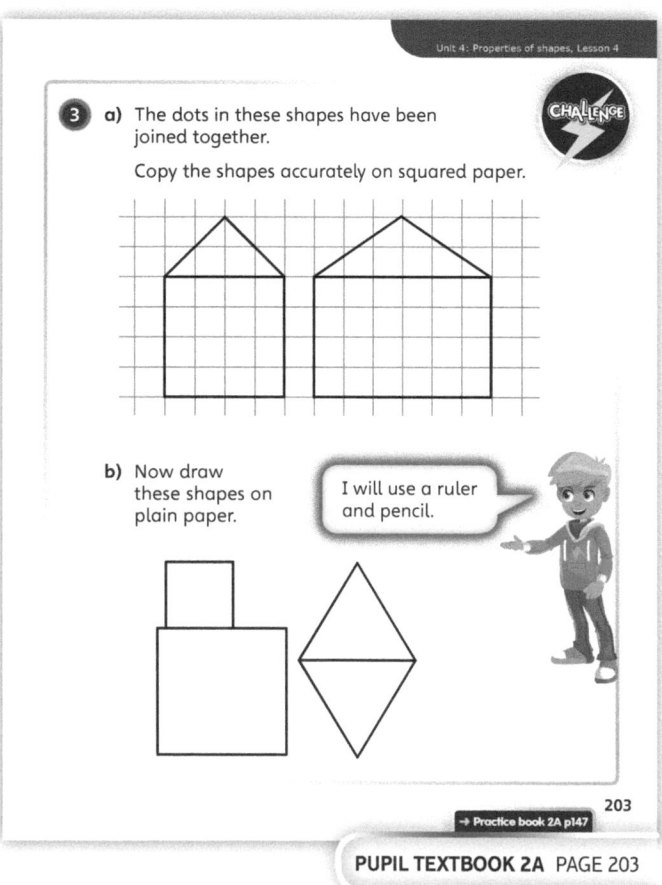

❸ a) The dots in these shapes have been joined together.

Copy the shapes accurately on squared paper.

CHALLENGE

b) Now draw these shapes on plain paper.

I will use a ruler and pencil.

203

→ Practice book 2A p147

PUPIL TEXTBOOK 2A PAGE 203

Practice

WAYS OF WORKING Independent thinking

IN FOCUS Question ③ asks children to copy triangles without squared paper, so they may need support with accurately drawing the angles and distances. Question ④ asks children to draw different squares. The squares have to be different sizes. Some children may think that positioning the square in a different place on the grid will create a different square. You might want to explore this misconception as a class.

It *is* possible to create other squares by joining diagonally opposite dots. This will give you a 'rotated' square, whose sides are not vertical and horizontal.

Point out that these squares are different because the distance between two diagonally opposite dots is greater than the distance between two dots that are vertically or horizontally adjacent.

STRENGTHEN Provide 2D shapes as a reference for children. Distribute laminated versions of the questions, so children can attempt them using dry-wipe markers and can correct and adapt what they do easily.

DEEPEN Provide children with a similar version of question ④, but with isometric paper. Challenge children to create different triangles, rectangles and other quadrilaterals. If children say they have found a square, help them to check – it is more likely to be a rectangle.

ASSESSMENT CHECKPOINT Question ① will determine whether children are able to identify a polygon when presented with only the vertices. In question ②, children must add the final vertex in order to complete each rectangle. Question ③ will determine whether children are able to accurately measure and copy triangles. Question ④ will determine whether children are able to accurately draw squares of different sizes and orientations.

ANSWERS Answers for the **Practice** part of the lesson can be found in the *Power Maths* online subscription.

Reflect

WAYS OF WORKING Pair work

IN FOCUS This section requires children to think carefully about how a shape is drawn in order to give clear instructions, which they can then try on a partner. The limited number of instructions prompts children to think about how to accurately locate each vertex of the shape.

ASSESSMENT CHECKPOINT This section will determine whether children are able to accurately identify the location of each vertex in relation to another.

ANSWERS Answers for the **Reflect** part of the lesson can be found in the *Power Maths* online subscription.

After the lesson

- Were children able to accurately draw the required shapes?
- Were children accurate in drawing the sides and angles of the shapes?
- What challenges did children face when drawing the shapes and how can you help them overcome these challenges?

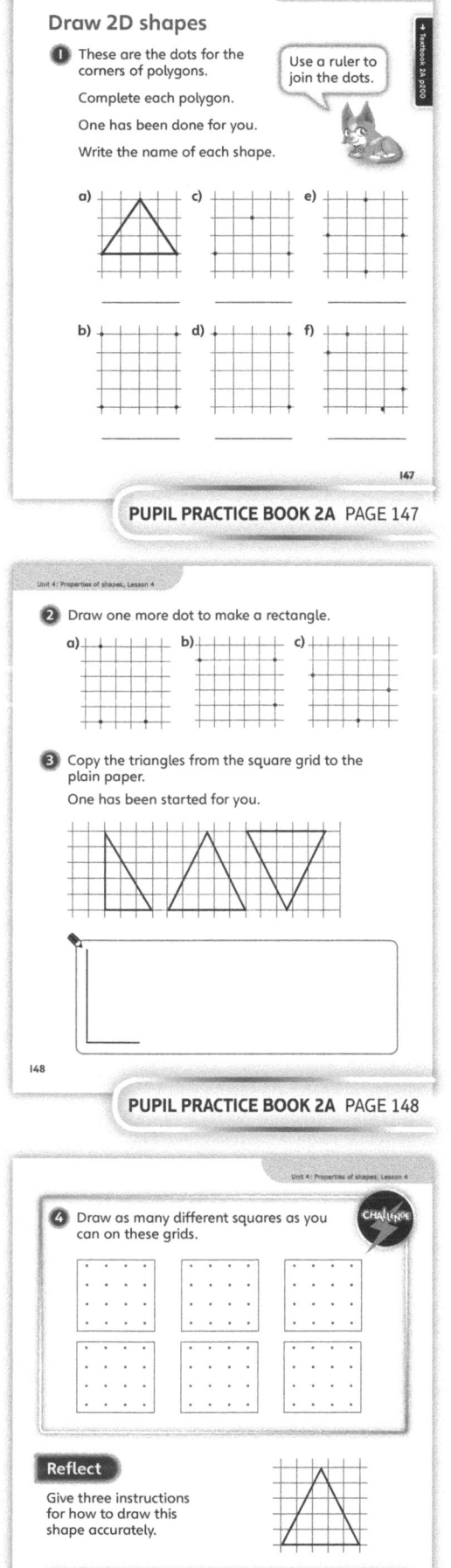

PUPIL PRACTICE BOOK 2A PAGE 147

PUPIL PRACTICE BOOK 2A PAGE 148

PUPIL PRACTICE BOOK 2A PAGE 149

Lines of symmetry on shapes

Learning focus

In this lesson, children will explore reflective symmetry. They will learn to identify shapes and images that have reflective symmetry and identify where the line of symmetry lies.

Before you teach

- How will you explain reflective symmetry?
- How will you support children who may struggle to understand the concept of reflective symmetry?
- What resources and images could you have available to reinforce understanding of reflective symmetry?

NATIONAL CURRICULUM LINKS

Year 2 Geometry – properties of shape

Identify and describe the properties of 2D shapes, including the number of sides and line symmetry in a vertical line.

ASSESSING MASTERY

Children can identify lines of symmetry in images and shapes. Children can understand that the halves either side of the line of symmetry are mirror images of each other and they can identify what the symmetrical shape will be when only one-half is shown.

COMMON MISCONCEPTIONS

Children may think that by drawing a line to halve a shape, they have found the line of symmetry, even if the two halves are not mirror images of each other. Show children an oblong. Ask:

- *Is this symmetrical? Where is the line of symmetry?*

Now draw a line from corner to corner. Ask:

- *Have I drawn the line of symmetry?*

STRENGTHENING UNDERSTANDING

A nice way to introduce reflective symmetry is to create symmetrical butterflies. Provide children with an outline of a butterfly that has a vertical line of symmetry. Children can then use different coloured paints and place blobs of paint on one side of the butterfly. They then fold it in half and smooth it down, so that the paint is printed on the opposite side. When children open it up, they reveal a symmetrical pattern. Discuss how the colours match up when the butterfly is folded in half. You could compare these with pictures of actual butterflies and discuss the symmetrical patterns.

GOING DEEPER

Provide children with some Rangoli-style patterns. These usually have a number of lines of symmetry. Can children find them all? Children could try creating their own symmetrical patterns with more than one line of symmetry using peg boards or colouring in squared paper.

KEY LANGUAGE

In lesson: line of symmetry, symmetrical

Other language to be used by the teacher: symmetry, vertical, horizontal, reflective

RESOURCES

Mandatory: mirrors

Optional: peg boards, squared paper, paint, butterfly templates, butterfly pictures, tracing paper, paper cutouts of regular polygons, Rangoli-style patterns

 In the eTextbook of this lesson, you will find interactive links to a selection of teaching tools.

Quick recap

Ask children to draw four different examples of 2D shapes on squared dotted paper or squared paper. Then ask them to name each shape and write the number of sides and vertices that it has.

Discover

Unit 4: Properties of shapes, Lesson 5

Lines of symmetry on shapes

Discover

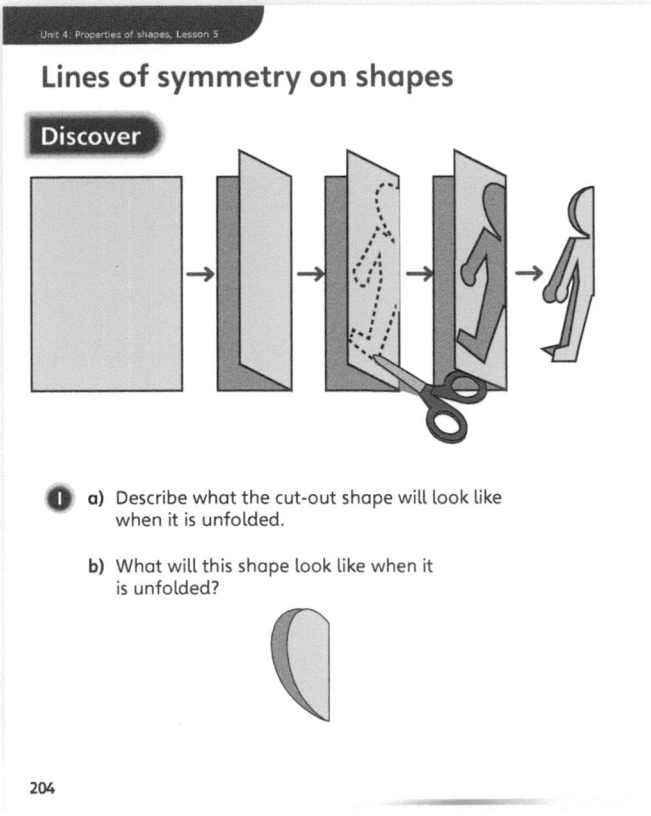

WAYS OF WORKING Pair work

ASK

- Questions ① a) and b): *How did you know what the shape would be?*
- Questions ① a) and b): *What can you tell me about the two halves of the image? Are they the same?*

IN FOCUS This section asks children to visualise a symmetrical image when only half of it is presented. This introduces reflective symmetry as the idea that when the image is folded along the line of symmetry, the two halves match.

ANSWERS

Question ① a): The shape will look like a person when it is unfolded.

Question ① b): The shape will look like a heart when it is unfolded.

① a) Describe what the cut-out shape will look like when it is unfolded.

   b) What will this shape look like when it is unfolded?

204

PUPIL TEXTBOOK 2A PAGE 204

Share

Share

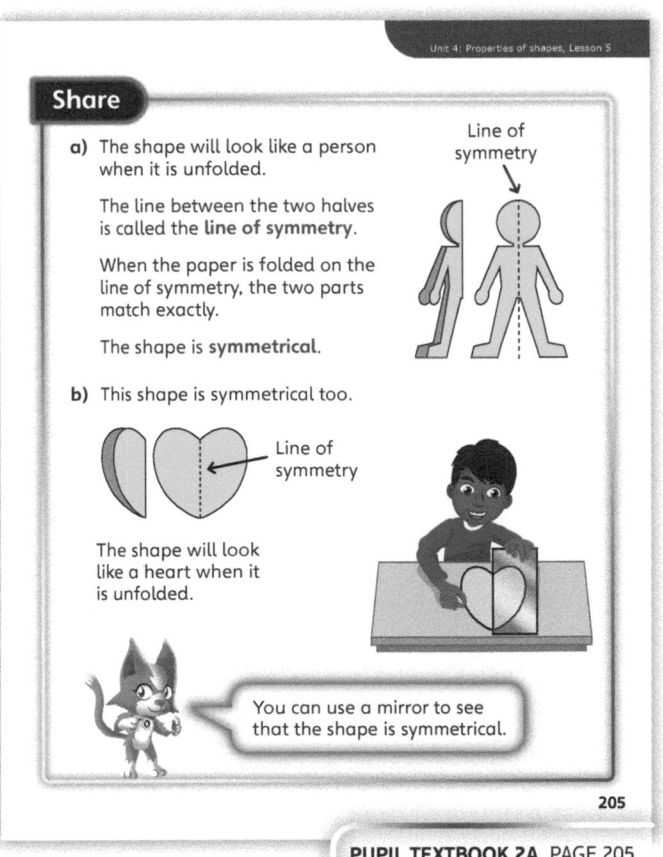

a) The shape will look like a person when it is unfolded.

   The line between the two halves is called the **line of symmetry**.

   When the paper is folded on the line of symmetry, the two parts match exactly.

   The shape is **symmetrical**.

b) This shape is symmetrical too.

   The shape will look like a heart when it is unfolded.

Line of symmetry

You can use a mirror to see that the shape is symmetrical.

WAYS OF WORKING Whole class teacher led

ASK

- *Can you see anything in the room that has a line of symmetry?*
- *Can you think of any 2D shapes that have a line of symmetry?*

IN FOCUS Sparks's comment tells children that a mirror can be used to test for lines of reflective symmetry. Provide some mirrors for children and ask them to try to find objects that have reflective symmetry.

205

PUPIL TEXTBOOK 2A PAGE 205

Think together

Whole class teacher led (I do, We do, You do)

ASK

- Question **2**: *How can you work out where to draw each vertex?*
- Question **2**: *Can you identify what the shape will be before you draw it?*
- Question **3**: *Are there any shapes that do not have a line of symmetry?*

IN FOCUS Question **2** provides children with half a symmetrical image and asks them to complete it. This will highlight whether children understand that with reflective symmetry, each side of the line of symmetry is a mirror image of the other. The use of squared paper provides children with a reference when trying to complete the image, enabling them to count the number of squares from the line of symmetry to vertices of the shape. Children can check their responses with a mirror, as suggested by Astrid.

STRENGTHEN For question **2**, children could trace over the half shape using tracing paper and then flip the paper over in order to see how the second half should be positioned.

Provide children with paper cutout versions of the shapes in question **3**. Children can then fold them in half in order to identify the line of reflective symmetry.

DEEPEN When children have completed question **2**, ask them if any of the shapes have more than one line of symmetry. This will be true for the squares and oblongs. Provide children with paper cutouts of regular polygons. Can they identify all the lines of symmetry by folding them?

ASSESSMENT CHECKPOINT Question **1** will determine whether children can visualise the whole image when presented with only half. Question **2** will determine whether children can draw the missing half of a symmetrical 2D shape. Do children understand that the half they need to draw is a mirror image of the first half? Question **3** will determine whether children can identify lines of reflective symmetry in 2D shapes.

ANSWERS

Question **1**: Circle, star, smiley face, house.

Question **2**: Check the shapes are completed correctly.
Top row: a square, an isosceles triangle.
Bottom row: a square, an oblong.

Question **3**: From left to right, top to bottom: no, no, yes, yes, no.

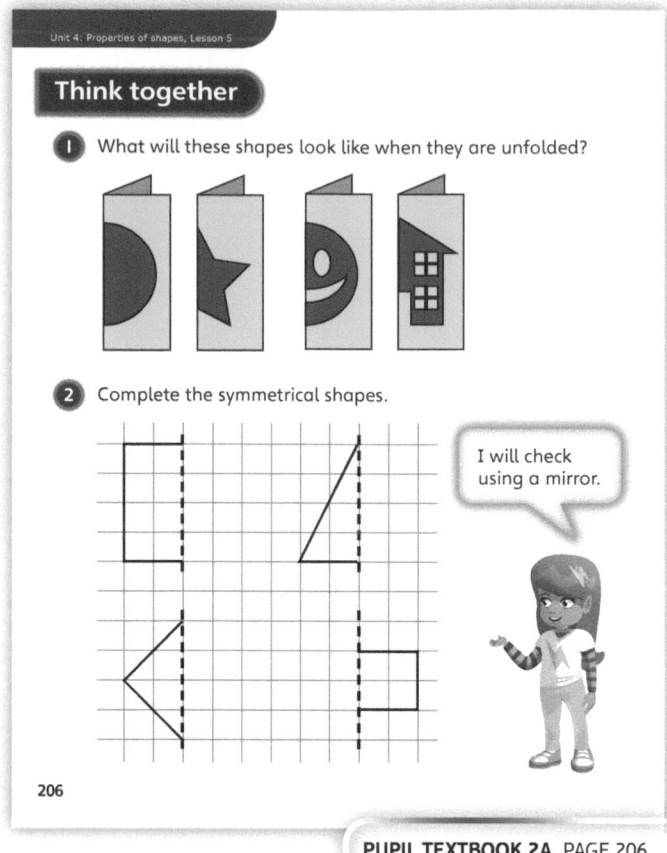

PUPIL TEXTBOOK 2A PAGE 206

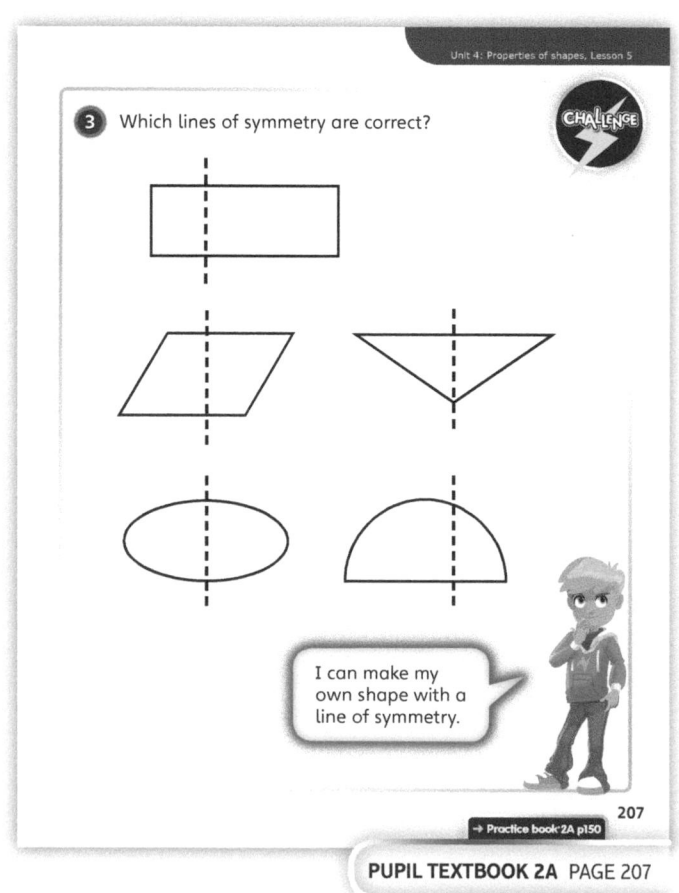

PUPIL TEXTBOOK 2A PAGE 207

Practice

Independent thinking

IN FOCUS Question ① requires children to draw the line of reflective symmetry on given 2D shapes that they may encounter in real-life contexts. In question ②, children are asked to use the line of symmetry to draw in the missing half of given 2D shapes. Children need to understand that the two halves of the shape are mirror images of each other along the line of symmetry and be able to visualise what the complete shape will look like.

STRENGTHEN For question ②, provide children with tracing paper so that they can trace the first half and then flip the paper to identify where to position the second half. For question ③, provide children with matching paper shapes. They can then fold them in half to identify the lines of symmetry, making sure that the two halves match when folded.

DEEPEN Provide children with peg boards or squared paper. Can children produce a pattern that has both vertical and horizontal lines of symmetry? What about a diagonal line of symmetry?

THINK DIFFERENTLY In question ④, children use what they know about lines of reflective symmetry to identify which given lines of symmetry are incorrect. Encourage children to explain how they know. They could use paper folding or mirrors to demonstrate this.

This question encourages children to think differently as it explores the position of the line of symmetry within the shape.

ASSESSMENT CHECKPOINT Use question ⑤ to assess children's understanding. Children should draw a shape with reflective symmetry, ensuring that the line of symmetry is drawn correctly.

ANSWERS Answers for the **Practice** part of the lesson can be found in the *Power Maths* online subscription.

Reflect

WAYS OF WORKING Pair work

IN FOCUS In this section, children will need to apply what they have learned so far in this unit. As the shape has fewer than five vertices, it can only be a quadrilateral or a triangle. Children have to think carefully about how they describe the shape; they should use what they have already learned about the properties of the shapes they draw.

ASSESSMENT CHECKPOINT This section will determine whether children are able to visualise shapes with reflective symmetry and whether they can use correct mathematical vocabulary to describe a given shape's properties.

ANSWERS Answers for the **Reflect** part of the lesson can be found in the *Power Maths* online subscription.

After the lesson ⏸

- Are children confident in identifying lines of reflective symmetry in 2D shapes?
- How does this build on the previous lessons on properties of shapes?
- What practical opportunities can you provide for children who may still not be secure in their understanding of reflective symmetry?

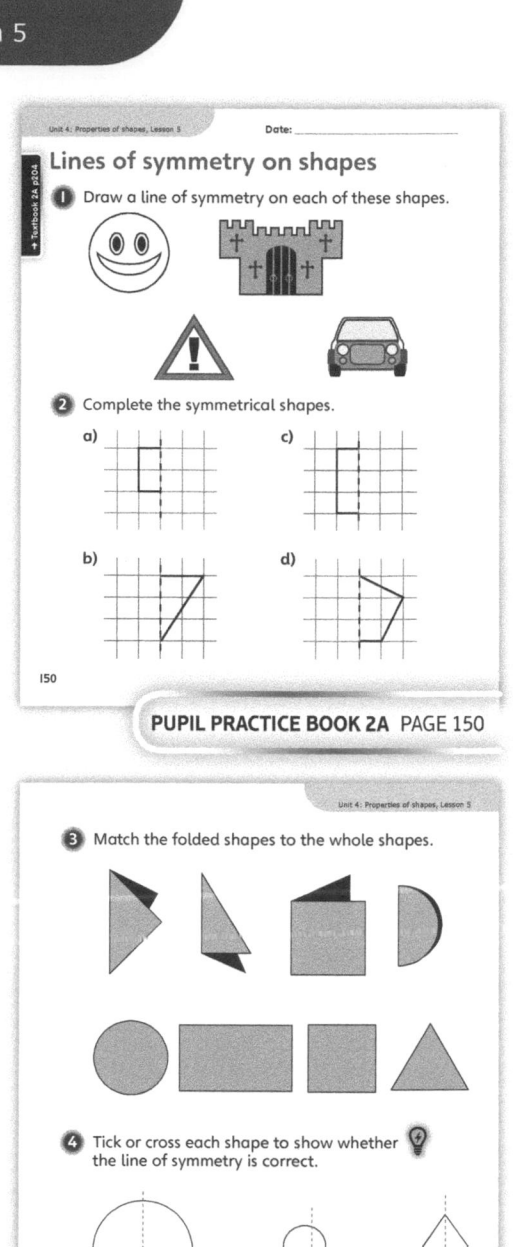

PUPIL PRACTICE BOOK 2A PAGE 150

PUPIL PRACTICE BOOK 2A PAGE 151

PUPIL PRACTICE BOOK 2A PAGE 152

Sort 2D shapes

Learning focus

In this lesson, children will draw on their previous learning about properties of 2D shapes in order to sort polygons by different criteria. This will include focusing on number of sides, number of vertices and reflective symmetry.

Before you teach

- Are children secure in identifying the properties of 2D shapes?
- How could you display the vocabulary used in this lesson for children to refer to?

NATIONAL CURRICULUM LINKS

Year 2 Geometry – properties of shape

Compare and sort common 2D and 3D shapes and everyday objects.

ASSESSING MASTERY

Children can sort 2D shapes based on a variety of given criteria. Children are also able to choose their own criteria to sort shapes based on the mathematical properties of each shape.

COMMON MISCONCEPTIONS

Children may be unsuccessful in counting the number of sides or vertices, particularly when the shape is irregular and has an internal reflex angle. Show children a variety of irregular polygons. Ask:

- *How many sides does this shape have? How many vertices does this shape have? How can you make sure that you count them all correctly?*

Children may focus on properties such as colour rather than mathematical properties. Provide children with a variety of 2D shapes and ask them to sort them into groups. If they sort them by colour, ask:

- *Can you sort them a different way?*

STRENGTHENING UNDERSTANDING

Wherever possible, provide children with shapes and sorting hoops so that they can physically sort the shapes. To begin with, you may need to provide labels for the different groups before children start to write their own. Ensure that you encourage children to talk about their decisions as they sort the shapes.

GOING DEEPER

Provide children with shapes and sorting criteria where some shapes belong either in both groups or in neither. Ask them where they would place those shapes. Some children may be ready to sort shapes using a Carroll diagram in which each group is determined by two criteria.

KEY LANGUAGE

In lesson: most, vertices, more than, triangle, hexagon, octagon, greater than, **polygon**, 2D, side, circle, oval, semicircle, fewest, symmetry

Other language to be used by the teacher: vertex, symmetrical, square, oblong, rectangle, pentagon, less than, fewer than, odd, even, sort, group

STRUCTURES AND REPRESENTATIONS

Sorting circles

RESOURCES

Mandatory: 2D shapes, labels

 In the eTextbook of this lesson, you will find interactive links to a selection of teaching tools.

Quick recap ↻

Give children manipulatives of rectangles, triangles, circles, squares, pentagons and hexagons so that they can practise handling and naming 2D shapes. Discuss the features of each shape together.

Discover

WAYS OF WORKING Pair work

ASK

- Question ① : *What do you notice about the shapes?*
- Question ① : *What is the same and what is different about two of the shapes?*

IN FOCUS This section requires children to identify the properties of 2D shapes and to recognise which properties are shared and which are different in order to categorise and sort shapes into sorting circles with given headings. They should observe that not all of the shapes are sorted correctly.

ANSWERS

Question ① a): An inverted kite is not a triangle.

Question ① b): A trapezium is not a rectangle.

Sort 2D shapes

Discover

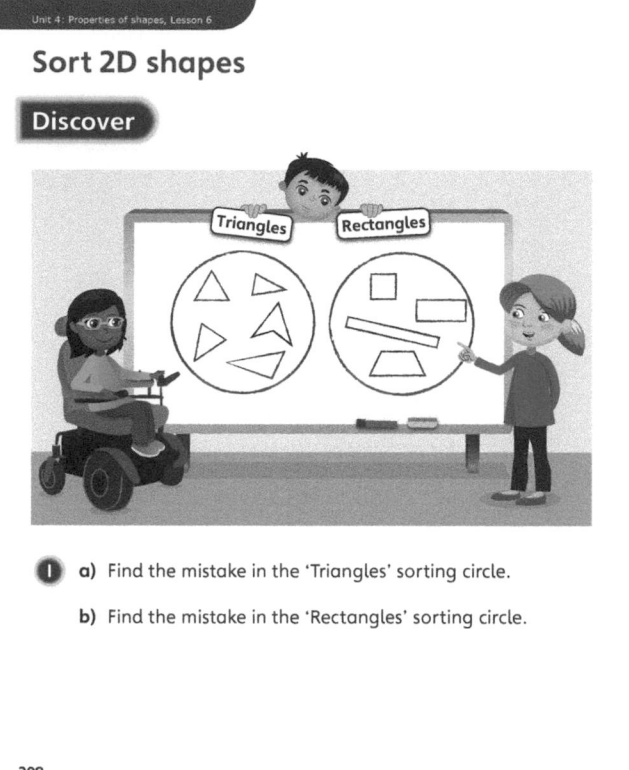

① a) Find the mistake in the 'Triangles' sorting circle.

b) Find the mistake in the 'Rectangles' sorting circle.

208

PUPIL TEXTBOOK 2A PAGE 208

Share

WAYS OF WORKING Whole class teacher led

ASK

- Question ① a): *What do you notice about the shape that is incorrect?*
- Question ① a): *Why might someone think this is a triangle?*
- Question ① b): *What is the same and what is different about the rectangles?*
- Question ① b): *How could you convince me that this shape is not a rectangle?*

IN FOCUS Question ① a) explores the properties of triangles. Children are presented with some shapes that closely resemble triangles and are asked to explain how they know these shapes are not triangles. Question ① b) explores a quadrilateral which is not a rectangle. Encourage children to discuss properties of rectangles and identify why this shape does not belong in the sorting circle with the other rectangles.

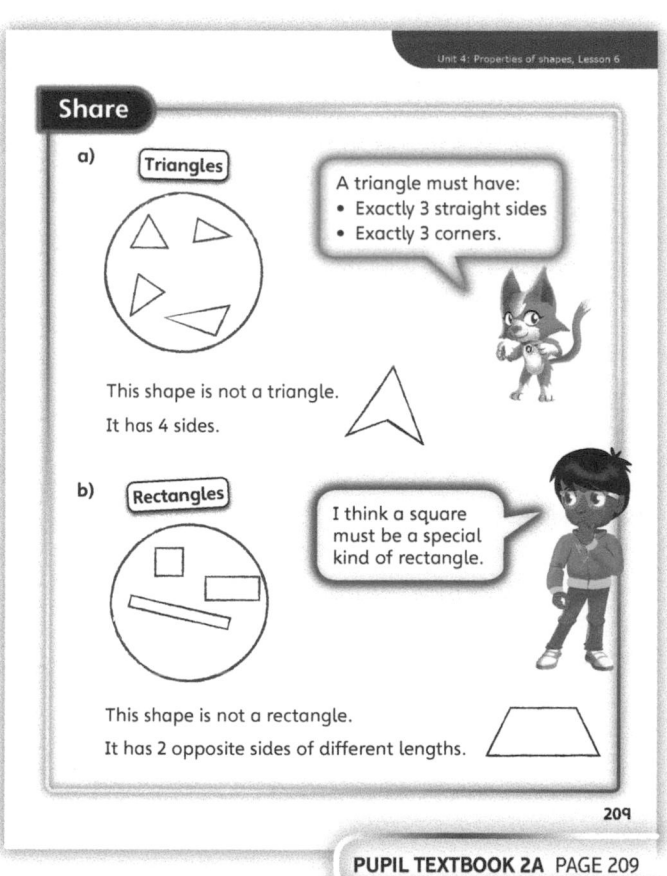

Think together

WAYS OF WORKING Whole class teacher led (I do, We do, You do)

ASK

- Question ❶: *What different properties can you look at when sorting shapes?*
- Question ❶: *Is there more than one way you could label the boxes?*
- Question ❸: *Can you sort the shapes into two groups, three groups or four groups?*

IN FOCUS Question ❸ requires children to come up with their own criteria for sorting the shapes. As Dexter says, there are a number of different ways to sort them. Encourage children to come up with more than one solution and discuss them with a partner. Astrid talks about using symmetry to sort the shapes, drawing on the learning from the previous lesson. However, the quarter circle has a line of symmetry that may not be very obvious to children as it is not vertical.

STRENGTHEN Provide children with 2D shapes and sorting circles so that they can physically sort the shapes. Begin by sorting by type of shape and then move towards sorting them by number of sides and vertices. Use the vocabulary 'more than', 'greater than', 'less than' and 'fewer than'.

DEEPEN Focus on Astrid's comment in question ❸. Provide children with a variety of shapes and some sorting circles, and ask them to sort the shapes based on symmetry. Children will most likely begin by sorting them into two groups: 'Symmetrical' and 'Not symmetrical'. Ask: *Can you have more than two groups?* Prompt children to sort shapes by how many lines of symmetry each one has. For example: 'No lines of symmetry', 'One line of symmetry' and 'More than one line of symmetry'.

ASSESSMENT CHECKPOINT Question ❶ will determine whether children can identify sorting criteria for shapes that have already been sorted. Question ❷ will determine whether children can identify and count the number of sides of irregular polygons and order them based on the number of sides. Question ❸ will determine whether children can identify different criteria for sorting shapes and then sort them accordingly.

ANSWERS

Question ❶: There is more than one possible answer. For example, the headings could be 'Polygons with 3 sides', 'Polygons with more than 3 sides' and 'Not polygons'.

Question ❷: A, D, E and F (same number of sides), C, B.

Question ❸: There is more than one possible way. An example would be '4 vertices' and 'Fewer than 4 vertices'.

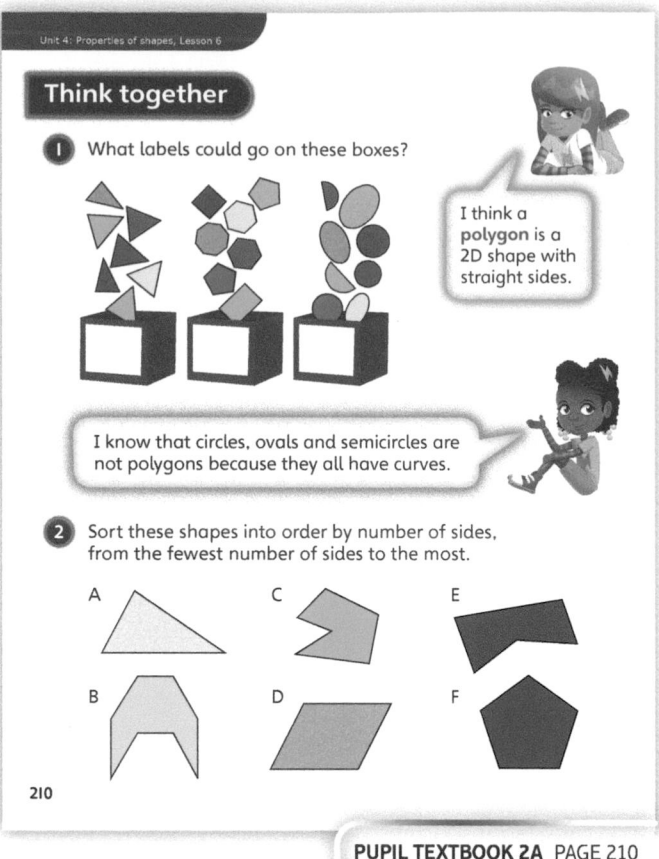

Unit 4: Properties of shapes, Lesson 6

Think together

❶ What labels could go on these boxes?

I think a **polygon** is a 2D shape with straight sides.

I know that circles, ovals and semicircles are not polygons because they all have curves.

❷ Sort these shapes into order by number of sides, from the fewest number of sides to the most.

A   C   E

B   D   F

210

PUPIL TEXTBOOK 2A PAGE 210

Unit 4: Properties of shapes, Lesson 6

❸ Think of your own ways of sorting these shapes into two groups.

CHALLENGE

I thought of sorting them by symmetry.

There must be lots of different ways to sort them. I am going to find some more.

→ Practice book 2A p153

211

PUPIL TEXTBOOK 2A PAGE 211

Practice

WAYS OF WORKING Independent thinking

IN FOCUS Question ④ asks children to identify and draw shapes with given numbers of vertices. Ask children to think of as many different shapes that would fit in each group as possible and remind them to include irregular shapes if necessary.

STRENGTHEN Have 2D shapes available for children to manipulate and sort. When they are counting the vertices, it may help children to mark off each vertex as they count it. When determining criteria for sorting, ask: *What is the same about the shapes? What is different about them?* This will help children to identify common properties that they can use as sorting criteria.

DEEPEN Provide children with two overlapping sorting circles. Explain that in the space where the circles overlap, shapes have to follow the rules for both groups. Provide children with a range of 2D shapes. Can they think of a way to sort the shapes so that there is at least one shape in the overlap?

THINK DIFFERENTLY In question ③, children match sorting labels for given groups of shapes. This prompts children to think of suitable labels about the properties of the shapes, and not simply the shape names. For example, 'has fewer than 5 vertices' rather than just 'triangles and quadrilaterals'.

ASSESSMENT CHECKPOINT Question ① will determine whether children can distinguish between polygons and non-polygons. Question ② will determine whether children can identify the number of vertices within a 2D shape and order the shapes accordingly. Question ③ will determine whether children can identify possible criteria for how shapes have been sorted. Question ④ will determine whether children can identify shapes with an odd number of vertices and shapes with an even number of vertices. Question ⑤ will determine whether children can identify possible criteria for how shapes have been sorted and suggest other possible shapes that meet those criteria.

ANSWERS Answers for the **Practice** part of the lesson can be found in the *Power Maths* online subscription.

Reflect

WAYS OF WORKING Pair work

IN FOCUS This section requires children to think about how to sort the shapes into two groups of three. There are a number of ways to do this; these will most likely involve numbers of sides or vertices. Children should compare what criteria they used and whether or not different criteria resulted in different groupings.

ASSESSMENT CHECKPOINT Are children able to identify common properties between shapes? Can they select criteria that include some shapes but eliminate others? Do the criteria ensure that all shapes can be sorted into two equal groups of three shapes?

ANSWERS Answers for the **Reflect** part of the lesson can be found in the *Power Maths* online subscription.

After the lesson ⏸

- How secure are children with the vocabulary relating to sorting?
- Was the use of practical activities successful in supporting children whose understanding in this area needed strengthening?
- Were children able to apply their knowledge of the properties of 2D shapes in order to sort them?

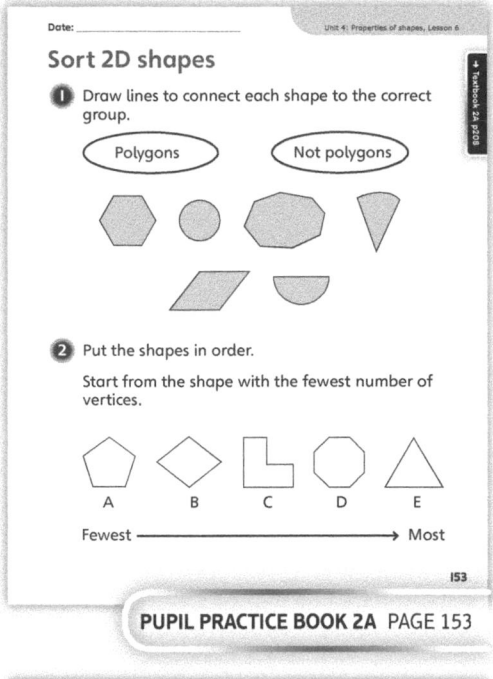

PUPIL PRACTICE BOOK 2A PAGE 153

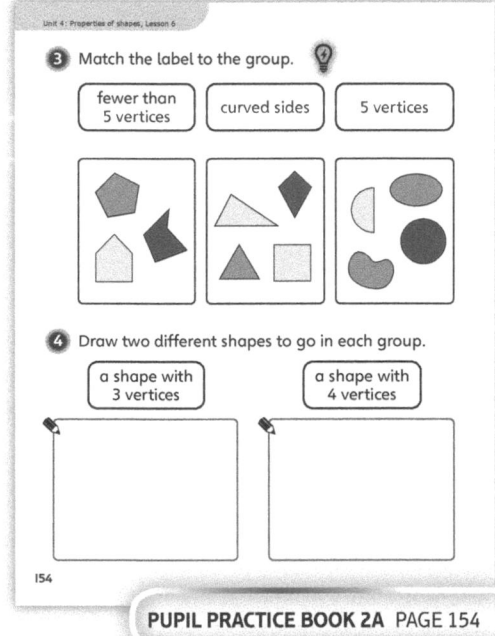

PUPIL PRACTICE BOOK 2A PAGE 154

PUPIL PRACTICE BOOK 2A PAGE 155

Make patterns with 2D shapes

Learning focus

In this lesson, children will identify patterns involving 2D shapes. By isolating the pattern core, children will be able to identify missing terms.

Before you teach

- Are children secure in identifying the 2D shapes that appear in this lesson?
- How will you support children as they explain their reasoning when identifying a given term within a pattern?

NATIONAL CURRICULUM LINKS

Year 2 Geometry – properties of shape

Order and arrange combinations of mathematical objects in patterns and sequences.

ASSESSING MASTERY

Children can identify the core of a pattern, using this to find missing terms and to make generalisations in order to find a given term, such as the 10th or 20th.

COMMON MISCONCEPTIONS

Children may struggle to identify the pattern core, particularly when a shape is repeated within the core. Ask:
- *What is the repeating part of the pattern? How many shapes are in the repeating part?*

STRENGTHENING UNDERSTANDING

Provide children with tracing paper. When they think they have identified the pattern core, ask them to trace over it and move their tracing across to test whether they have correctly identified the core.

Children may need time to create their own repeating patterns. Provide children with 2D shapes so they can use tracings of them to create a pattern. Ask children to highlight the pattern core.

GOING DEEPER

Ask children to create a repeating pattern of more than one row. Can they create a pattern that has a core both horizontally and vertically? Can they use this to predict what the 10th row will look like? How about the 100th?

KEY LANGUAGE

In lesson: pattern, repeating, triangle, circle

Other language to be used by the teacher: term, core, square, kite, pentagon, hexagon

RESOURCES

Mandatory: 2D shapes

Optional: tracing paper

 In the eTextbook of this lesson, you will find interactive links to a selection of teaching tools.

Quick recap

Challenge children to copy and continue repeating AB AB AB patterns. You could use objects or actions, for example: pen, pencil, pen, pencil, pen, pencil ... jump, clap, jump, clap, jump, clap ...

Discover

Unit 4: Properties of shapes, Lesson 7

WAYS OF WORKING Pair work

ASK

- Question ① a): *What is the repeating part of the pattern?*
- Question ① a): *How do you know which shapes complete the pattern?*
- Question ① b): *How did you work out what the 20th term would be?*

IN FOCUS Question ① b) requires children to generalise in order to identify the 20th term in the pattern. Children need to reason that every other term is always a circle and that 20 is an even number. Even though children may not have come across the terminology of even numbers yet, they should still be able to recognise this pattern. Therefore, the 20th term must be a circle. Children have to draw on their knowledge of number as well as make careful observations about the pattern. Some children may begin by continuing the pattern to the 20th term. If this happens, support them by asking: *What shape appears at every other term? Can you use this to help you work out the 20th term?*

ANSWERS

Question ① a): C is the correct option to complete the pattern.

Question ① b): The 20th shape must be a circle.

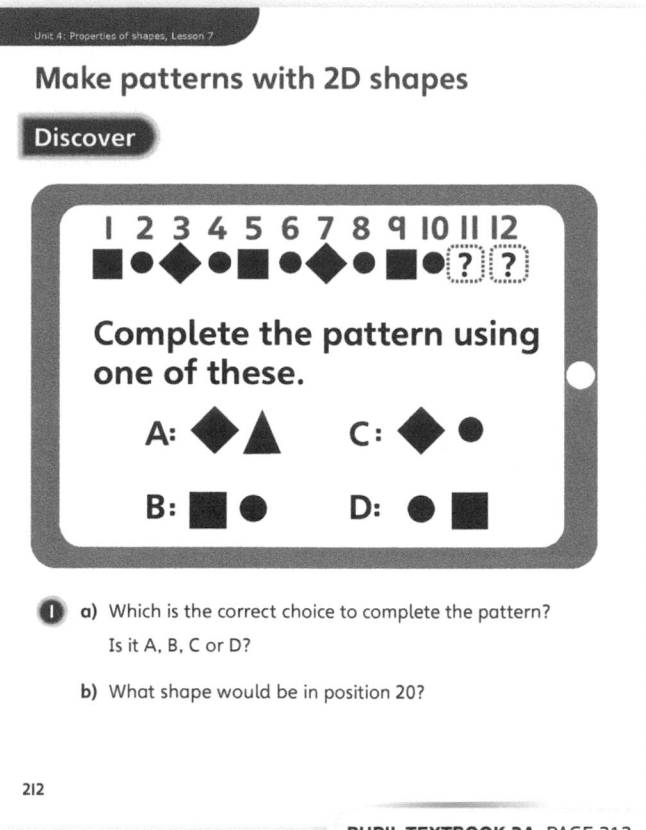

Make patterns with 2D shapes

Discover

Complete the pattern using one of these.

A: ◆ ▲   C: ◆ ●

B: ■ ●   D: ● ■

① a) Which is the correct choice to complete the pattern? Is it A, B, C or D?

b) What shape would be in position 20?

212

PUPIL TEXTBOOK 2A PAGE 212

Share

WAYS OF WORKING Whole class teacher led

ASK

- Question ① a): *Can you describe the pattern core?*
- Question ① a): *How many shapes are there in the pattern core?*

IN FOCUS Question ① a) highlights how the repeating part, or pattern core, needs to be identified in order to work out what the missing shapes are. Isolating the pattern core is an important skill that enables children to not only describe the pattern but to make generalisations and find missing terms.

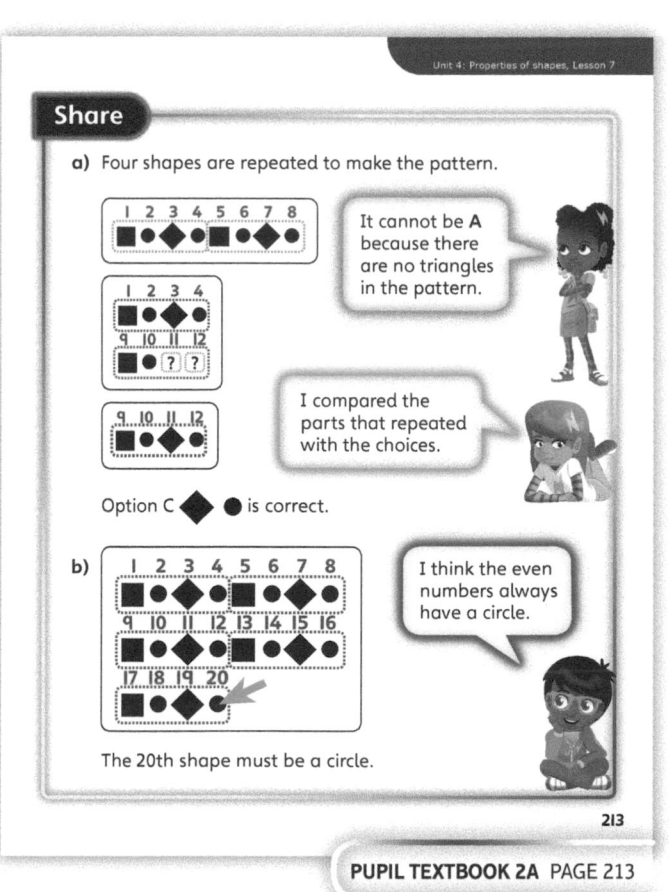

Share

a) Four shapes are repeated to make the pattern.

It cannot be **A** because there are no triangles in the pattern.

I compared the parts that repeated with the choices.

Option C ◆ ● is correct.

b) The 20th shape must be a circle.

I think the even numbers always have a circle.

213

PUPIL TEXTBOOK 2A PAGE 213

Think together

WAYS OF WORKING Whole class teacher led (I do, We do, You do)

ASK

- Questions ❶, ❷ and ❸: *Can you identify the pattern core for the pattern?*
- Question ❶: *How did you work out the missing shapes?*
- Question ❸: *Can you use what you know about the patterns to work out the 30th term?*

IN FOCUS Question ❷ asks children to find the 15th term. Here the pattern has a core of 5 shapes. By identifying the core, children may be able to rationalise that any term that is a multiple of 5 will be the last shape in the pattern core. They may not have come across the terminology 'multiple of 5' yet, but they may still be able to recognise that each 5th term is the same.

STRENGTHEN Provide children with tracing paper so that they can trace the pattern cores. They can then use this to determine the missing shapes or continue the patterns. Discuss the shapes that are in the repeating part and how many shapes there are in the core.

DEEPEN Focus on question ❸ and ask children to work out the position of each shape. Ask children to choose one shape from the pattern. In what positions does that shape appear? When will it appear next? When will it appear for the 10th time? 20th time? 100th time?

ASSESSMENT CHECKPOINT Question ❶ will determine whether children can identify the pattern core and can identify the missing shapes in a pattern by using their knowledge of the pattern core. Question ❷ will determine whether children can use their knowledge of the pattern core to work out a given term. Question ❸ will determine whether children can use their knowledge of the pattern core to continue a pattern.

ANSWERS

Question ❶: ▲ ▼

Question ❷: ▲

Question ❸ a):

Question ❸ b):

Question ❸ c):

PUPIL TEXTBOOK 2A PAGE 214

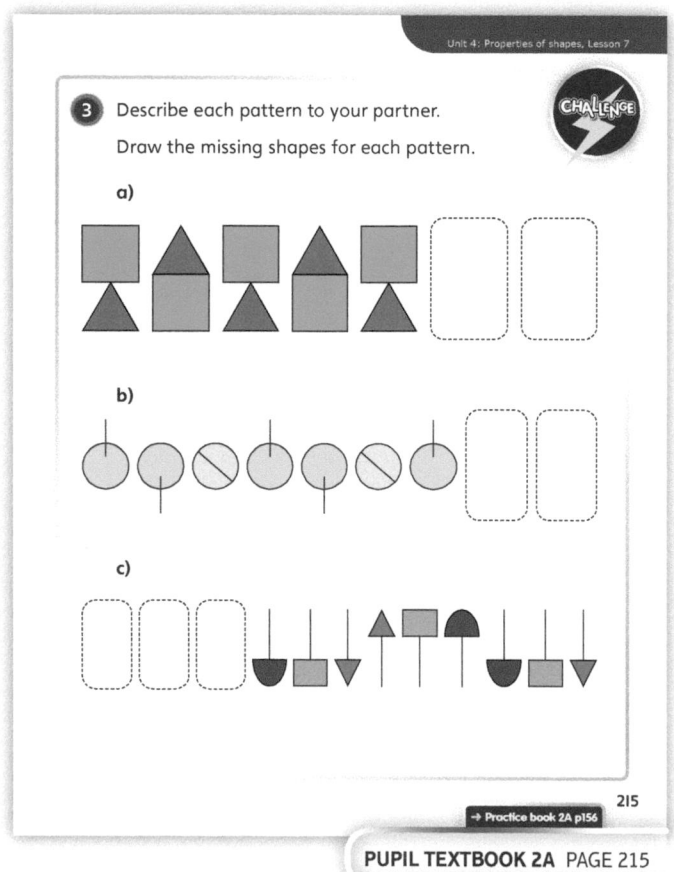

PUPIL TEXTBOOK 2A PAGE 215

Practice

Independent thinking

IN FOCUS Question **5** has patterns that work both horizontally and vertically. There is also no repeating part, so children will not be able to identify the pattern core. Instead, children will have to reason about how the shape changes as it goes from left to right as well as top to bottom in order to fill in the missing terms. In order to do this, they will have to look at completed rows and columns to identify the pattern. Instead of asking children to identify the repeating part, ask them to describe how the shape changes as they go across and down.

STRENGTHEN Asking children to say the name of each shape as they look at the patterns from left to right can help them to identify the missing part. It may also help if an adult or a partner says the name of each shape so children can listen for the repeating part. Once the repeating part has been identified, ask children to circle the pattern core each time throughout the pattern to help them see it in isolation.

DEEPEN Extend question **5** by asking children to draw one more row and one more column. Can children explain why they chose the shapes they did? Did anyone come up with a different possibility? Can children try to convince a partner that their answer is correct?

THINK DIFFERENTLY In question **3**, children are given the first six terms of some patterns and are required to use this to find the 7th and the 8th term of each pattern. They also need to find the 20th term of pattern a). They will need to identify the pattern core and work out how this can be applied in order to continue the pattern and find later terms.

ASSESSMENT CHECKPOINT Question **1** will determine whether children are able to identify the pattern core. Question **2** will determine whether children can identify the pattern core and use this to complete the pattern. Question **3** will determine whether children can rationalise about the pattern in order to work out the 7th, 8th and 20th terms. Question **4** will determine whether children can identify the pattern core and use this to continue a pattern. Question **5** will determine whether children can identify how the shape changes in each column and thus complete the pattern.

ANSWERS Answers for the **Practice** part of the lesson can be found in the *Power Maths* online subscription.

Reflect

Pair work

IN FOCUS This question requires children to come up with their own pattern. You may wish to turn this into a game. Children can choose how many shapes are missing, allowing them to make the problem simple or more challenging. Once they have finished designing their problem, they should test it out on a partner to see if it works.

ASSESSMENT CHECKPOINT Can children create a pattern with a repeating part? Can they identify the information needed to solve the problem?

ANSWERS Answers for the **Reflect** part of the lesson appear in the *Power Maths* online subscription.

After the lesson

- Were children confident in identifying the pattern core?
- Were children secure in describing the pattern?
- Do children need further support in applying what they know about the pattern in order to identify missing terms?

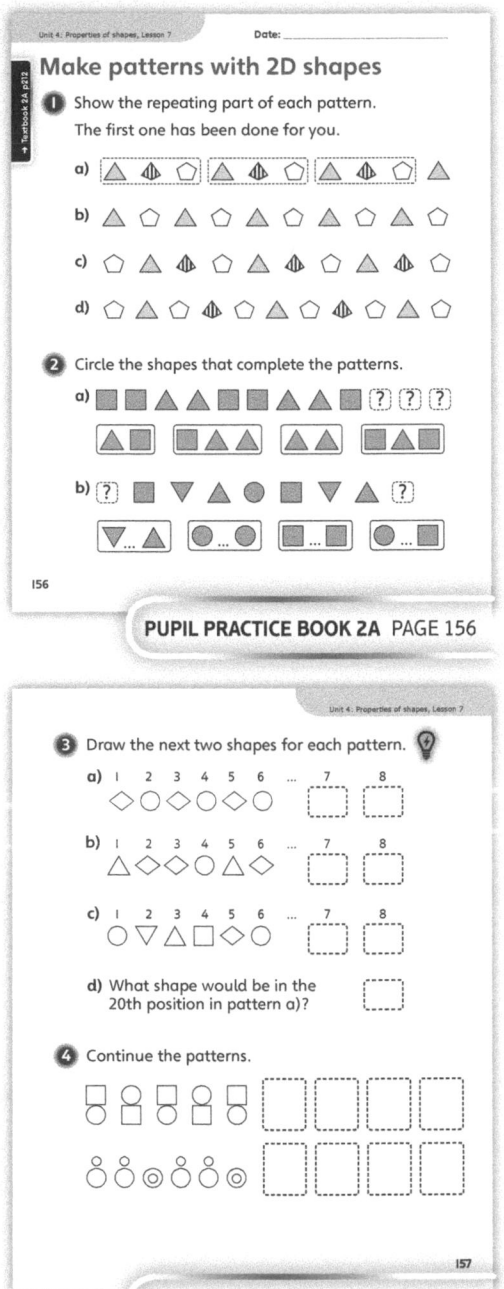

PUPIL PRACTICE BOOK 2A PAGE 156

PUPIL PRACTICE BOOK 2A PAGE 157

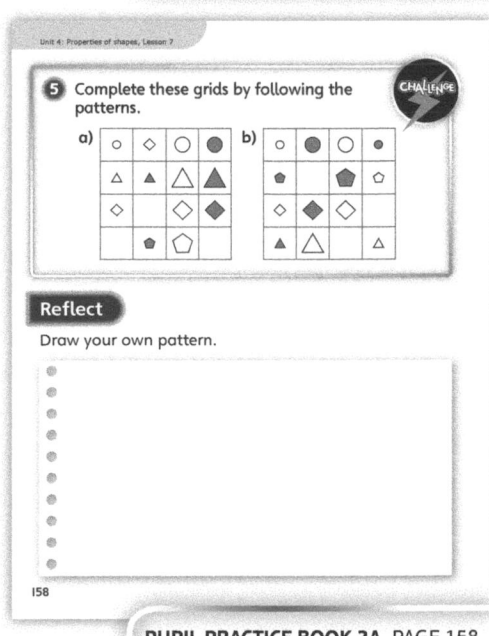

PUPIL PRACTICE BOOK 2A PAGE 158

Count faces on 3D shapes

Learning focus

In this lesson, children will count and describe the faces of 3D shapes. They will learn that a curved surface on a 3D shape is not classed as a face.

Before you teach

- Are children secure in identifying the 2D shapes that occur in this lesson when describing faces?
- Can you provide the vocabulary needed for this lesson for children to refer to?
- What concrete apparatus can you provide for children in order to reinforce their learning?

NATIONAL CURRICULUM LINKS

Year 2 Geometry – properties of shape

Identify and describe the properties of 3D shapes, including the number of edges, vertices and faces.

ASSESSING MASTERY

Children can identify how many faces there are on a range of 3D shapes and are able to describe them based on their knowledge of 2D shapes.

COMMON MISCONCEPTIONS

Children may think that a curved surface is considered a face. Show children a cylinder. Ask:
- *How many faces does this cylinder have? What 2D shapes can you get from printing with a cylinder?*

For some shapes, children may miscount the number of faces as they struggle to keep track of which ones they have counted. Show children a cuboid. Ask:
- *How many faces does this cuboid have?*

Observe how children count the faces and whether they have a systematic approach.

STRENGTHENING UNDERSTANDING

Give children some 3D shapes and different coloured paint. Ask children to paint each face a different colour or a different pattern and make a print of each face. Children can then label each 2D shape they have produced and count how many shapes they have made with their chosen 3D shape. This will help them to identify the number of faces and to describe them.

GOING DEEPER

Provide children with nets of familiar 3D shapes made with construction materials. Ask children to predict what they will make and then test their predictions by making the shape. Children could also challenge each other by placing a variety of 3D shapes in a feely bag. They take it in turns to feel a shape and describe its faces to a partner. Their partner then has to work out what the shape is.

KEY LANGUAGE

In lesson: 3D, 2D, face, flat, surface, cuboid, rectangle, square, square-based pyramid, triangle, cone, **hemisphere**, **curved surface**, sphere, circle

Other language to be used by the teacher: 3D shape, cube, cylinder, oblong, ovoid

RESOURCES

Mandatory: variety of 3D shapes, construction materials to create polyhedrons

Optional: paint of various colours, feely bag

 In the eTextbook of this lesson, you will find interactive links to a selection of teaching tools.

Quick recap

Give children manipulatives of cubes, spheres, cuboids, pyramids and cylinders so that they can practise handling and naming 3D shapes. Discuss the features of each shape together.

Discover

Count faces on 3D shapes

Discover

Will

1 **a)** Will wants to paint every face of the box a different colour.

How many colours will he need?

b) Describe the shape of each face.

216

WAYS OF WORKING Pair work

ASK

- Question 1 a): *How many faces can you see? How many faces are hidden?*
- Question 1 a): *Can you think of another shape that has six faces?*
- Question 1 b): *How do you know what shape each face is?*

IN FOCUS Question 1 a) requires children to visualise the shape in order to identify the number of faces. Some children may struggle with this and could benefit from having a physical cuboid, so that they could count each face. Encourage children to see that each face that is visible has a similar shaped face opposite it.

ANSWERS

Question 1 a): Will needs 6 different colours.

Question 1 b): The shape of each face is a rectangle.

PUPIL TEXTBOOK 2A PAGE 216

Share

Share

a) Will's box is a cuboid.

A face is a flat surface on a 3D shape. Each face is a 2D shape.

A cuboid has 3 pairs of faces.

A cuboid has 6 faces in total.

⌐1⌐ ⌐2⌐ ⌐3⌐ ⌐4⌐ ⌐5⌐ ⌐6⌐

Ben will need 6 different colours.

b) Each face is a rectangle.

A cuboid can have 2 square faces.

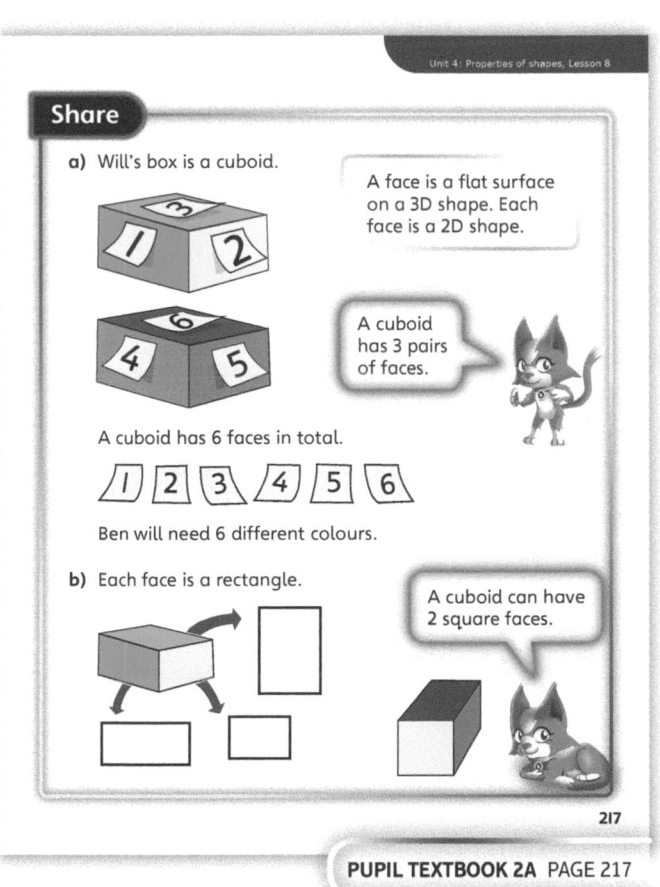

217

WAYS OF WORKING Whole class teacher led

ASK

- Question 1 a): *Is Sparks's first comment always true?*
- Question 1 a): *How did you make sure that you counted all the faces correctly?*
- Question 1 b): *Do you know of any other shapes that have rectangular faces (square or oblong)?*

IN FOCUS Being able to describe what type of faces a 3D shape has is an essential part of describing its properties. Two different 3D shapes can have the same number of faces, so describing the shape of each face enables children to distinguish between them. For example, a triangular prism and a square-based pyramid both have five faces. However, the prism has two triangular faces and three rectangular faces, whereas the pyramid has one square face and four triangular faces.

PUPIL TEXTBOOK 2A PAGE 217

Think together

Think together

WAYS OF WORKING Whole class teacher led (I do, We do, You do)

ASK

- Question ❷: *What faces will you need to make a cube? How about a triangular prism?*
- Question ❸: *What shapes could you use to print a square, a triangle and a circle?*

IN FOCUS Question ❸ looks at the misconception that a curved surface is a face. Children may see the hemisphere as having two faces or the cylinder as having three faces. It is important to explain that, by definition, a face is flat. Thinking about whether they can print with it can help children to make the distinction between a face and a curved surface. It is important to ensure that children use the term 'curved surface' when describing the types of shape in this question.

STRENGTHEN Provide children with 3D shapes that they can manipulate. Encourage them to mark off each face to help them count the faces reliably. Provide children with paint and paper so that they can print with the 3D shapes to help identify the shapes of the faces.

DEEPEN Ask children to think about combining two different 3D shapes so that the faces fit together exactly. What would happen to the total number of faces? For example, place a square-based pyramid on top of a cube to make a new 3D shape. Ask: *How many faces did you have to start with? How many faces are there now? Can you explain why the total number of faces changes in the way that it does?*

ASSESSMENT CHECKPOINT Question ❶ will determine whether children are able to count the number of faces on a pictorial representation of a polyhedron. Question ❷ will determine whether children are able to identify the number of faces and types of face of a given polyhedron. Question ❸ will determine whether children can distinguish between faces and curved surfaces.

ANSWERS

Question ❶: The cube has 6 faces; the square-based pyramid has 5 faces; the cuboid has 6 faces.

Question ❷ a): Anna will need 1 square face.

Question ❷ b): Anna will need 4 triangular faces.

Question ❷ c): She will need 5 faces in total.

Question ❸: The sphere has no faces; the cylinder has 2 circular faces; the cone has 1 circular face; the ovoid has no faces; the hemisphere has 1 circular face.

Think together

❶ How many faces does each shape have?

❷ Anna wants to make a square-based pyramid from construction materials.

> I know that some pyramids have a square base and some have a triangular base.

a) How many square faces does Anna need?
b) How many triangular faces does she need?
c) How many faces will she need in total?

218

PUPIL TEXTBOOK 2A PAGE 218

❸ How many faces does each shape have?

What shapes are the faces?

CHALLENGE

> A cone and a hemisphere have a circular face and a curved surface.

> Does a sphere have a face? If I try to print using a sphere, I don't get a circle.

> Remember, a face is a flat surface.

219

→ Practice book 2A p159

PUPIL TEXTBOOK 2A PAGE 219

Practice

WAYS OF WORKING **WAYS OF WORKING** Independent thinking

IN FOCUS Question **5** requires children to apply their understanding of multiplication, as well as their knowledge of 3D shapes, in order to solve the problem. In question **5** a), children have to calculate 2 × 6. They could do this by repeated addition or by drawing on their multiplication facts. In question **5** b), they have to identify the shape with two faces, recognising that the cone has one circle face and one curved surface.

STRENGTHEN Before children start, provide them with concrete representations of the 3D shapes in this section. Prompt children to match name labels to the shapes, count the faces and identify the shape of each face. They could then use their notes as a reference point when answering the questions.

DEEPEN Ask children to set up a 3D shape shop. Provide them with a range of 3D shapes to sell. They have to price the shapes following these rules: curved surfaces cost 2p; triangular faces cost 5p; square and oblong faces cost 10p and circular faces cost 20p. Can they work out the total cost of each shape?

THINK DIFFERENTLY In question **3**, children use what they have learned to fill in the missing words for sentences about the faces of 3D shapes that all have some curved surfaces.

ASSESSMENT CHECKPOINT Questions **1** and **2** determine whether children can identify the number of faces and the shapes of faces of a variety of polyhedrons. Question **3** will determine whether children can apply their learning from this lesson to find the combined total of a number of 3D shapes. Question **4** will determine whether children can distinguish between faces and curved surfaces. Question **5** will determine whether children can apply their understanding of counting in 2s, along with properties of 3D shapes, to solve a problem.

ANSWERS Answers for the **Practice** part of the lesson can be found in the *Power Maths* online subscription.

Reflect

WAYS OF WORKING Pair work

IN FOCUS This part of the lesson requires children to visualise a 3D shape, then describe its faces. Children need to have a secure understanding of what they have learned in this lesson in order to do this. This could be extended by asking children to think of a shape and describe it to a partner. Their partner then has to work out what the shape is.

ASSESSMENT CHECKPOINT This section will determine whether children can correctly count and name the faces of a 3D shape without having a pictorial or concrete representation to refer to.

ANSWERS Answers for the **Reflect** part of the lesson can be found in the *Power Maths* online subscription.

After the lesson

- How can you further support children who were not secure in identifying the number of faces on a given 3D shape?
- Can children create a display of the properties of 3D shapes that they can add to as they learn more shapes?

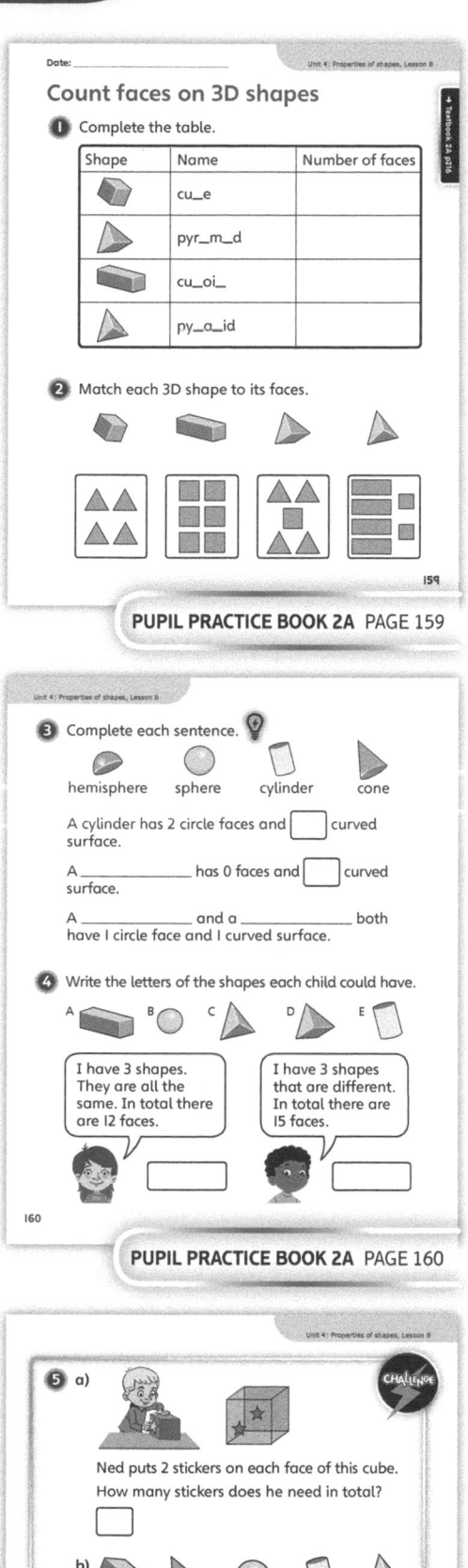

PUPIL PRACTICE BOOK 2A PAGE 159

PUPIL PRACTICE BOOK 2A PAGE 160

PUPIL PRACTICE BOOK 2A PAGE 161

Count edges on 3D shapes

Learning focus

In this lesson, children will identify edges of a 3D shape as the line where two faces meet. They will learn to use the property of number of edges to describe 3D shapes.

Before you teach

- What practical opportunities can you provide to support learning in this lesson?
- How can you link learning from the last lesson to this lesson?
- How will you reinforce the new vocabulary in this lesson?

NATIONAL CURRICULUM LINKS

Year 2 Geometry – properties of shape

Identify and describe the properties of 3D shapes, including the number of edges, vertices and faces.

ASSESSING MASTERY

Children can identify and count the edges of 3D shapes and use this knowledge when comparing and describing 3D shapes.

COMMON MISCONCEPTIONS

Children may make the mistake of thinking the point where a face meets a curved surface is an edge. Give children a cylinder. Ask:
- *How many edges does this cylinder have?*

Children may miscount the number of edges on 3D shapes. Provide children with a cube. Ask:
- *How many edges does this cube have?*

Observe how children count them.

STRENGTHENING UNDERSTANDING

Using concrete 3D shapes, prompt children to mark off the edges with a dry-wipe marker as they count them. Discuss how they can be systematic in the way they count the edges.

Where possible, provide children with straws to make the 3D shapes. As each straw represents an edge, this will support children in identifying where the edges are on a 3D shape.

GOING DEEPER

Provide children with straws and challenge them to make shapes with different numbers of edges. Can children make a 3D shape with two edges? How about three, four or five edges? Why? What different number of edges can they use to make 3D shapes?

KEY LANGUAGE

In lesson: cube, face, **edges**, 3D, **prism**, triangular prism, pyramid

Other language to be used by the teacher: 3D shape, cuboid, triangle-based pyramid, square-based pyramid, pentagon-based pyramid, pentagonal prism, hexagonal prism

RESOURCES

Mandatory: construction straws or art straws, 3D shapes

Optional: dry-wipe marker

 In the eTextbook of this lesson, you will find interactive links to a selection of teaching tools.

Quick recap

Play 'Guess my shape'. Say: *I am thinking of a shape. It has 6 faces. What could it be?*

Discover

WAYS OF WORKING Pair work

ASK

• Question ① a): *Can you describe the faces of the cube?*
• Question ① b): *How many straws are needed to make two cubes? What about three cubes?*

IN FOCUS Question ① b) highlights how, if the type of shape does not change, then the number of faces and edges remains the same regardless of the size of the shape. Some children may want to count the edges of all the cubes in this problem as they may not see that the number of edges is constant as the size of the shape changes. Use this as an opportunity to reinforce that all cubes have 12 edges. This is one of the properties of a cube.

ANSWERS

Question ① a): Hassan needs 12 straws.

Question ① b): Each cube has 6 square faces and 12 edges, and this fact stays the same. The length of each edge and the size of each face change between the cubes.

Count edges on 3D shapes

Discover

Look at my cube!

Molly   Bob   Hassan

① **a)** Hassan is making his own cube to dunk in the bubble mixture.
How many straws does he need?

b) There are three different cubes: small, medium and large.

What is the same? What is different?

220

Share

WAYS OF WORKING Whole class teacher led

ASK

• Question ① a): *What other shape has 12 edges?*
• Question ① a): *How would you describe what an edge is?*
• Question ① b): *How can you be sure that you have counted all the edges only once?*

IN FOCUS Question ① a) defines an edge. Children need to understand that an edge is only created where two faces meet. This is a good opportunity to allow children to make a cube from straws, so that they can see how each straw they use represents an edge. This will provide a practical context for children to refer back to when identifying edges on 3D shapes.

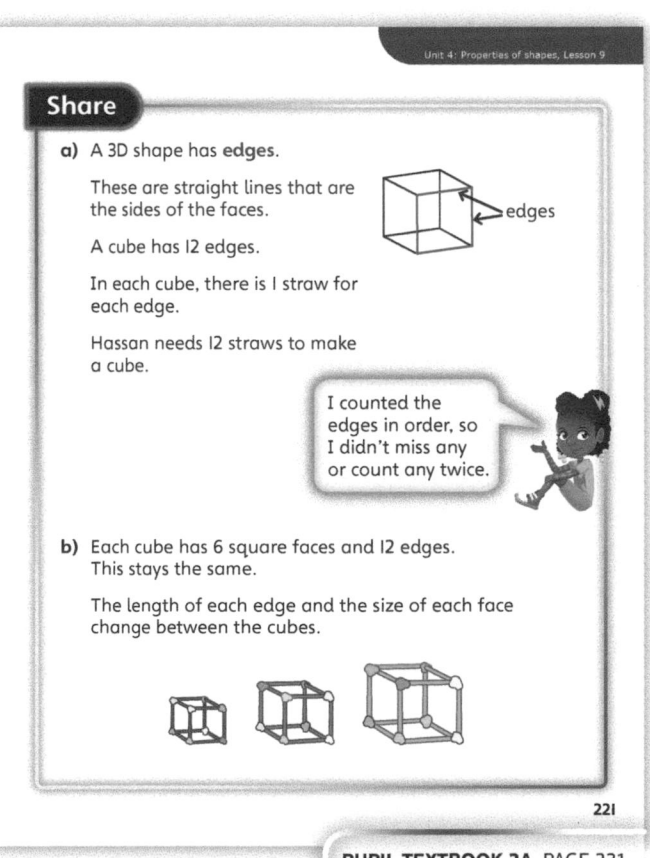

Share

a) A 3D shape has **edges**.

These are straight lines that are the sides of the faces.

A cube has 12 edges.

edges

In each cube, there is 1 straw for each edge.

Hassan needs 12 straws to make a cube.

I counted the edges in order, so I didn't miss any or count any twice.

b) Each cube has 6 square faces and 12 edges. This stays the same.

The length of each edge and the size of each face change between the cubes.

221

Think together

WAYS OF WORKING Whole class teacher led (I do, We do, You do)

ASK

- Question ②: *Can you put the shapes in order from the fewest edges to the most?*
- Question ②: *Can you describe the faces of the different shapes?*
- Question ②: *If the triangular prism and the square-based pyramid have the same number of faces, why do they not have the same number of edges?*

IN FOCUS Question ③ asks children to investigate the relationship between the number of faces and the number of edges of 3D shapes. For all polyhedrons, there will be more edges than faces. However, this is not the case for hemispheres and cylinders, which do not have any edges.

STRENGTHEN For questions ① and ②, provide children with concrete representations of the shapes. Children can then physically count the edges. Encourage children to mark off the edges with a dry-wipe marker, so that they can keep track of which edges they have counted.

DEEPEN Ask children to make a cube and a square-based pyramid with straws. Tell them to place the pyramid on top of the cube so that the square face of the pyramid sits exactly on a square face of the cube. What has happened to the total number of edges? Can children explain why that is? What happens if they join two other shapes in a similar way?

ASSESSMENT CHECKPOINT Questions ① and ② will determine whether children can accurately count the number of edges of a 3D shape. Question ③ will determine whether children can accurately identify and count the faces and edges of 3D shapes.

ANSWERS

Question ①: 6, 12, 9

Question ②: The cube has 12 edges. The triangular prism has 9 edges. The square-based pyramids each have 8 edges.

Question ③: All 3D shapes with flat faces have more edges than faces. Cylinders and hemispheres do not.

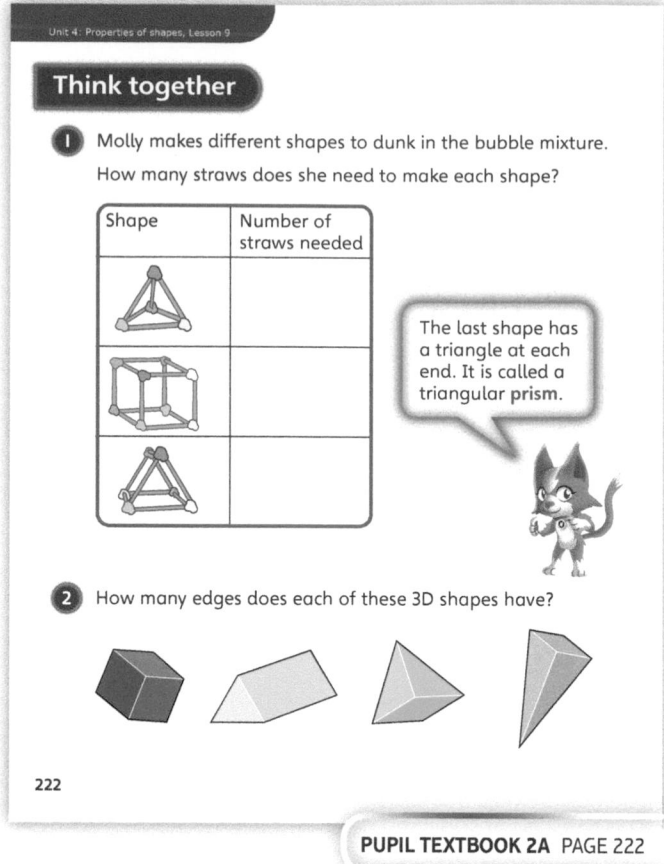

Unit 4: Properties of shapes, Lesson 9

Think together

① Molly makes different shapes to dunk in the bubble mixture. How many straws does she need to make each shape?

Shape	Number of straws needed

The last shape has a triangle at each end. It is called a triangular **prism**.

② How many edges does each of these 3D shapes have?

222

PUPIL TEXTBOOK 2A PAGE 222

③ Kendi and Abbie are making shapes from construction materials.

CHALLENGE

My cube has more edges than faces.

Does my pyramid have more faces or more edges?

Kendi

Abbie

Does a 3D shape always have more edges than faces?

I will investigate other shapes.

223

→ Practice book 2A p162

PUPIL TEXTBOOK 2A PAGE 223

Practice

WAYS OF WORKING Pair work

IN FOCUS Question **5** requires children to apply their multiplication knowledge and what they have learned about edges of 3D shapes in order to find the solution. Children have to first identify how many edges each shape has, then determine the calculation that is needed. Children should find that they have to calculate 4 × 6.

STRENGTHEN Ask children to help contribute towards the class display of 3D shapes. Present some shapes and ask children to count the number of faces and edges for each. Children then label each shape accordingly. Display this so that children are able to refer to it when answering the questions.

DEEPEN Extend question **4** by asking children to explore the relationship between the number of sides of the end face and the number of edges on the prism. For example, a triangle has 3 sides and a triangular prism has 9 edges; a pentagon has 5 sides and a pentagonal prism has 15 edges. Can children see a relationship? Can they explain it?

ASSESSMENT CHECKPOINT Question **1** will determine whether children can accurately count the number of edges for different 3D shapes. Question **2** will determine whether children can count and compare the number of edges for different polyhedrons. Question **3** will determine whether children can accurately identify and count the number of faces and edges for different polyhedrons. Question **4** will determine whether children can accurately identify and count the number of edges for unfamiliar polyhedrons. Question **5** will determine whether children can combine their learning of properties of 3D shapes with their addition or multiplication skills in order to solve problems.

ANSWERS Answers for the **Practice** part of the lesson can be found in the *Power Maths* online subscription.

Reflect

WAYS OF WORKING Pair work

IN FOCUS This section asks children to think carefully about the definition of a face and an edge for 3D shapes. By explaining the distinction between the two to a partner, children help to secure their understanding of the concepts.

ASSESSMENT CHECKPOINT This section will determine whether children are secure in their understanding of what edges and faces are in relation to 3D shapes.

ANSWERS Answers for the **Reflect** part of the lesson can be found in the *Power Maths* online subscription.

After the lesson ⏸

- Did children confuse faces and edges? If so, how can you further support them?
- Were children able to apply their learning from the previous lesson?
- Were the practical opportunities successful in reinforcing children's understanding of the properties of 3D shapes?

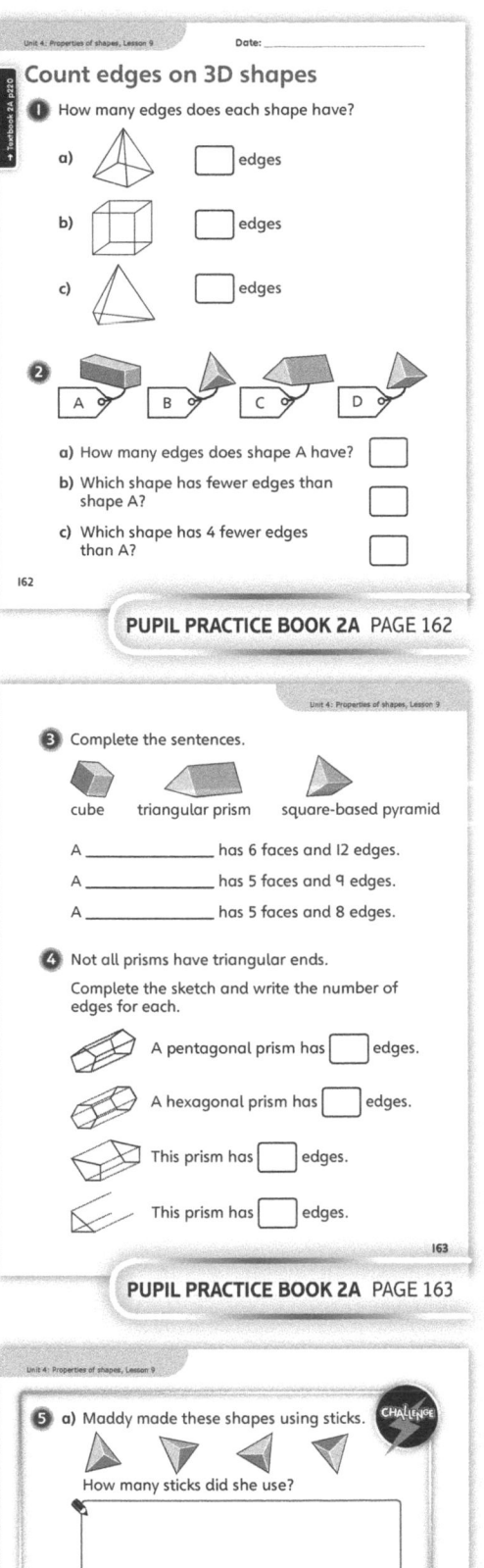

PUPIL PRACTICE BOOK 2A PAGE 162

PUPIL PRACTICE BOOK 2A PAGE 163

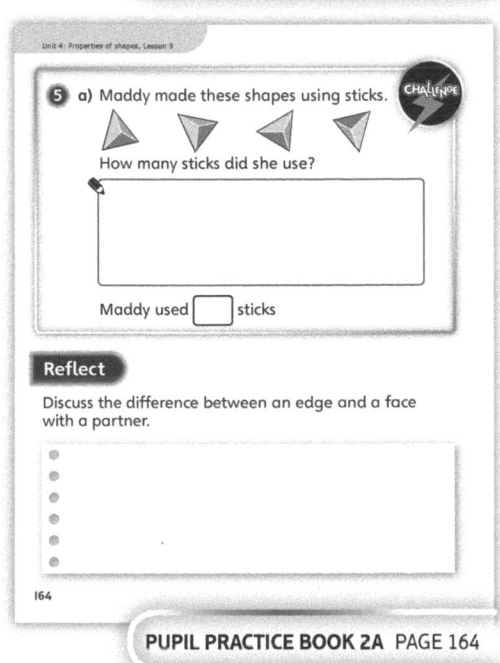

PUPIL PRACTICE BOOK 2A PAGE 164

Count vertices on 3D shapes

Learning focus

In this lesson, children will learn that vertices on a 3D shape are where three or more edges meet. They will then use this knowledge to help describe the properties of 3D shapes.

Before you teach

- What resources can you provide to help secure children's understanding of properties of 3D shapes?

NATIONAL CURRICULUM LINKS

Year 2 Geometry – properties of shape

Identify and describe the properties of 3D shapes, including the number of edges, vertices and faces.

ASSESSING MASTERY

Children can correctly identify and count the vertices of a variety of 3D shapes. They can describe and compare shapes by the number of vertices.

COMMON MISCONCEPTIONS

Children may miscount the number of vertices or they may confuse vertices with edges. Show a cube. Ask:
- *How many faces does this cube have? How many edges does this cube have? How many vertices?*

Observe how children count. Are they being systematic? Are they counting the correct property?

STRENGTHENING UNDERSTANDING

If possible, provide children with marshmallows and straws, so that they can make their own 3D shapes. They could then count how many marshmallows they have used to determine how many vertices there are.

GOING DEEPER

Ask children to explore what happens to the number of vertices when two different shapes are combined. For example: *A cube and a square-based pyramid have a total of 13 vertices. If you place the pyramid on top of the cube so that the square faces match, what happens to the number of vertices? Can you explain it? What about if you combine two other shapes?*

KEY LANGUAGE

In lesson: triangle-based pyramid, vertex, vertices

Other language to be used by the teacher: face, edge, 3D shapes, square-based pyramid, pentagon-based pyramid, hexagon-based pyramid, cube, cuboid, triangular prism, pentagonal prism, hexagonal prism, cone, cylinder, sphere, hemisphere, circle, triangle, square, oblong, rectangle, pentagon, hexagon, curved surface

RESOURCES

Mandatory: 3D shapes

Optional: marshmallows, joining tubes, straws

 In the eTextbook of this lesson, you will find interactive links to a selection of teaching tools.

Quick recap 🔍

Play 'Guess my shape'. Say: I am thinking of a shape. It has 12 edges. What could it be?

Discover

Unit 4: Properties of shapes, Lesson 10

WAYS OF WORKING Pair work

ASK

• Question **1** a): *How many sticks are there in each joining tube?*

• Question **1** a): *What part of the shape do the sticks represent?*

IN FOCUS This section draws children's attention to another property of 3D shapes. Using joining tubes to join the edges of the 3D shape makes the vertices easier to identify. It also helps to communicate the idea that vertices are where the edges meet.

ANSWERS

Question **1** a): Mia needs 4 joining tubes for this pyramid.

Question **1** b): Mia needs 5 joining tubes for a pyramid with a square base.

Count vertices on 3D shapes

Discover

I am using sticks and joining tubes.

1 a) Mia is making a triangle-based pyramid. How many joining tubes does she need?

b) How many joining tubes does she need to make a pyramid like this?

224

PUPIL TEXTBOOK 2A PAGE 224

Share

WAYS OF WORKING Whole class teacher led

ASK

• Question **1** a): *How would you describe a vertex?*
• *How many joining tubes would Mia need if she made a cube instead of a square-based pyramid?*
• Question **1** b): *Can you explain why the square-based pyramid has one more joining cube than the triangle-based pyramid?*

IN FOCUS Question **1** a) explains to children that the joining cubes represent the vertices of the shape. Children need to be secure in the distinction between edges and vertices and how the vertices are created. It is important that the correct terminology is used as some children may call the vertices 'corners'. Encourage children to use the correct vocabulary to avoid confusion or misinterpretation.

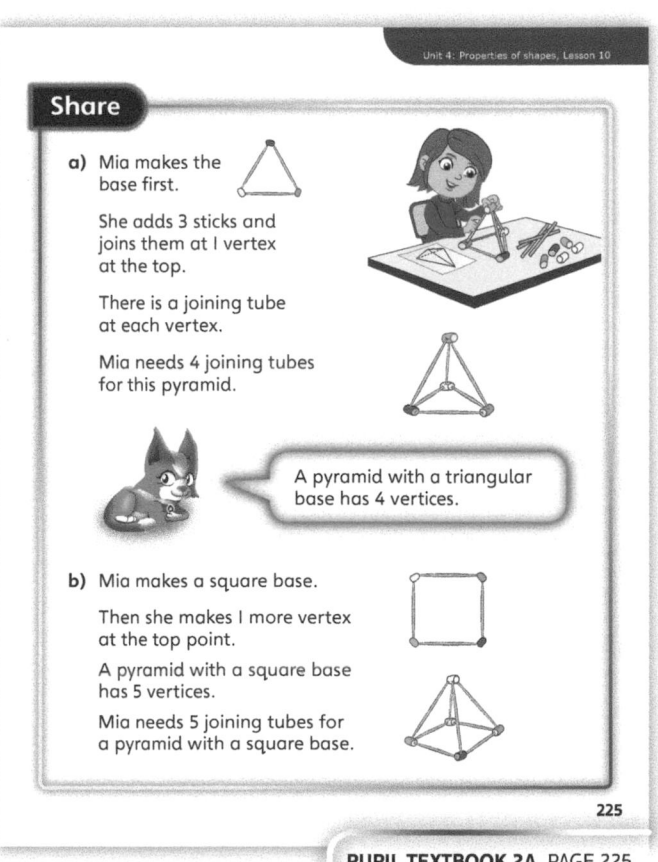

Share

a) Mia makes the base first.

She adds 3 sticks and joins them at 1 vertex at the top.

There is a joining tube at each vertex.

Mia needs 4 joining tubes for this pyramid.

A pyramid with a triangular base has 4 vertices.

b) Mia makes a square base.

Then she makes 1 more vertex at the top point.

A pyramid with a square base has 5 vertices.

Mia needs 5 joining tubes for a pyramid with a square base.

225

PUPIL TEXTBOOK 2A PAGE 225

Think together

WAYS OF WORKING Whole class teacher led (I do, We do, You do)

ASK

- Question **2**: *Can you see a relationship between the shape of the base and the number of vertices? What do you think that relationship is?*
- Question **2**: *Are there always more edges than vertices? Why do you think that is?*
- *Can you order the shapes from questions **1** and **2** from fewest vertices to most?*

IN FOCUS Question **3** tackles a misconception about vertices. Some children may think, incorrectly, that a vertex is created where two cubes join together. In order to find a solution to this question, children need to recognise that a cuboid has eight vertices and that different types of cuboids can be made using the eight cubes.

STRENGTHEN Ask children to create their own versions of the shapes in questions **1** and **2** using joining tubes and straws. Children can then use these shapes to physically count the number of vertices on each shape.

DEEPEN Extend question **3** by asking children: *What is the greatest number of vertices you can create by combining three cubes? (12) What is the fewest number of vertices? (8) What about combining four cubes? (16 and 8) Or five cubes (20 and 8)?*

ASSESSMENT CHECKPOINT Question **1** will determine whether children can identify and count the vertices of different 3D shapes. Question **2** will determine whether children can visualise and count the vertices of 3D shapes. Question **3** will determine whether children can count the vertices of unfamiliar irregular polyhedrons and if they recognise that cuboids have 8 vertices.

ANSWERS

Question **1**: 8, 4, 8

Question **2**: Pentagon-based pyramid needs 6 joining tubes.
Hexagon-based pyramid needs 7 joining tubes.
Triangle-based pyramid needs 4 joining tubes.
Oblong-based pyramid needs 5 joining tubes.

Question **3**: Yes, she would have to make a cuboid either 1 × 1 × 8 or 1 × 2 × 4 or 2 × 2 × 2.

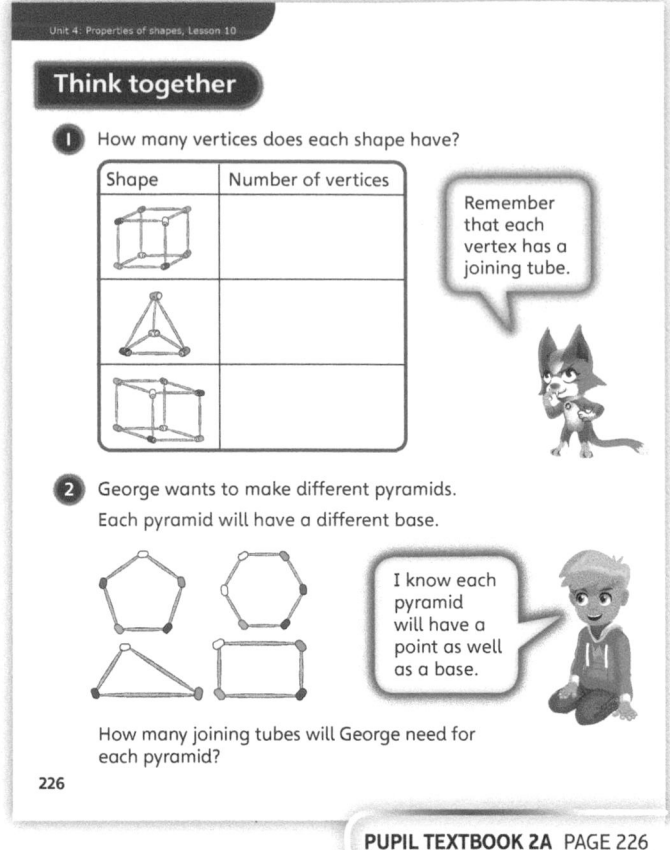

PUPIL TEXTBOOK 2A PAGE 226

PUPIL TEXTBOOK 2A PAGE 227

Practice

Pair work

IN FOCUS Question **5** combines the three types of properties of 3D shapes that children have learned to identify in the last three lessons. It encourages children to look for patterns between the number of faces, edges and vertices of different types of pyramids, as well as the relationship between faces, edges and vertices of different types of pyramids, as well as the relationship between faces, edges and vertices.

STRENGTHEN Provide children with 3D shapes that correspond to those in the questions to enable them to physically count the vertices. This will support them in answering the questions.

DEEPEN Provide children with a range of polyhedrons and ask them to count and write down the number of faces, edges and vertices of each shape. They could create a simple table to do this. Ask: *Can you see any link between the number of faces, edges and vertices that is true for all the shapes? What about if you start by adding the number of faces and vertices together? What happens if you now subtract the number of edges?*

Children should discover that faces + vertices − edges = 2. Ask: *Is this the same for all the shapes?*

THINK DIFFERENTLY Question **4** challenges children to use their addition skills in order to explore which shapes have a given number of vertices.

ASSESSMENT CHECKPOINT Questions **1** and **2** will determine whether children can correctly identify and count the number of vertices of 3D shapes. Question **3** will determine whether children can compare and group shapes based on their number of vertices. Question **4** will determine whether children can apply their knowledge of vertices and addition in order to find solutions. Question **5** will determine whether children are able to identify the number of faces, edges and vertices of 3D shapes.

ANSWERS Answers for the **Practice** part of the lesson can be found in the *Power Maths* online subscription.

Reflect

WAYS OF WORKING Group work

IN FOCUS This section requires children to visualise and either recall or count the number of vertices on a 3D shape. Children should start to remember how many vertices there are on familiar shapes such as cuboids and pyramids.

ASSESSMENT CHECKPOINT This section will determine whether children can recall or visualise the number of vertices on familiar 3D shapes.

ANSWERS Answers for the **Reflect** part of the lesson can be found in the *Power Maths* online subscription.

After the lesson

- Were children secure on the distinction between faces, edges and vertices?
- Are children able to describe 3D shapes by the number of faces, vertices and edges?
- Is there opportunity to display the definitions and properties of 3D shapes to reinforce learning?

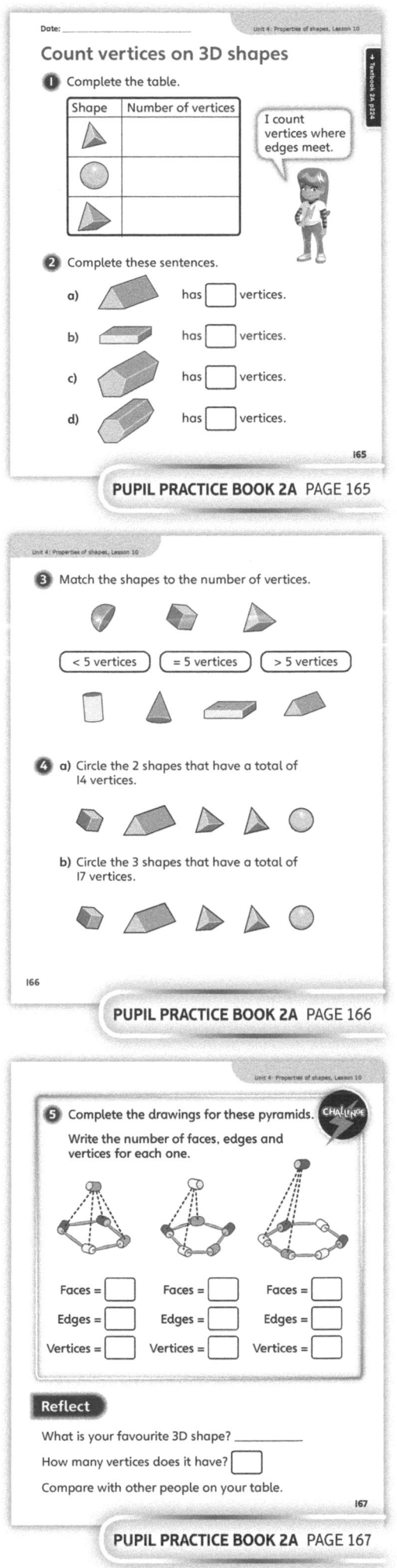

PUPIL PRACTICE BOOK 2A PAGE 165

PUPIL PRACTICE BOOK 2A PAGE 166

PUPIL PRACTICE BOOK 2A PAGE 167

Sort 3D shapes

Learning focus

In this lesson, children will sort 3D shapes based on their properties. Children will apply their learning from the previous three lessons about 3D shapes.

Before you teach ⏸

- Do children still need support when identifying properties of 3D shapes?
- How will you encourage children to explain their reasoning with regard to sorting 3D shapes?
- How will you make the links to the previous lessons clear?

NATIONAL CURRICULUM LINKS

Year 2 Geometry – properties of shape

Compare and sort common 2D and 3D shapes and everyday objects.

ASSESSING MASTERY

Children can sort 3D shapes based on given criteria, identifying similarities and differences in the properties of shapes in order to choose their own criteria for sorting 3D shapes. Children will be able to use groups (sets), both separate and overlapping, to sort 3D shapes.

COMMON MISCONCEPTIONS

Children may confuse properties of shapes, such as mistaking edges for vertices. Give children a 3D shape. Ask:
- *Can you point to an edge? A vertex? A face?*

Children may make the mistake of describing a curved surface as a face. Hold up a cylinder. Ask:
- *How many faces does this shape have?*

STRENGTHENING UNDERSTANDING

Provide children with a range of 3D shapes and sorting hoops. Discuss the properties of each shape and note down the number of faces, vertices and edges for each. Children can then work in pairs to sort the shapes into two or more groups. Ask children to label each group and then compare with another pair to see if they sorted the shapes in a different way.

GOING DEEPER

Provide children with a Carroll diagram, with the headings 'Pyramid', 'Prism', 'Six or fewer vertices' and 'More than six vertices'. Can children think of a shape to go in each group?

KEY LANGUAGE

In lesson: faces, pyramid, cuboid, sphere, curved surface, pentagon, edges, 2D

Other language to be used by the teacher: 3D shapes, groups, overlapping groups, vertex, vertices, hemisphere, triangle-based pyramid, square-based pyramid, pentagon-based pyramid, hexagon-based pyramid, cube, triangular prism, pentagonal prism, cylinder, cone, circle, oblong, rectangle

STRUCTURES AND REPRESENTATIONS

Sorting circles

RESOURCES

Mandatory: 3D shapes

Optional: a bag

 In the eTextbook of this lesson, you will find interactive links to a selection of teaching tools.

Quick recap 🔍

Reveal a 3D shape slowly from behind a screen or inside a bag. Prompt children to guess what the shape is before it is fully revealed.

Discover

WAYS OF WORKING Pair work

ASK

- Question ❶ a): *What is a pyramid?*
- Question ❶ a): *Which shapes do you find tricky to sort?*
- Question ❶ a): *Why do you think the children put the cone in the wrong group?*

IN FOCUS Question ❶ a) requires children to look carefully at the shapes and determine whether they meet the criteria for the group in which they are placed. The cone looks similar to a pyramid as it has an apex, so some children may think that it is in the correct group. However, it is not a polyhedron; it has a curved surface and, therefore, is not a pyramid.

Some of the shapes have a different orientation from how they are usually presented and this may confuse some children. Tell children to look carefully at each shape to count the faces. Having corresponding concrete shapes may help some children.

ANSWERS

Question ❶ a): ◯, △ and ⬟ are in the wrong groups.

They belong in the 'Other' group as they are not pyramids and they do not have 6 faces.

Question ❶ b): A pentagon-based pyramid could go in both groups as it is a pyramid and has 6 faces.

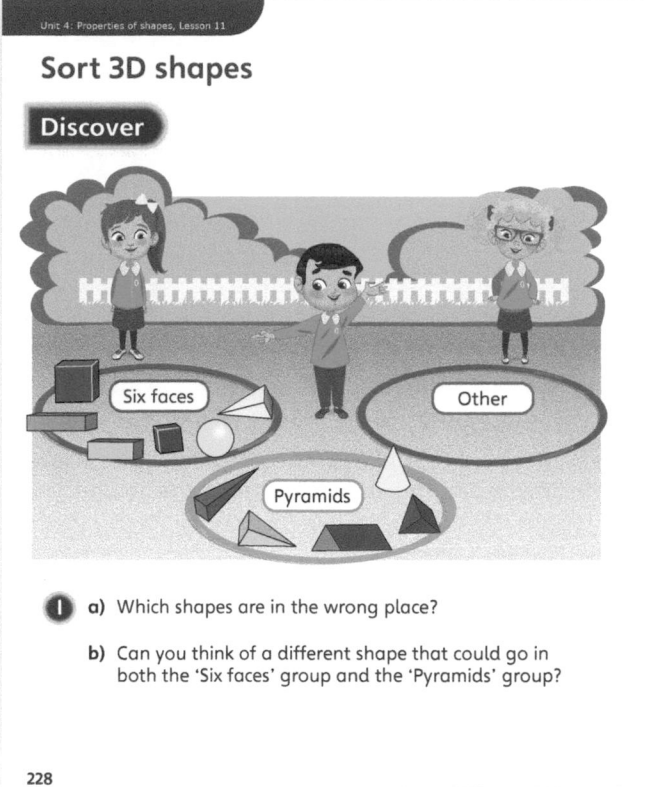

Sort 3D shapes

Discover

❶ a) Which shapes are in the wrong place?

b) Can you think of a different shape that could go in both the 'Six faces' group and the 'Pyramids' group?

228

PUPIL TEXTBOOK 2A PAGE 228

Share

WAYS OF WORKING Whole class teacher led

ASK

- Question ❶ a): *What group could you have for the cone and sphere? Can you think of another shape that would go in this group?*
- Question ❶ a): *What could you rename the 'Six faces' group so that the triangular prism could go into it as well as the cuboids?*
- Question ❶ a): *Can you think of a different way of sorting the shapes?*

IN FOCUS Question ❶ b) introduces the idea of overlapping (intersecting) groups. In order for a shape to go in the intersection, it has to meet the criteria for both the groups. Some children often struggle to place an item if it belongs in more than one group, so it is important to show how this can sometimes occur.

Share

a) This is how the shapes should be sorted.

I know that all cuboids have 6 faces. A sphere and a cone have a curved surface, so I put these in the 'Other' group.

b) A pyramid with a pentagon base has 6 faces in total.

It could go in both groups.

I looked for a pyramid that had 6 faces.

229

PUPIL TEXTBOOK 2A PAGE 229

Think together

Whole class teacher led (I do, We do, You do)

ASK

- Question **1**: *Could you sort these in a similar way but by number of faces?*
- Question **2**: *Would the order be the same if it was number of edges or vertices?*
- Question **3**: *Can you think of more than one way of sorting the shapes?*

IN FOCUS Question **3** presents children with 3D shapes as everyday objects. First, children have to look beyond the real-life context and focus on the mathematical properties. Astrid helps children by giving them a criterion for one of the groups. Children then have to decide how to sort the remaining shapes into two groups. There is more than one possible solution to this and it would be valuable to ask children to share and justify how they sorted the shapes.

STRENGTHEN Provide children with concrete representations of the shapes in questions **1** and **2** to help them count the edges and faces, so that they can physically sort and order them. Encourage them to talk about their thinking as they do it and to explain their decisions.

DEEPEN Look at question **1** and ask: *What would happen if the groups were '< 12 edges' and '> 12 edges'?* After children have had time to think and discuss, ask them where they would put the cuboids. Some children may think that cuboids are in both groups, but a shape cannot have both more than 12 edges and less than 12 edges. The cuboids need to be outside the circles. Draw a rectangle that encompasses the two circles and ask where the cuboids would go. Explain that the cuboids belong in the rectangle, but not in either of the circles. Ask: *What are you sorting? You are sorting 3D shapes, so you can call the universal group '3D shapes'.*

ASSESSMENT CHECKPOINT Question **1** will determine whether children can sort 3D shapes by the number of edges. Question **2** will determine whether children can compare and order shapes by the number of faces. Question **3** will determine whether children can identify properties of 3D shapes and sort them by their own chosen criteria.

ANSWERS

Question **1**: The sphere, hemisphere, cylinder and triangle-based pyramid all have fewer than 10 edges. The cube, cuboids and pentagonal prism all have more than 10 edges.

Question **2**: The order of fewest faces to most faces is C, A and D, E, B. The order would be the same if they were sorted by edges or vertices.

Question **3**: There is more than one way to solve this. Examples might be 'curved surfaces', 'flat surfaces', 'number of edges', 'number of faces' or 'number of vertices'.

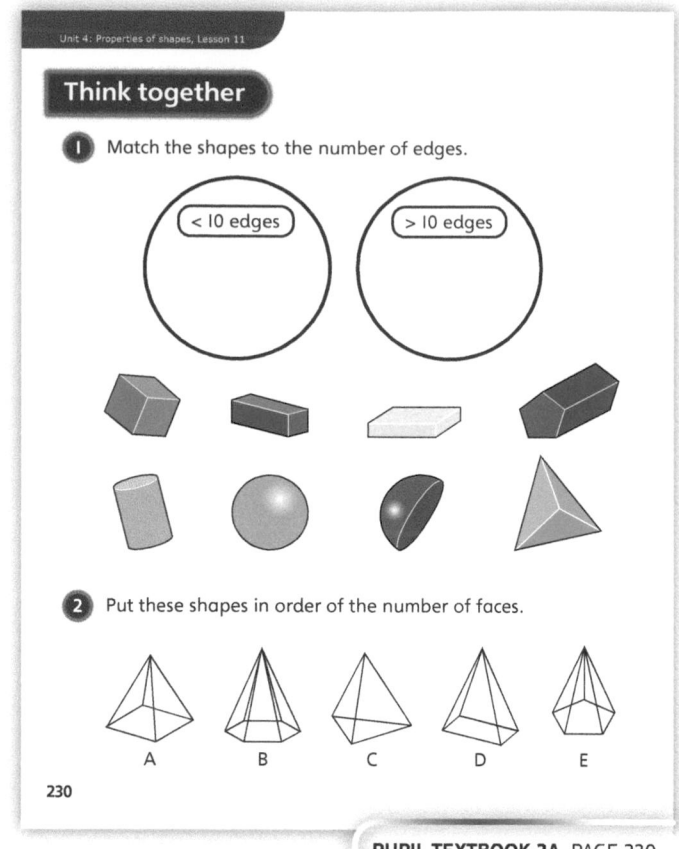

PUPIL TEXTBOOK 2A PAGE 230

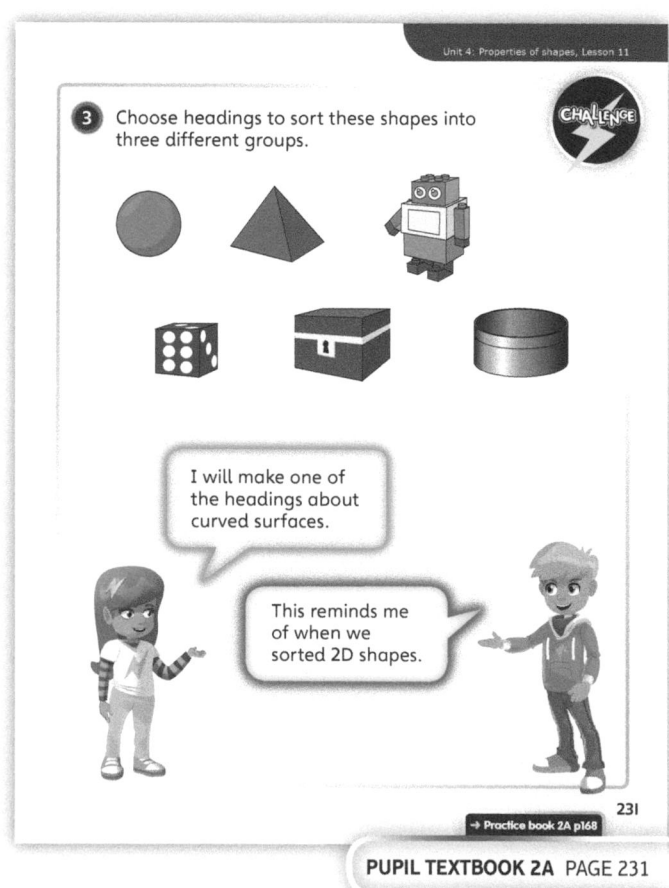

PUPIL TEXTBOOK 2A PAGE 231

Practice

WAYS OF WORKING Pair work

IN FOCUS Question ④ requires children to choose their own criteria for sorting shapes into two groups. The most straightforward sorting criteria would be a 'true' or 'not true' approach, such as 'curved surface' and 'no curved surface'. But children could try a 'less than' or 'more than' approach and consider the number of vertices, edges or faces.

STRENGTHEN If children are finding it difficult to sort 3D shapes, provide them with a variety of concrete 3D shapes and ask them to select all shapes with a given property. For example, ask: *Can you find all the shapes with a circular face? A curved surface?* You could then extend this by selecting several shapes and then asking children: *What do all these shapes have in common?* Children could then carry out the same activity in pairs.

DEEPEN Extend question ④ by asking children to sort the shapes so that at least one shape has to go in the overlap between the groups. Could children sort them into three groups, with at least one shape belonging to all three groups? For example, 'Has a triangular face', 'Has a rectangular face', 'Has fewer than eight vertices'.

THINK DIFFERENTLY In question ④, children need to identify what the headings could be for two sorting circles so that they can sort a set of given 3D shapes into each of the groups without having any left over.

ASSESSMENT CHECKPOINT Question ① will determine whether children can identify shapes that do not match a given criterion. Question ② will determine whether children can identify several shapes that all meet a given criterion. Question ③ will determine whether children can identify shapes that meet more than one criterion. Question ④ will determine whether children can devise their own criteria for sorting 3D shapes. Question ⑤ will determine whether children can compare and order 3D shapes by the number of vertices and edges.

ANSWERS Answers for the **Practice** part of the lesson can be found in the *Power Maths* online subscription.

Reflect

WAYS OF WORKING Pair work

IN FOCUS This question prompts children to think of more than one way to sort the shapes. In order to do this, they have to consider more than one type of property. The added challenge of having to sort them in two different ways means that children have to think carefully about the shapes' similarities and differences.

ASSESSMENT CHECKPOINT This section will determine whether children can identify similarities and differences between 3D shapes and whether they can sort them in more than one way by considering different types of properties.

ANSWERS Answers for the **Reflect** part of the lesson can be found in the *Power Maths* online subscription.

After the lesson

- Did the discussions between children help them to reason more deeply about properties of 3D shapes?
- Did any children struggle due to gaps in their understanding of properties of 3D shapes?

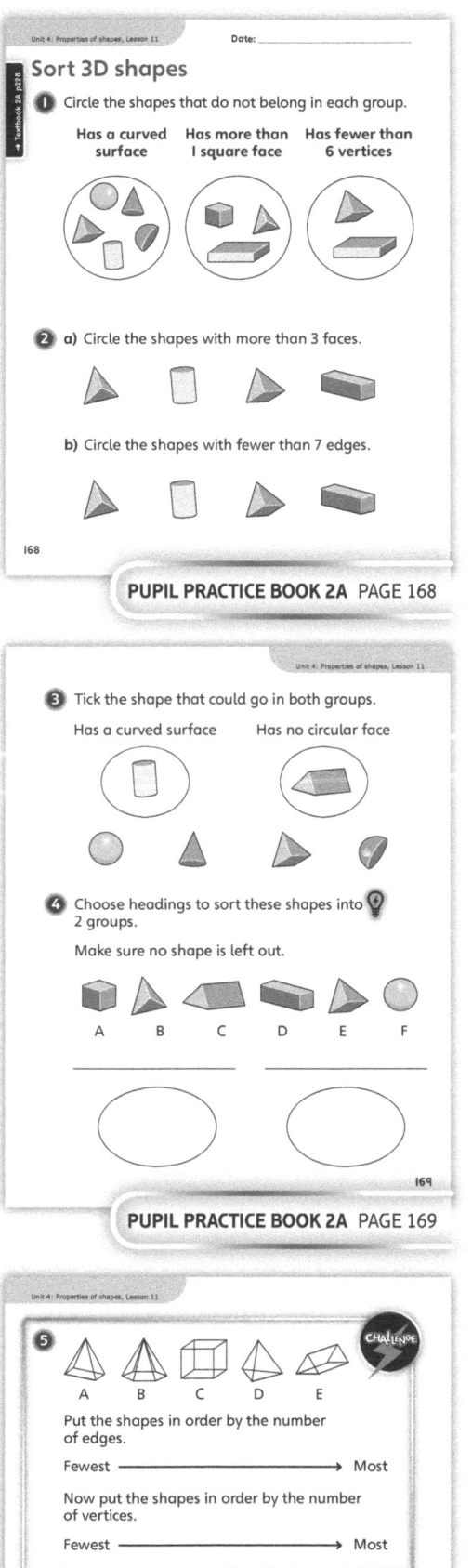

PUPIL PRACTICE BOOK 2A PAGE 168

PUPIL PRACTICE BOOK 2A PAGE 169

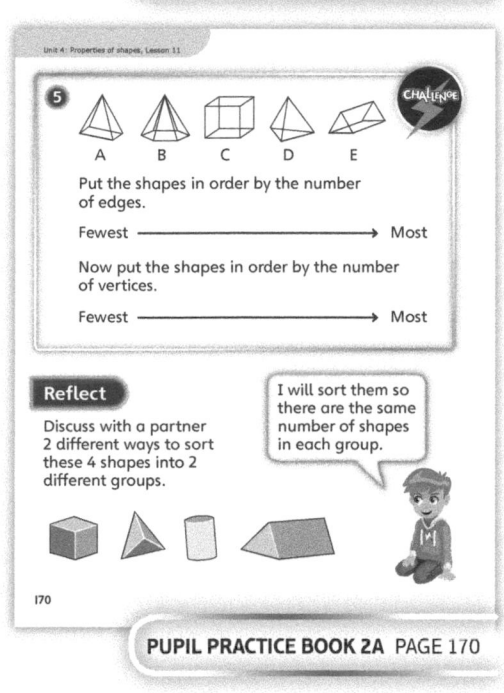

PUPIL PRACTICE BOOK 2A PAGE 170

Make patterns with 3D shapes

Learning focus

In this lesson, children will make symmetrical patterns with 3D shapes. They will use what they know about symmetrical patterns to identify missing shapes and create their own.

Before you teach

- Are children secure in naming the 3D shapes in this lesson?
- How can you make links from this lesson to children's learning in previous lessons?
- How will you promote discussions of thinking and strategies when children complete and create patterns?

NATIONAL CURRICULUM LINKS

Year 2 Geometry – properties of shape

Order and arrange combinations of mathematical objects in patterns and sequences.

ASSESSING MASTERY

Children can identify the symmetry in patterns and use this to find missing shapes. Children can create their own symmetrical patterns from a given set of 3D shapes and apply their knowledge of properties of 3D shapes when designing their own patterns.

COMMON MISCONCEPTIONS

Children may assume that the pattern is a repeating pattern when there is no repeating core. Children may try to repeat the whole sequence of shapes in order to create a repeating pattern or reverse the mirror image so that the pattern repeats. Show children a symmetrical pattern made from 3D shapes. Ask:
- *Can you name all the shapes? Are there any parts of the pattern that repeat?*

STRENGTHENING UNDERSTANDING

Allow children to create their own symmetrical patterns with 3D shapes. Start with a single shape for the centre of the pattern, then ask children to pick up two of the same shape, one in each hand. Next, tell them to place these shapes on either side of the centre shape. Repeat this process of placing two identical shapes on either side to create a symmetrical pattern. Working from the centre out, ask children to point and name each shape.

GOING DEEPER

Ask children to explore making a symmetrical pattern with 3D shapes that has both a vertical and horizontal line of symmetry. How will children go about doing this? Where will they start? How can they ensure that their pattern has horizontal and vertical symmetry?

KEY LANGUAGE

In lesson: 3D, symmetrical, cylinder, repeating pattern

Other language to be used by the teacher: sphere, cone, triangle-based pyramid, square-based pyramid, cube, cuboid, faces, edges, vertices, symmetry, vertical, horizontal, reflective

RESOURCES

Mandatory: 3D shapes, mirrors

Optional: pictures of 3D shapes

 In the eTextbook of this lesson, you will find interactive links to a selection of teaching tools.

Quick recap 🔎

Play 'Guess my shape'. Say: *I am thinking of a shape. It has fewer than 6 vertices. What could it be?*

Discover

WAYS OF WORKING Pair work

ASK

- Question ➊ a): *Which shape is not repeated in the pattern?*
- Question ➊ a): *Are you able to continue the pattern?*
- Question ➊ b): *Which shape is there only one of? How can this help you?*

IN FOCUS Question ➊ a) asks children to identify the pattern in the sequence of shapes. There is no repeating core in the pattern and this may puzzle children as they have looked at repeating patterns in Lesson 7 of this unit. Children may try to repeat the whole sequence or think that the repeating core is cube, cylinder, pyramid. Ask children to look at the centre shape first and work out from both sides rather than starting from the end if they struggle to see the pattern.

ANSWERS

Question ➊ a): It is a symmetrical pattern with the line of symmetry running down the centre of the blue cuboid.

Question ➊ b): The cylinder has to be in the centre, with two identical shapes either side and the final two identical shapes at either end.

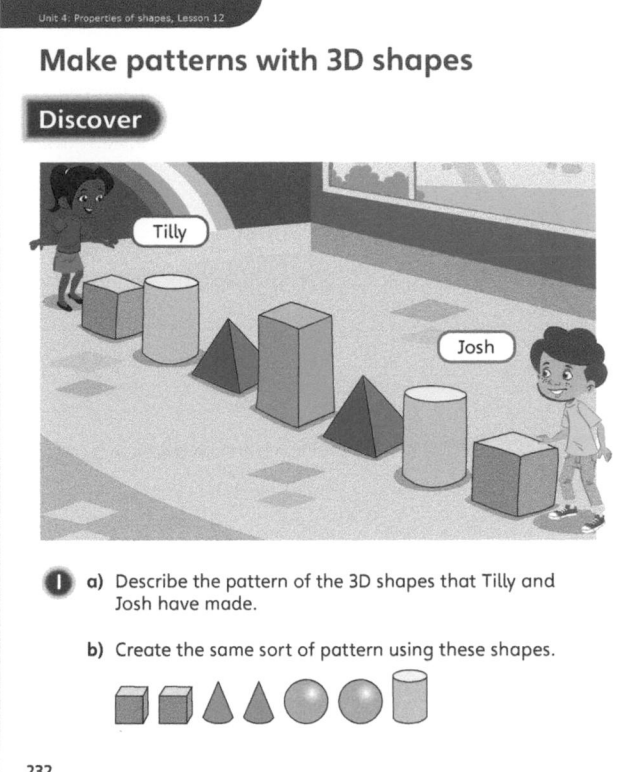

Make patterns with 3D shapes

Discover

➊ a) Describe the pattern of the 3D shapes that Tilly and Josh have made.

b) Create the same sort of pattern using these shapes.

232

PUPIL TEXTBOOK 2A PAGE 232

Share

WAYS OF WORKING Whole class teacher led

ASK

- Question ➊ a): *Could you rearrange these shapes and still have a symmetrical pattern?*
- Question ➊ a): *If you removed the centre cuboid would it still be symmetrical?*
- Question ➊ b): *How many different ways are there to create a symmetrical pattern with these shapes?*
- Question ➊ b): *Why do you think symmetrical patterns start with the middle shape?*

IN FOCUS Question ➊ b) shows children how a symmetrical pattern can be created. By starting with a shape in the middle, symmetry can be assured by placing identical shapes either side. Although in the pictures the characters place the spheres next to the cylinder, they could have placed the cubes or the cones next to the cylinder. This can lead to a discussion about there being more than one possibility and how children could find all the possibilities.

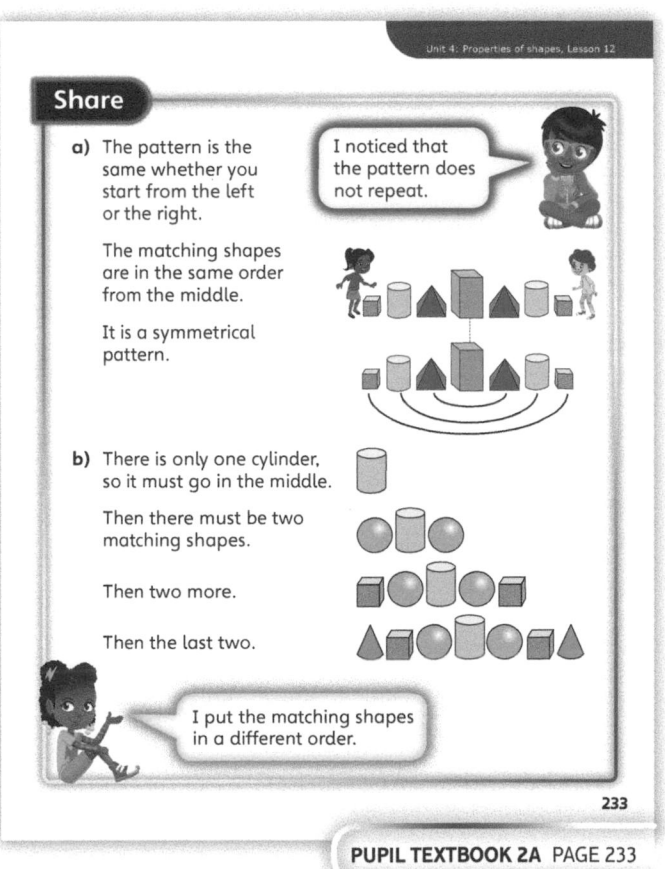

Share

a) The pattern is the same whether you start from the left or the right.

I noticed that the pattern does not repeat.

The matching shapes are in the same order from the middle.

It is a symmetrical pattern.

b) There is only one cylinder, so it must go in the middle.

Then there must be two matching shapes.

Then two more.

Then the last two.

I put the matching shapes in a different order.

233

PUPIL TEXTBOOK 2A PAGE 233

Think together

WAYS OF WORKING Whole class teacher led (I do, We do, You do)

ASK

- Question **1**: *Which shape will you look at first?*
- Question **2**: *Which shape or shapes will you start with?*
- Question **2**: *Is your pattern different from a partner's pattern? Explain to your partner why you chose to do it your way.*
- Question **3**: *Is it possible to use these shapes to create a pattern that is repeating and symmetrical?*

IN FOCUS Question **3** requires children to design their own symmetrical pattern. Children have to choose how many of each shape they have. They could have one shape in the centre or two identical shapes, so the line of reflective symmetry runs between them. This question allows children to explore different ways of creating symmetrical patterns. This question also draws on children's learning from Lesson 7 by asking them to create a repeating pattern. This helps children understand the distinction between repeating and symmetrical patterns.

STRENGTHEN Provide children with a mirror for question **1** so that they can use it to identify what shapes are missing. If they struggle to find the line of reflective symmetry, this can be drawn for them.

For questions **2** and **3**, provide children with the corresponding 3D shapes or pictures of the shapes so that they can explore arranging them. Children can then use a mirror to test whether their design is symmetrical.

DEEPEN Extend question **2** by asking children to find as many ways as they can to use the shapes to create a symmetrical pattern. Can they work systematically to ensure that they have them all? Ask them to check their solutions with a partner to see if they have missed any out.

ASSESSMENT CHECKPOINT Question **1** will determine whether children can identify missing shapes in a symmetrical pattern. Question **2** will determine whether children can create a symmetrical pattern with given shapes. Question **3** will determine whether children can design their own symmetrical patterns and whether they understand the difference between symmetrical and repeating patterns.

ANSWERS

Question **1**: In the first pattern, the missing shapes are cube and sphere. In the second pattern, the missing shapes are sphere and cube.

Question **2**: There are various solutions to this problem. Check children's answers to ensure the patterns are symmetrical.

Question **3**: There are various solutions to this problem. Check children's answers to ensure they have created both a symmetrical and a repeating pattern.

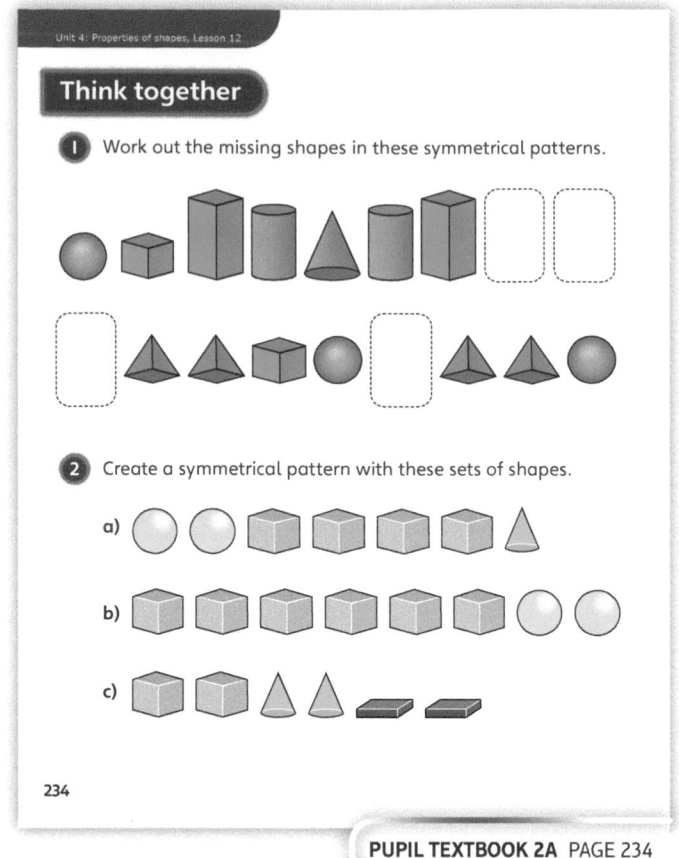

PUPIL TEXTBOOK 2A PAGE 234

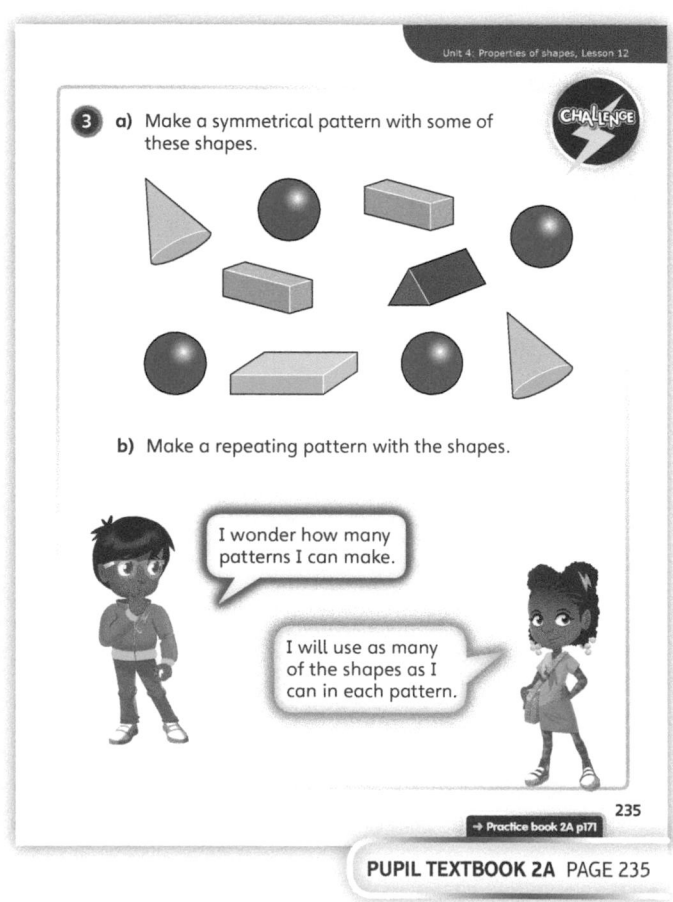

PUPIL TEXTBOOK 2A PAGE 235

Practice

WAYS OF WORKING Pair work

IN FOCUS Question **5** requires children to apply what they have learned about properties of shape in order to find the solution. As there are to be only three shapes, two of them have to be the same. Children must think of a shape, count the faces and double it. They then subtract this from the total number of faces in order to identify the centre shape. This is a multi-step problem that requires a high level of reasoning. Therefore, children will benefit from working in pairs or small groups to promote discussion.

STRENGTHEN Provide children with mirrors to identify missing shapes and to check their own symmetrical patterns. Where possible, provide children with 3D shapes so that they can manipulate them and easily adapt their ideas.

DEEPEN Provide children with a range of cutouts of 3D shapes. Children work in pairs. Each child takes it in turn to add two identical shapes to create a joint symmetrical pattern. Each time they add two shapes, they count the number of faces, edges and vertices, and add that number to the running total for the pattern. The first pair to reach or pass 100 wins. This can be further extended by asking children to create a pattern that has both a vertical and a horizontal line of reflective symmetry.

THINK DIFFERENTLY In question **3**, children are given a series of 3D shapes and are asked to use them to make a pattern. They should consider what shape will be in the middle of the pattern and how they will need to arrange the other shapes around it in order to maintain reflective symmetry.

ASSESSMENT CHECKPOINT Question **1** will determine whether children can identify missing shapes from a symmetrical pattern. Questions **2** and **3** will determine whether children can create their own symmetrical patterns from given 3D shapes. Question **4** will determine whether children can create their own symmetrical patterns, devise their own problems and identify patterns to find missing shapes. Question **5** will determine whether children can carry out multi-step problems by applying their understanding of properties of shape and symmetrical patterns.

ANSWERS Answers for the **Practice** part of the lesson can be found in the *Power Maths* online subscription.

Reflect

WAYS OF WORKING Pair work

IN FOCUS This question tests children's understanding of symmetrical and repeating patterns. In order for children to give a concise and accurate explanation of the difference, they need to be secure in their understanding. Ask children to discuss and compose their answer to this problem with a partner and to use a whiteboard, so that they can edit and improve their answer before writing it in their practice books.

ASSESSMENT CHECKPOINT This section will determine whether children have a secure understanding of what constitutes a repeating pattern and a symmetrical pattern and what the differences are between them.

ANSWERS Answers for the **Reflect** part of the lesson can be found in the *Power Maths* online subscription.

After the lesson

- Were children able to use their knowledge of the names and properties of 3D shapes fluently in this lesson?
- How confident were children in identifying, completing and creating symmetrical patterns?

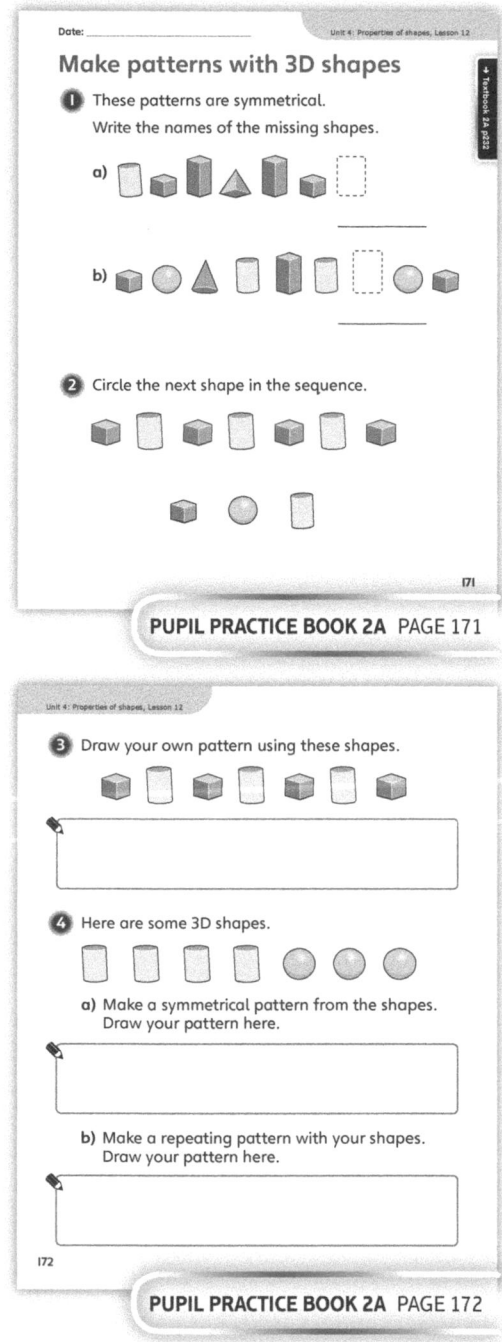

PUPIL PRACTICE BOOK 2A PAGE 171

PUPIL PRACTICE BOOK 2A PAGE 172

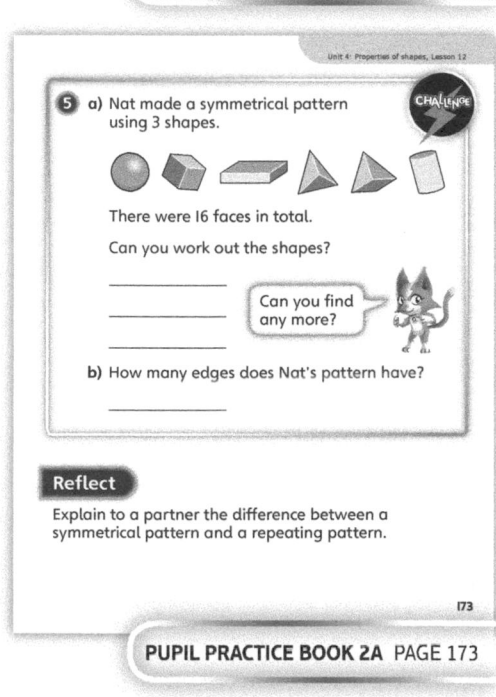

PUPIL PRACTICE BOOK 2A PAGE 173

End of unit check

> **Don't forget the unit assessment grid in your *Power Maths* online subscription.**

WAYS OF WORKING Group work adult led

IN FOCUS This end of unit check requires children to draw on their understanding of the properties of 2D and 3D shapes and how shapes are categorised based on specific properties that they share.

Think!

WAYS OF WORKING Pair work or small groups

IN FOCUS This question requires children to apply their understanding of the vertices of 2D shapes in order to solve the problem. By working with a partner, children can discuss and justify their strategies. Provide children with squares of paper to encourage a 'trial and improve' approach.

Key vocabulary in this question includes: vertices, sides, pentagon, hexagon, triangle, quadrilateral.

Encourage children to think through or discuss the number of vertices for a variety of shapes before writing their answer in **My journal**.

ANSWERS AND COMMENTARY Children will be able to recognise the key vocabulary used in this section and reliably count each property. Children can recognise similarities and differences between shapes and use the correct vocabulary to describe these. Children will confidently name a variety of 2D and 3D shapes and recognise patterns involving shapes.

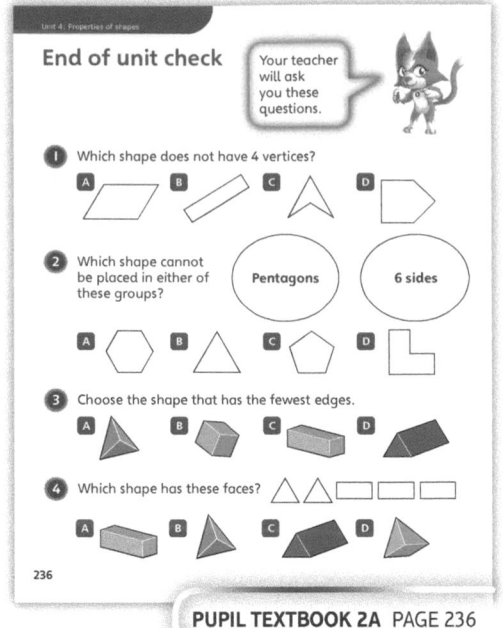

PUPIL TEXTBOOK 2A PAGE 236

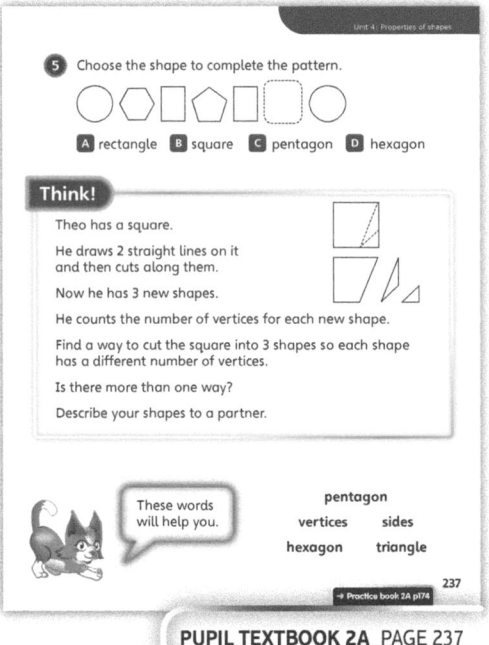

PUPIL TEXTBOOK 2A PAGE 237

Q	A	WRONG ANSWERS AND MISCONCEPTIONS	STRENGTHENING UNDERSTANDING
1	D	A, B or C suggests that children have miscounted or do not understand the meaning of 'vertices'. C may also suggest that children did not include the concave vertex.	Provide children with concrete 2D and 3D shapes so that they can practise recognising and counting the different properties. Children can also use dry-wipe markers to mark off properties as they count them.
2	B	A suggests that children have miscounted. C suggests that children have miscounted or do not recognise a pentagon. D suggests that children have miscounted or have not recognised all sides.	
3	A	B, C or D suggests that children have miscounted or do not understand the meaning of 'edges'.	
4	C	A, B or D suggests that children do not recognise the shapes of faces.	
5	D	A, B or C suggests that children do not recognise the names of 2D shapes or have failed to recognise the symmetrical pattern.	

My journal

WAYS OF WORKING Independent thinking

ANSWERS AND COMMENTARY

If children cut off one of the square's corners, they produce a pentagon and a triangle. Children could then cut off a corner from the triangle to create a smaller triangle, a quadrilateral and a pentagon.

Alternatively, children could cut the square from side to side to produce two quadrilaterals. By cutting a corner off from either quadrilateral, children will end up with a pentagon, a quadrilateral and a triangle.

Provide children with squares of paper so that they can physically cut the square and then reassemble it, so that they can see where the cut lines are. Encourage children to count the sides or vertices and then refer to a classroom display of the properties of shape to help children identify the shapes they have created.

Power check

WAYS OF WORKING Independent thinking

ASK

- *Do you think you are better at naming and describing 2D shapes and 3D shapes than before?*
- *Do you think you can sort shapes in different ways?*

Power puzzle

WAYS OF WORKING Pair work or small groups

IN FOCUS Use this **Power puzzle** to assess whether children are able to identify and create different types of cuboid. Working in pairs will encourage children to work collaboratively to find all the solutions.

ANSWERS AND COMMENTARY With 24 cubes, children could create a 1 × 1 × 24 cuboid, a 1 × 2 × 12 cuboid, a 1 × 3 × 8 cuboid, a 1 × 4 × 6 cuboid, a 2 × 2 × 6 cuboid or a 2 × 3 × 4 cuboid. In order to find all the possibilities, children need to understand that cuboids need to have six faces and that the faces can be square or oblong.

With 27 cubes, children can create a 1 × 1 × 27 cuboid, a 1 × 3 × 9 cuboid or a 3 × 3 × 3 cuboid. To find all three, children need to understand that a cube is a special type of cuboid.

If children are unsuccessful in identifying all the cuboids, or if they create shapes that are not cuboids, provide opportunities for them to create their own cuboids using a range of construction equipment, straws and multilink cubes. Encourage children to produce a range of cuboids, including cubes, and to check the shapes they make against the key mathematical properties.

After the unit ⏸

- Were children able to use the key vocabulary accurately, fluently and confidently when naming, sorting and describing 2D and 3D shapes?
- Can you find opportunities to reinforce this learning in other areas of the curriculum?

Strengthen and **Deepen** activities for this unit can be found in the *Power Maths* online subscription.

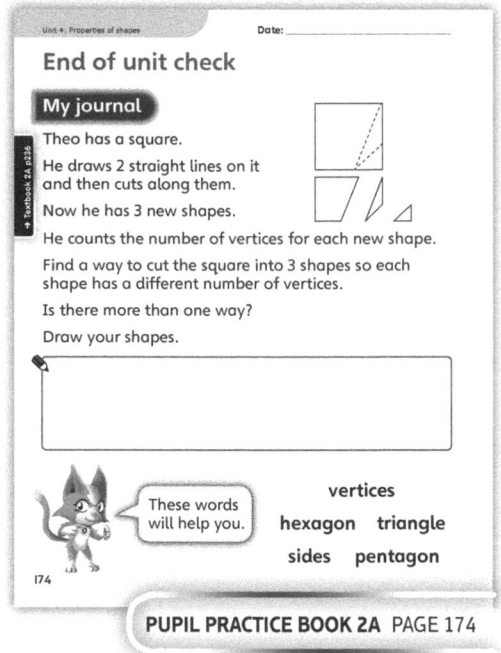

PUPIL PRACTICE BOOK 2A PAGE 174

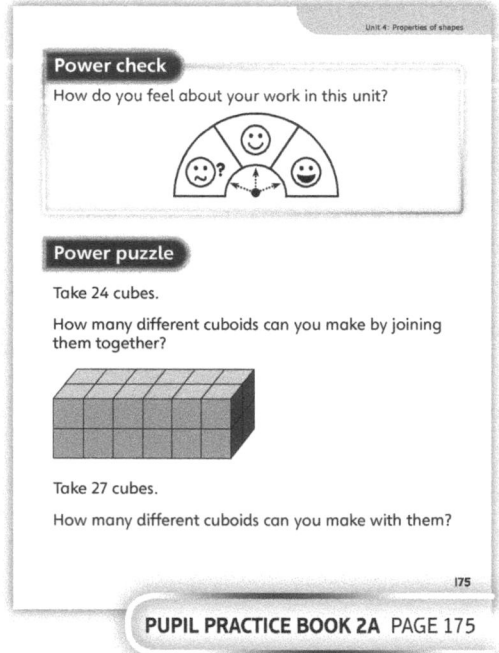

PUPIL PRACTICE BOOK 2A PAGE 175

Published by Pearson Education Limited, 80 Strand, London, WC2R 0RL.

www.pearsonschools.co.uk

Text © Pearson Education Limited 2018, 2022
Edited by Pearson and Florence Production Ltd
First edition edited by Pearson, Little Grey Cells Publishing Services and Haremi Ltd
Designed and typeset by Pearson and Florence Production Ltd
First edition designed and typeset by Kamae Design
Original illustrations © Pearson Education Limited 2018, 2022
Illustrated by Laura Arias, Fran and David Brylewski, Nigel Dobbyn, Adam Linley, Nadene Naude
and Dusan Pavlic at Beehive Illustration, Kamae Design and Florence Production Ltd
Back cover illustration © Will Overton at Advocate Art and Nadene Naude at Beehive Illustration
Cover design by Pearson Education Ltd
Series editor: Tony Staneff; Lead author: Josh Lury
Authors (first edition): Tony Staneff, Natasha Dolling, Jonathan East, Julia Hayes, Neil Jarrett
and Timothy Weal
Consultants (first edition): Professor Jian Liu

The rights of Tony Staneff and Josh Lury to be identified as authors of this work have been
asserted by them in accordance with the Copyright, Designs and Patents Act 1988.

First published 2018
This edition first published 2022

26 25 24 23 22
10 9 8 7 6 5 4 3 2 1

British Library Cataloguing in Publication Data
A catalogue record for this book is available from the British Library

ISBN 978 1 292 45050 6

Printed in the UK by Ashford Press Ltd

For Power Maths online resources, go to:
www.activelearnprimary.co.uk

Note from the publisher
Pearson has robust editorial processes, including answer and fact checks, to ensure the
accuracy of the content in this publication, and every effort is made to ensure this publication
is free of errors. We are, however, only human, and occasionally errors do occur. Pearson is
not liable for any misunderstandings that arise as a result of errors in this publication, but it is
our priority to ensure that the content is accurate. If you spot an error, please do contact us at
resourcescorrections@pearson.com so we can make sure it is corrected.